Rana P. B. Singh · Olimpia Niglio · Pravin S. Rana
Editors

Placemaking and Cultural Landscapes

Springer

Editors
Rana P. B. Singh
Department of Geography
Institute of Science
Banaras Hindu University
Varanasi, Uttar Pradesh, India

Pravin S. Rana
Tourism Management
Faculty of Arts
Banaras Hindu University
Varanasi, Uttar Pradesh, India

Olimpia Niglio
Department of Civil Engineering
and Architecture
University of Pavia
Pavia, Italy

Research Center for Edo-Tokyo Studies
Hosei University
Tokyo, Japan

ISSN 2198-3542 ISSN 2198-3550 (electronic)
Advances in Geographical and Environmental Sciences
ISBN 978-981-19-6273-8 ISBN 978-981-19-6274-5 (eBook)
https://doi.org/10.1007/978-981-19-6274-5

This Springer imprint is published by the registered company Springer Nature Singapore Pte Ltd.
The registered company address is: 152 Beach Road, #21-01/04 Gateway East, Singapore 189721, Singapore

This anthology of essays is dedicated to
Professor Sung-Kyun Kim
(18 February 1956–19 June 2020)

Photo by Rana P. B. Singh, 16 October 2013, SNU GH, Seoul

Of Kim's mind and his manners shines,
In his village-world, and land-filled lines.
In each of space, he seems to shake a lance,
Touching heart, and giving love to each incidence,
Accompanying him, one learns making pathway,

Thus, reaching to destination, all the stresses go sway.

—Rana P. B. Singh

Foreword: Landscapes of Hope and Care

It is no exaggeration to say that the essays in this volume offer a wealth of insights into *Placemaking and Cultural Landscapes*. Each of their core concepts—"place" and "landscape"—has rich supporting theoretical and regional literatures. One can think of this pairing, between places and landscapes, as a dialogue between two closely-related ideas. These studies suggest a set of relationships where the activities of placemaking collectively produce larger cultural landscapes. Other studies draw attention to differences and tensions between concepts of place and landscape (e.g., Malpas 2011). However, it seems fair to say that this volume emphasizes their interrelationships and complementarities.

It is thus exciting to consider the intersections among various chapters in the book. On one level, their relationships are manifestly plural, with many types of places producing varied cultural landscapes. This pattern brings to mind the philosopher William James' (1909) *Pluralistic Universe*. At the same time, the essays individually and collectively suggest a unity of thought and a universe of many diverse but well-composed parts. Several integrative themes come to mind, which are shared below for the reader's consideration.

Meanings of Place and Landscape

From the opening chapter by editors Rana P. B. Singh and Olimpia Niglio onwards, the volume shows concern for the construction of meaning in placemaking and cultural landscapes (Singh and Niglio). Its chapters explore the many meanings of meaning. They arise in contexts ranging from linguistic study to literary imagination, sacred ritual and cosmological symbolism. Some have deep historical roots in embodied experience while others arise in new modes of virtual media. The latter has been greatly amplified by the COVID-19 pandemic, which several chapters consider in compelling new ways (e.g., Francini and Montacchini; and Olsen).

Explorations of meaning in placemaking and cultural landscapes extend beyond philosophical approaches to meaning in language and psychology (e.g., Ogden and

Richards 1938; Speaks 2021; Wolf 2010). The locative domains of meaning seek to discern and share meanings "out there" in the world, beyond one's own activities and thoughts, in places and cultural landscapes produced by others. It is a challenging goal and a generous one.

Place and landscape have diverse languages. They have evolved in widely varying ways in different cultures and times. And still, their meanings are comprehensible and communicable, in some measure, to readers in the present moment. We learn, for example, that the Korean spatial construct of *pung-su* at traditional villages like Hahoe bears comparison with spatial orientations and principles of *feng-shui* in China (Kim and Singh). The authors note that both traditions evoke "air-and-water," which brings to mind a comparison with *ab-hawa* (water-air) features of Persianate architecture in southern and southwest Asia.

The language and meanings of landscape present challenges in this regard, as some languages do not have precise synonyms for landscape or parallel historical developments in landscape theory with those of northern Europe and the United States (Olwig 2019; Spirn 2000). It may be no accident that landscape conservation *as such* may be less salient in regions that do not have close cognates with that concept. Landscape is an emergent construct, for example, in areas where Arabic, Persian, and Turkish are spoken, and where other concepts may be useful and enriching to the field (Wescoat 2016). In those contexts, scholarship should strive to discern and promote the vernacular languages of land and place that have developed in the region, and to compare them with denotations of what landscapes are said to be and connotations of what they purportedly mean. Several chapters offer philosophically-based guides into these complex questions of meaning, some of them grounded in religious thought, while others draw upon phenomenological conceptions of lifeworlds and lifeways (Seamon).

Scope: Sacralities, Settlements, and Ecologies

The scope of placemaking and cultural landscapes as a field of study in these essays is truly extensive. Deep in psychology, extending in all directions, and to the heights of cosmology, they encompass many scales of human understanding. Three major substantive themes stand out, though readers will no doubt discern others as well.

Many chapters focus on sacred places and landscapes. They range from sacred gardens in eastern Europe (Fekete) to sacred groves in tribal India (Borde), monasteries in Korea (Lee-Niinioja), cloisters (Francini and Montacchini), convents (Volzone et al.), temple towns in India (Ramineni et al.), pilgrimage paths (Rodriguez), and sacred capital cities in China (Biatta, Magli, Giambruno, and Zhou). They proceed from origin myths to rituals, symbols, and eschatologies—that is, across the full spectrum of religious imagination, revelation, and experience. Religious experience is often featured in placemaking and cultural landscapes. It is thus interesting to consider the *varieties of religious experience*, to borrow another phrase from William James (1990). This volume contributes to our understanding of the loci of such experience, which has been less fully explored in religious studies than the psychology of religious experience, though that may be changing (e.g., Eck 2011; Smith 1987; Tweed 2006). This volume foregrounds the geographic character of the sacred.

In the temporal context, several essays explore rural and urban landscapes, along with the eroding boundary between them in the twenty-first century. Some underscore the living heritage of rural places in the context of modernizing trends (Faraji and Masoumi; Kowkabi). Others draw attention to cities as a category of valued places (Singh and Kumar) and to the expanding rubric of historic urban landscapes since its adoption by the World Heritage Commission in 2011 (Bandarin and Van Oers 2012; Silva 2020). Eroding distinctions between the urban and rural are increasingly evident worldwide, leading to growing interest in rurban, suburban, and peri-urban placemaking (Hui and Wescoat 2019).

A third major substantive theme in this volume treats cultural landscapes as inherently natural and ecological (O'Donnell). This framing renews early twentieth-century cultural geography (Sauer 1963). It was eclipsed to some extent by more exclusively social scientific and humanistic conceptions of landscape and by their critique of various forms of oppression in landscape history and (mis)representation (Adams et al. 2001; Mitchell 2002; Olwig and Mitchell 2009). Ironically, these latter developments occurred around the same time that the discipline of landscape ecology began to flourish as a jointly natural and social science (Nassauer 1997). It is thus exciting to follow the integrated cultural ecological in this volume. For example, the roles of water in cultural landscapes range from floodplains and canals of Zambia (Chaiwa) to Sundarbans mangroves and riparian islands in Assam (Kumar).

Several chapters engage all three themes—sacrality, rurbanity, and ecology—from the introduction (Singh and Niglio) to the eco-religious ritual landscapes of adivasi tribes in India (Borde) and hill people of Himachal Pradesh (Pethe and Shinde).

Structures of Experience and Inquiry

How are placemaking and cultural landscapes to be understood? The essays offer insights into experiential and scholarly modes of inquiry. Each chapter proceeds in its own way and with its own voice to communicate the depth of experience and feeling at a place, using methods that can be organized along a temporal scale.

They begin with archaeological studies and methods, excavating layers of early historical settlement and material culture (Akture; Niglio). These studies remind us that the discipline of landscape archaeology employs detailed surface surveys to reconstruct long-term processes of human-environment relations. Other chapters deal with the more recent past of modern nineteenth-century and twentieth-century landscape change. In between lie medieval and early modern landscapes that developed prior to colonization and receive somewhat less attention than ancient and recent places in modern scholarship.

The studies of contemporary landscapes are structured in interesting ways. Some are organized thematically by level or scale from the small site to region and cosmos, employing analogies to connect each level with those above and below. Others deal with levels of inquiry from perception to cognition, behavior, and social life that remind one of Norberg-Schultz and other phenomenologists' levels of spatial experience. Analogies between microcosm and macrocosm feature prominently in the studies of sacred landscapes.

A few essays employ comparative methods that interest me deeply, including two that compare sacred institutional complexes—convents and monasteries (Lee-Niinioji; and Volzone et al.). Comparative methods in the history of religions came under criticism for their association with orientalist projects (Eliade and Kitagawa 1959), but the "magic" of comparative inquiry clearly endures (Smith 1982; Patton and Ray 2000). That sense of magic pervades this volume, and there are endless comparisons that might be made among its chapters.

Another prominent line of inquiry in landscape heritage conservation is "problem-driven." It begins with concerns about contemporary issues and threats, the complexity of which gives rise to inquiry and analysis. The search deeper understanding of the problem leads to alternatives that help clarify and may help resolve the problem. This pragmatic approach is inherent in conservation studies and implicit in others as well.

Cultural Institutions, Economies, and Quality of Life

Engagement with landscape heritage draws one into related issues of political economy, which involve cultural policies, institutions, politics, values, economics, and so on. These topics can themselves sometimes pose threats to the integrity or authenticity of a place or landscape. But this volume and others remind us that political economy is an inherent aspect of placemaking and valued landscapes. Even remote places have become prosperous temple towns through jointly cultural and economic processes (Gaur; Ramineni, Bhardwaj, and Bhaskar).

It is interesting to consider that placemaking, place attachment, and the resultant "sense of place" are major topics of scholarly research, but they have not been institutionalized in the way that "cultural landscapes" have become a specific UNESCO category for World Heritage conservation since 1992, followed by "historic urban landscapes" in 2011. Authors in this volume have decades, or perhaps centuries, of experience working with heritage institutions, helping design them as well as apply them in emerging fields of practice.

Several chapters discuss how UNESCO World Heritage policies survey, assess, and conserve valued landscapes (Akture; Elizbarashvili et al.; Jain). The broader literature in this field recognizes that heritage institutions can add to the commodification or distortion of valued places (Lowenthal 1998). In what situations does heritage listing help conserve cultural values and places, and when does it adversely affect them? In what context does designation of heritage sites in danger make a difference for conservation on the ground? The cultural landscape approach helps ensure that these questions are carefully addressed.

It is interesting to consider the politics as well as institutions of landscape heritage conservation (Chalana and Ashima 2021). The early twenty-first century has witnessed increased attention to explicitly political, vis-à-vis institutional, aspects of heritage conservation. Cultural places have complex politics some of which, for better or worse, do not jibe with the contemporary politics of societies in which they are situated, or cannot be regarded as equitable or sustainable in international conservation practice.

The political economy of placemaking and cultural landscapes is thus a fascinating topic. Religious tourism is arguably an age-old phenomenon, but it is expanding rapidly and taking new forms which bring financial resources to sacred heritage sites while altering their experience (Pusalkar). Some historic sites have become the focus of violent conflict or dispossessing development proposals. To anticipate and transform such hazards, heritage organizations are advancing strategies for socio-economically inclusive conservation in heritage cities, and exploring ways to assess the economic value of unique cultural phenomena (Licciardi and Amirtahmasebi 2013; World Bank 2014). Although working in different ways and contexts, cultural landscape historians like J.B. Jackson (1984) and his followers have written persuasively about the inherent dignity of ordinary and vernacular landscapes.

Some exemplary long-term conservation projects include a core commitment to community-led development. In the past, tourists to the World Heritage Site of Humayun's Tomb-Garden in Delhi had little interaction with pilgrims or residents at the nearby Sufi shrine of Nizamuddin Auliya, notwithstanding the close historical connection between those two places. That has changed dramatically with the award-winning conservation initiatives in both places guided by the Aga Khan Trust for Culture (2021) working closely with the Nizamuddin community and Archaeological Survey of India.

If we draw together politics and economics into political economy, and focus that field on improving the quality of life and quality of place, what can we learn? Researchers on quality of life distinguish between subjective and objective aspects of well-being (cf. Nussbaum and Sen 1993). Research on quality-of-life helps ensure that many dimensions of well-being are included in placemaking and cultural landscape conservation. Conversely, research on placemaking and cultural landscapes helps integrate diverse aspects of human and environmental well-being in ways that deserve further study.

Hope and Care in Placemaking and Cultural Landscapes

By the end of this volume, one arrives with a broad sense of hope. The places and landscapes presented are hopeful, even those marked by troubled histories. The authors are not naïve about conflict and pain in the world, but they focus in larger measure on positive values of cultural landscapes that peace scholar Elise Boulding (2000) once described as the most pervasive but often hidden side of history.

Hope is more than passive acceptance. It is cultivated through cultural and academic means, and through practices of care for valued places and cultural landscapes (Lawson 2007). Placemaking generally involves process of care, as does conservation of natural and cultural landscapes to an even greater extent. These practices and products of care are evident throughout the case studies and chapters in this volume.

James L. Wescoat Jr.
Aga Khan Professor Emeritus
of Landscape Architecture
and Geography
Massachusetts Institute of Technology
Cambridge, MA, USA
wescoat@mit.edu

References

Adams PC, Hoelscher S, Till KE (eds) (2001) Textures of place: exploring humanistic geographies. University of Minnesota Press, Minneapolis

Aga Khan Trust for Culture (2021) Nizamuddin urban renewal initiative. https://www.nizamuddinrenewal.org/. Accessed 23 Dec 2021

Bandarin F, Van Oers R (2012) The historic urban landscape: managing heritage in an urban century. Wiley-Blackwell, Chichester UK

Boulding E (2000) Cultures of peace: the hidden side of history. Syracuse University Press, Syracuse

Chalana M, Ashima K (2021) Heritage conservation in postcolonial india, approaches and challenges. Routledge, New Delhi

Eck D (2011) India: a sacred geography. Harvard University Press, Cambridge

Eliade M, Kitagawa J (1959) The history of religions: essays in methodology. University of Chicago Press, Chicago

Hui R, Wescoat JL Jr. (2018) Visualizing peri-urban and rurban water conditions in Pune District, Maharashtra, India. Geoforum 102:255–266

Jackson JB (1984) Discovering the vernacular landscape. Yale University Press, New Haven

James W (1909) A pluralistic universe. Longmans, New York

James W (1982 reprint) The varieties of religious experience. Penguin, New York

Lawson V (2007) Geographies of care and responsibility. Ann Assoc Am Geogr, 97:1–11

Licciardi G, Amirtahmasebi R (2013) The economics of uniqueness: investing in historic city cores and cultural heritage assets for sustainable development. World Bank, Washington DC

Lowenthal D (1999) The heritage crusade and the spoils of history. Cambridge University Press, Cambridge

Malpas J (ed) (2011) The place of landscape: concepts, contexts, studies. MIT Press, Cambridge MA

Mitchell, WJT (ed) (2002) Landscape and power, 2nd ed. University of Chicago Press, Chicago

Nassauer JI (1997) Placing nature: culture and landscape ecology. Island Press, Washington DC

Nussbaum MC, Sen A (eds) (1993) The quality of life. Clarendon Press, Oxford

Ogden CK, Richards IA (1938) The meaning of meaning: a study of the influence of language upon thought and of the science of symbolism. Harcourt, Brace, New York

Olwig K (2019) The meanings of landscape: essays on place, space, nature and justice. Routledge, London
Olwig K, Mitchell D (eds) (2009) Justice, power and the political landscape. Routledge, London
Patton KC, Ray B (2000) A magic still dwells: comparative religion in the postmodern age. University of California Press, Berkeley
Sauer C (1963) Land and life. Ed. John Leighly. University of California Press, Berkeley
Smith JZ (1982) In Comparison a Magic Dwells. In: Imagining Religion: from Babylon to Jonestown, pp 19–35. University of Chicago Press, Chicago
Smith JZ (1987) To take place: toward theory in ritual. University of Chicago Press, Chicago
Speaks J (2021) Theories of Meaning The stanford encyclopedia of philosophy, Edward N. Zalta (ed), https://plato.stanford.edu/archives/spr2021/entries/meaning. Accessed 20 Dec 2021
Spirn AW (2000) Language of landscape. Yale University Press, New Haven
Tweed TA (2006) Crossing and dwelling: a theory of religion. Harvard University Press, Cambridge
Wescoat JL Jr. (1999) Mughal gardens: the re-emergence of comparative possibilities and the wavering of practical concerns. In: Conan M (ed) Perspectives on garden histories. Dumbarton Oaks, Washington, DC, pp 107–126
Wescoat JL Jr. (2016) Between Garden and Geography: Landscape as an Emergent Construct in the Wider Middle East. In: Gharipour M (ed) Contemporary urban landscape design in the Middle East. Routledge, London, pp 22–44
Wolf S (2010) Meaning in life and why it matters. Princeton University Press, Princeton
World Bank (2014) Inclusive urban revitalization guidelines for Indian cities. Report. Washington DC

Preface and Acknowledgments

The idea of interlinkages and reciprocity between placemaking and cultural landscapes had been initiated under the aegis of the Asian Cultural Landscape Association (ACLA) under the leadership of (late) Sung-Kyun Kim (SNU Seoul, Korea) and Rana P. B. Singh (Varanasi, India) in the IFLA—Asia Pacific Region Congress of Landscape Architecture: 5–7 December 2011, at Seoul National University, Seoul, Korea, and followed up in the successive symposia organized under the aegis of ACLA (for the discourses on this development, see Chapter 2 in this volume). In course of this journey, friends and fellow practitioners from different parts of the world joined our walkways and actively performed their expositions and interactive synthesis in different ways and different contexts, illustrating examples from different corners of the globe. At one stage, it has been realized that only directly focused papers and also invited papers on the themes to be put together in preparing this anthology—a tribute and remembrance of Prof. Sung-Kyun Kim (1956–2020), and including one of his experiential posthumous papers that is shared with Rana P. B. Singh, i.e., '*Pung-su*: Evolving Cultural Landscapes and Placemaking in Korea' (Chapter 3 in this volume). After having the contract for this book, the world faced the pandemic situation of COVID-19, which hurdled the project at different levels over time. In the meantime, the then series editor of the Springer Nature's *Advances in Geographical and Environmental Sciences* Series (ISSN: 2198-3542), Prof. R. B. Singh (1955–2021) passed away. However, the project moved on slowly and reached its destination, with the support of the Series Editor Ms. Yosuke Nishida; we appreciate her sense of humankindness and help.

This book aims to examine the role of contemporary social-cultural issues in the context of UN-SDGs and associated cultural and sacred landscapes with an emphasis on awakening the deeper cultural sense in harmonizing the world and the role of society and spiritual system, drawing upon the multi-disciplinary and cross-cultural interfaces—all in the frame of 'FoE, Future of the Earth' and 'RWYC, Reconnecting With Your Culture', exemplifying from diverse examples. These papers will add a new dimension of cultural understanding in the broad domain of the emerging Human Geoscience, considered a key policy science for contributing toward sustainability/survivability science together with future earth initiatives. The chapters are arranged into three broad sections, i.e., the Framing, Placemaking: Identity and Exposition, and Cultural Landscapes: Emergence and Scenario. In the first introductory chapter, an appraisal and review of the trends of studying and practicing placemaking are presented, along with the perspectives for the visioning future.

The first and foremost thanks are due to Prof. James L. Wescoat Jr. (Emeritus Aga Khan Professor of Landscape Architecture, MIT, USA), who has been kind enough to write the 'foreword,' presenting a review of the framework and contents of the chapters and an analytical appraisal of the concept and use of the placemaking and cultural heritage. We have been fortunate enough to receive feedback, comments and suggestions, and guidance from all the contributors. However, let us mention the special support we received from the beginning of the frame in 2011 till today, from Ar Patricia M. O'Donnell, President of the ICOMOS IFLA International Scientific Committee on Cultural Landscapes; many of her concepts and constructs are tested and elaborated in different contexts.

Altogether there are twenty chapters, representing different case studies by in-dweller scholars and researchers concerned with their field studies. The countries represented include Australia, China, Korea, Finland, Georgia, India, Iran, Italy, Portugal, Turkey, the UK, the USA, and Zambia. In organizing thematic symposia on the broad theme of concern, we received support from the three international organizations, viz. International Geographical Union, IGU, Commissions C16.25 "Landscape Analysis and Landscape Planning", and C16.08 "Cultural Approach in Geography", and also from the ACLA (Asian Cultural Landscape Association). At the final stage, we have received support from the newly formed international movement RWYC, Reconnecting With Your Culture. Many of their inputs and visions are apprised and synthesized in the introductory Chapter (1). For a global and deeper level of synthesis and interfaces between culture and heritage, we are grateful to one of our senior guides D. Paul Schafer, Founder and Director of the World Culture Project (Canada). We are grateful to all the personnel and scholars for helping us in different capacities.

For the final setting and production, we are grateful to the production team at Springer Nature, especially Rini Christy Xavier Rajasekaran, the project coordinator. We are grateful to Ms. Yosuke Nishida, *Editor*: Earth Sciences & Geography, who has been a rational critic and a source of inspiration, and supported us in all the ways during the last two years of this project. Without their constant assistance and support, it would have not been possible to get the book released in the stipulated time.

Varanasi, India — Rana P. B. Singh
Pavia, Italy — Olimpia Niglio
Varanasi, India — Pravin S. Rana
October 2022, Vijaya Dashmi

Contents

Part I The Framing

1 **Interfaces Among Placemaking and Cultural Landscapes: Review and Appraisal** 3
Rana P. B. Singh, Olimpia Niglio, and Pravin S. Rana

2 **Vision and Exposition of Placemaking Under ACLA: Homage and Memorial Tribute to Sung-Kyun Kim** 23
Rana P. B. Singh and Olimpia Niglio

3 **Pung-su: Evolving Cultural Landscapes and Placemaking in Korea** 39
Sung-Kyun Kim and Rana P. B. Singh

4 **Cultural Landscapes: Integrating Culture and Nature to Uplift Global Sustainability Through the Lenses of the UN SDGs 2030 Agenda** 61
Patricia M. O'Donnell

5 **Branding the Image of Religious Heritage in India** 79
Samidha Mahesh Pusalkar

6 **Cultural Landscapes: Essence and Application Perspectives in Georgia** 95
Nodar Elizbarashvili, Sandro Gogoladze, Gela Sandodze, Lado Grigolia, and Teona Qobalia

Part II Placemaking: Identity and Exposition

7 **Ayodhya, India: Placemaking and Transformation of Historic Urban Landscape** 113
Rana P. B. Singh and Sarvesh Kumar

8 Placemaking Approach in Revitalizing Cultural Tourism in Temple Towns: Case of Melukote, Karnataka ... 135
Surekha Ramineni, Monalisa Bhardwaj, and H. Tejaswini

9 The Challenges of Integrated Conservation and Development in Historic Rural Landscapes; Case Study: The Historic Villages of East Azerbaijan, Iran ... 153
Farnaz Faraji and Elham Masoumi

10 Chinese Cities as Sacred Landscapes: The Case of the Capitals of the Ming Dynasty ... 173
Norma Camilla Baratta, Mariacristina Giambruno, Giulio Magli, and Jing Zhou

11 Archetypal Characters of Sacred Places and Cultural Landscape: Sansa, Korean Buddhist Mountain Monasteries Comparing Cistercian Architecture ... 189
Hee Sook Lee-Niinioja

12 Placemaking of the Barotse Cultural Landscape, Zambia ... 209
Chilangwa Chaiwa

13 Pandemics, Situational Under-Tourism, and the Search for Sustainability ... 227
Daniel H. Olsen

14 Post-COVID-19 Strategies: Cloisters as Urban Oasis to Reconnect Heritage and Communities ... 239
Carlo Francini, Alessia Montacchini, and Tatiana Rozochkina

Part III Cultural Landscapes: Emergence and Scenario

15 Archaeological Sites in Northern Japan: Interfacing Landscape and Sacred Rituals ... 257
Olimpia Niglio

16 (Dis)Continuous Spiritual Significance and Ritual Use of Çatalhöyük and Ephesus Archaeological Landscapes: A Comparative Historical Overview ... 275
Zeynep Aktüre

17 Cultural Landscape Modelling: A Practical Approach for Evaluating Sustainability. The Historic Village "Maymand" in Iran ... 295
Leila Kowkabi

18 Reuse of Unoccupied Religious Monuments for Tourist Accommodation: Santa Maria da Ínsua (North of Portugal) ... 317
Rolando Volzone, Soraya Genin, João Estêvão, and Mafalda Patuleia

19 Survival of Heritage in Majuli in the Context of Climate Change 335
M. Satish Kumar

20 '*Pūch*' as an Institution for Maintaining the Cultural Landscape of the Kullu Valley 351
Ashwini Pethe and Kiran Shinde

Correction to: Reuse of Unoccupied Religious Monuments for Tourist Accommodation: Santa Maria da Ínsua (North of Portugal) C1
Rolando Volzone, Soraya Genin, João Estêvão, and Mafalda Patuleia

About the Author

The responsible series editor of this book is Prof. Yukio Himiyama (Emeritus, Hokkaido University, Japan).

Editors and Contributors

About the Editors

Rana P. B. Singh has been a professor of cultural landscapes and heritage studies in the Department of Geography, the Institute of Science, Banaras Hindu University, Varanasi, India. He is the president of the Asian Cultural Landscape Association (ACLA), the chief coordinator of the International Council on Monuments and Sites (ICOMOS) National Scientific Committee "Cultural Landscapes", contributing member of the ICOMOS-IFLA (the International Federation of Landscape Architects) International Scientific Committee "Cultural Landscapes", an expert and voting member of the ICOMOS International Scientific Committee "Places of Religion and Rituals" (PRERICO), and president (Asia) of the Reconnecting With Your Culture (RWYC). For more than four decades, he has been studying, lecturing, and promoting cultural and sacred landscapes and pilgrimages in India, also has done field studies in Japan, Italy, Sweden, Korea, and China, and delivered lectures in almost all parts of the world. His publications include 43 books and anthologies, and 340 research papers.

Olimpia Niglio is Professor of architectural restoration, Department of Civil Engineering and Architecture, University of Pavia, Italy. Since 2021, she is a permanent visiting professor at Hosei University, Tokyo, Japan. She is also Research Fellow at Kyoto University Graduate School of Human and Environmental Studies, Japan. She was previously Professor at Hokkaido University (Japan), Universidad de Bogotá Jorge Tadeo Lozano (Colombia), and at International School University of Ibagué (Colombia). She studied at the University of Naples "Federico II" where she also obtained the Ph.D., Executive Master at Business School Sole24ore and Post Ph.D. (Miur) in Conservation of Architectural Heritage. She is the director of the EdA Esempi di Architettura International Research Center, Vice President of ICOMOS PRERICO (International Committee Monuments and Sites), and also the Vice President of the Asian Cultural Landscape Association (ACLA). Her publications include 9 books and more than 300 research papers published in Italian and English.

Pravin S. Rana is an assistant professor in tourism management in the Faculty of Arts, Banaras Hindu University, Varanasi, India. He has teaching experience of more than 20 years and is specialized in religious tourism and pilgrimages. He is the secretary of the Asian Cultural Landscape Association (India). He has contributed essays in research *series* and *handbooks* published by Routledge (London), CABI International (Oxfordshire UK), and Springer Nature (Cham & New York, Singapore). He has published 3 books, including *Pilgrimage Tourism: A Study of Varanasi Region* (SRME Publishers, 2014) and *Banaras Region, A Spiritual and Cultural Guide* (Indica Books, 2020), and 28 research papers. From August 2005 to July 2009, he served as Manager of the Children Programme at the World Literacy of Canada India Office. He has delivered seminars in Norway, Canada, Singapore, Malaysia, and Korea. He is a member of Forum UNESCO - University and Heritage-International network (FUUH), Spain.

Contributors

Zeynep Aktüre Independent Scholar, Izmir, Turkey

Norma Camilla Baratta Department of Architecture and Urban Studies, Politecnico di Milano, Milan, Italy

Monalisa Bhardwaj Ramaiah Institute of Technology, Bangalore, Karnataka, India

Chilangwa Chaiwa National Heritage Conservation Commission, Lusaka, Zambia

Nodar Elizbarashvili Department of Regional Geography and Landscape Planning, Tbilisi State University, Tbilisi, Georgia

João Estêvão DINÂMIA'CET-Iscte | Iscte-University Institute of Lisbon, Lisbon, Portugal

Farnaz Faraji School of History, Classics and Archaeology, Newcastle University, Newcastle Upon Tyne, UK

Carlo Francini Florence World Heritage Office and Relations with UNESCO Office, Florence, Italy

Soraya Genin DINÂMIA'CET-Iscte | Iscte-University Institute of Lisbon, Lisbon, Portugal

Mariacristina Giambruno Department of Architecture and Urban Studies, Politecnico di Milano, Milan, Italy

Sandro Gogoladze Department of Regional Geography and Landscape Planning, Tbilisi State University, Tbilisi, Georgia

Lado Grigolia Department of Regional Geography and Landscape Planning, Tbilisi State University, Tbilisi, Georgia

Sung-Kyun Kim Landscape Architecture, Seoul National University, Seoul, Korea

Leila Kowkabi Faculty of Architecture and Urban Planning, University of Art, Tehran, Iran

M. Satish Kumar Department of Geography, Archaeology & Palaeoecology, School of Natural and Built Environment, Faculty of Engineering and Physical Sciences, Queen's University Belfast, Belfast, BT7 1NN Northern Ireland, UK

Sarvesh Kumar Department of Geography, D.D.U. Gorakhpur University, Station Road, Gorakhpur, UP, 273009 India

Hee Sook Lee-Niinioja Piritanaukio, Helsinki, Finland

Giulio Magli Department of Mathematics, Politecnico di Milano, Milan, Italy

Elham Masoumi Architecture Department, College of Fine Arts, Tehran University, Tehran, Iran

Alessia Montacchini Department of Architecture (DIDA), University of Florence, Florence, Italy

Olimpia Niglio Department of Engineering and Architecture, University of Pavia, Pavia, Italy;
Faculty of Engineering and Design, Hosei University, Tokyo, Japan;
Cultural Landscapes & Heritage Studies, Banaras Hindu University, Varanasi, India;
Architectural Restoration & Cultural Heritage, University of Pavia, 27100 Pavia, PV, Italy

Patricia M. O'Donnell ICOMOS IFLA International Scientific Committee on Cultural Landscapes, PO Box 321, 501 Lake Road, Charlotte, VT, 05445-0321 USA

Daniel H. Olsen Department of Geography, Brigham Young University, Provo, UT, USA

Mafalda Patuleia TRIE | Universidade Lusófona, Lisbon, Portugal

Ashwini Pethe School of Architecture, MIT ADT University, Pune, MR 412201 India

Samidha Mahesh Pusalkar Preservation of Architectural Heritage, 36° Cycle, Department of Architecture and Urban Studies, Politecnico di Milano, Milano, Italy

Teona Qobalia Department of Regional Geography and Landscape Planning, Tbilisi State University, Tbilisi, Georgia

Surekha Ramineni Ramaiah Institute of Technology, Bangalore, Karnataka, India

Pravin S. Rana Tourism Management, Faculty of Arts, Banaras Hindu University, Varanasi, UP, 221005 India

Tatiana Rozochkina Department of History, Archeology, Geography, Fine and Performing Arts (SAGAS), University of Florence, Florence, Italy

Gela Sandodze Department of Regional Geography and Landscape Planning, Tbilisi State University, Tbilisi, Georgia

Kiran Shinde Department of Social Inquiry, La Trobe University, Melbourne, VIC, VIC-3552 Australia

Rana P. B. Singh Cultural Landscapes & Heritage Studies, Banaras Hindu University, Varanasi, India;
Asian Cultural Landscape Association (ACLA), Varanasi, India;
Usha Apartment (Flat 101-102), Lane 5—Janaki Nagar Colony, Flyover BLW end—Badi Patia Road, PO: Bazardiha, Varanasi, 221106 UP, India

H. Tejaswini Ramaiah Institute of Technology, Bangalore, Karnataka, India

Rolando Volzone DINÂMIA'CET-Iscte I Iscte-University Institute of Lisbon, Lisbon, Portugal

Jing Zhou Architecture Department, Human Settlement and Civil Engineering School, Xi'an Jiaotong University, Xi'an, China

Part I
The Framing

Chapter 1
Interfaces Among Placemaking and Cultural Landscapes: Review and Appraisal

Rana P. B. Singh, Olimpia Niglio, and Pravin S. Rana

Abstract The concept of placemaking is used in comprehensive and complex ways in different contexts of disciplinary boundedness and application in prospective planning and pathways to sustainable development. There is no mutual agreement on a common definition or its attributes. This concept is also taken as an approach and procedure too. Among all the concerns, at the core is 'place', which synthesises the multidimensional system of interdependency and interconnectedness. Culture is discussed as a vital force in practising placemaking. In the recent era of the paradigm shift, culture is accepted as the fourth pill that envelops the three basic pillars of SDG, viz. society, environment, and economy, as provoked in the Mexico Declaration in 1982. Through the recently developed mission of Reconnecting With Your Culture (RWYC), a new march is in the move to enhance quality and deeply-rooted education and dialogues making a bridge between locality and universality. The attributes of cultural landscapes and cultural heritage are the essential ingredients of placemaking. Ultimately, it is accepted that placemaking belongs to everyone: its message and mission are bigger than any one person or organisation. These are illustrated here within a frame of appraisal and review of the recent trends in understanding and studying placemaking.

Keywords Placemaking · Cultural landscapes · Cultural heritage · Culture · Nature-culture interface · Sense of place · Spirit of place · Global understanding

---

R. P. B. Singh (✉) · O. Niglio
Cultural Landscapes & Heritage Studies, Banaras Hindu University, Varanasi, India
e-mail: ranapbs@gmail.com
URL: https://bhu-in.academia.edu/RanaPBSINGH/Papers

O. Niglio
Architectural Restoration & Cultural Heritage, University of Pavia, 27100 Pavia, PV, Italy
e-mail: olimpia.niglio@unipv.it

P. S. Rana
Tourism Management, Faculty of Arts, Banaras Hindu University, Varanasi, UP 221005, India
e-mail: psrana@bhu.ac.in

R. P. B. Singh et al. (eds.), *Placemaking and Cultural Landscapes*,
Advances in Geographical and Environmental Sciences,
https://doi.org/10.1007/978-981-19-6274-5_1

> In our everyday lives, places are not experienced as independent, clearly defined entities that can be described simply in terms of their location or appearance. Rather they are sensed in a chiaroscuro of setting, landscape, ritual, routine, other people, personal experiences, care and concern for home, and in the context of other places. (Relph 1976: p. 28)

1.1 Placemaking: A Vision and March

The interconnectedness between placemaking and cultural landscapes is a general concern in the multidisciplinary frame that consists of many points of view on the common impacts of social, economic, cultural, and rational jurisprudence planning, wellbeing, and overall development. Moreover, concerned with the composite interactions between development and environment, its concern is to analyse the ways, paths, and implications for framing sustainability in all the social and economic activities that are part of activating the advancement of economy and culture. Most of the human transformations are part of the activities of 'placemaking'—considered to be the intentional creation of a sense of place for networking and functioning of Nature-Culture interaction reciprocatively. Moreover, placemaking is an art engaging in creating people and getting themselves interconnected through a culture where 'engagement' is a process and 'cultural landscape' is the product. A comprehensive and intentional needs to consider the interaction, three key aspects of a destination to fully approach placemaking consists of: (1) its natural environment and associated built landscape and their reciprocate interdependency (its *habitatscape*); (2) the people who are a living culture in that landscape (its *peoplescape*), and constantly making them alive and revive through rituals, traditions, and performances; and (3) the mental image and belief systems that people hold of the destination (its *mindscape*), through manifestation, archetype representation, and imposed mythologies and folklores. Resultantly, any cultural landscape illustrates a holistic, complex, and interdependent mixed-up of Nature-Culture symbiosis and results in the formation of an 'identity' and 'belongingness'—represented with uniqueness and distinctiveness—however, also having the instinct of commonality and universality. This also "means that the[ir] identity may be more difficult to change. A placemaking approach can help to understand the existing identity of a place and how it came to be, and it can be used to reshape that identity to create new opportunities" through human interaction, interfaces, and imitations (Lew 2021: p. 126). Similarly, the 'place-framing approach harnesses socio-spatial relations and networks and 'socio-spatial positionality' (Martin 2003: p. 747).

In terms of existentiality and expositions, six major components of the concept of place are identified by Lukermann (1964) that reflect the whole spectrum of placemaking:

1. *The idea of the location*—as it relates to other things and places, is fundamental, and can be described in terms of internal characteristics (site) and external connectivity to other locations (situation); thus, places have spatial expansion possessing an inside and outside.

2. *Integration of elements of Nature and Culture*—each place involves its order, its special ensemble making the place a unique entity, which distinguishes it from the next place.
3. *Places are interconnected*—although every place is unique, however, they are interconnected by a system of spatial interactions and transfers; thus, they become part of the circulatory network.
4. *Places are localised*—they are parts of a larger territory and maintain focus in a system of localisation.
5. *Places are emerging or becoming*—over time historical and cultural changes add new elements superseding the older ones, of course having the essence in different forms, represented with a distinct historical component.
6. *Places have meaning*—they are characterised by the beliefs and faith system of human beings, developed through human consciousness, which resultantly makes the personality of the place.

Elaborating on the structure of understanding 'place', Lukermann (1964) has explained places as "complex integrations of nature and culture that have developed and are developing in particular locations, and which are linked by flows of people and goods to other places. A place is not just the 'where' of something; it is the location plus everything that occupies that location seen as an integrated and meaningful phenomenon" (Relph 1976: p. 3). However, he does not distinguish among similar concepts like 'region', 'area', 'location', and 'territory'; but, over time the acceptance of the concept of 'placemaking' is conceived comprehensively. Places certainly matter to us; therefore, "if we are at all concerned about the psychological consequences and moral issues in uprooting and increasing geographical mobility and placelessness, then we must explore the possibility of developing an approach for making places self-consciously and authentically" (Relph 1976: p. 147). All the place-related particularities affect place meanings including inherent (aesthetic) meanings, instrumental (goal-directed) meanings, cultural (symbolic) meanings, and individual (expressive) meanings, and over time together they make a complex whole (cosmality) that human beings conceive and perceive in the form of a sense of place (*genius loci*).

In the process of Nature-Culture interfaces, humans create places by imposing their inherent or possessed unique manifestation to certain parts of the geographic space and its environs. According to Tuan (1979: p. 410), "People demonstrate their sense of place when they apply their moral and aesthetic discernment to sites and locations. Modern man, it is often claimed, has lost this sensitivity. He transgresses against the *genius loci* because he fails to recognize it, and he fails to recognize it because the blandness of much modern environment combined with the ethos of human dominance has stunted the cultivation of place awareness". Humans receive and re-create geographic space through their imagery perceptions, primarily sight, sound, and their movement and process of settling down. In the process of transformation, as space is being re-created in the form of places, through the selection with suitability, based on preferences and adjustability, of certain items or locations, and thus resultantly giving them meanings, values, eulogies, and senses of goodness and badness. This is the process of placemaking, i.e. from top to down: Nature to Culture.

In a broad perspective, placemaking should be expressed as the holistic process of producing, designing, crafting, creating, or otherwise bringing into being the material and experiential elements of a place landscape. Says Lew (2021: p. 112), "Everyone does place-making (i.e. everyone is a participant in placemaking) through the continuing process of expressing their preferences in the geographic space they inhabit. This includes humans, non-human animals/wildlife, plants/vegetation, buildings, economic and business structures, and activities, recreation facilities and activities, and much more". In Japan, for rural landscapes that maintained historical-cultural continuity, the word *satoyama* is used, which refers to how every attribute of agricultural landscapes is interrelated to each other—forming a complex ecosystem, thus leading to "society in harmony with nature"; nevertheless, in the metaphysical sense, it also refers to the transition and coexistence of human-made landscapes and the natural world. However, there is no such expressive word for this in English. Of course, Massey (2005: pp. 149–152) suggested "*throwntogetherness*", which includes the idea that things that seemingly look unrelated to each other are related, though perhaps at levels that are not always obvious. This sense in the Oriental world is to be compared with archetypal representation or cosmic whole.

In the era of the cybernetic age, the world is rapidly shrinking under the expanding threat of economic, technological, and cultural globalisation. One of the characteristics of globalisation is the homogenisation (or flattening) of differences between places (Hall and Lew 2009). Globalisation is considered bad when local cultures and landscapes are replaced by global ones, resulting in 'placelessness', which describes both an environment without significant places and the underlying attitude which does not acknowledge significance in places. It reaches back into the deepest levels of place, cutting roots, eroding symbols, replacing diversity with uniformity and experiential order with conceptual order" (Relph 1976: p. 143). If a sociocultural landscape is indistinguishable from those found anywhere in the world, then it is considered 'placeless', lacking a 'sense of place', and inauthentic (Tuan 1977). Further, places may be conceived as "public symbols of the field of care, but the power of the symbols to create place depends ultimately on the human emotions that vibrate in a field of care" (Tuan 1979: p. 421). By exposing place "as a multifaceted phenomenon of experience and examining the various properties of place, such as location, landscape, and personal involvement, some assessment can be made of the degree to which these are essential to our experience and sense of place. In this way, the sources of meaning, or essence of a place can be revealed" (Relph 1976: p. 29).

Even after having the passage of a long march, adopting and being getting adapted by several attributes and supporting mechanisms, the study of "place" has maintained its identity as a central point, with a notion that 'place' refers to a synthesis of land and people. Under the umbrella of UNDESD (United Nations Decade of Education for Sustainable Development 2005–2015, cf. Higgitt 2009), the study of 'place' as an organic reality has been conceived as a core concern for the education of sustainable development (see Fig. 1.1). Place synthesises "varying dimensions of territorial concern, evolutionary change, landscape scenarios, society and its systems,

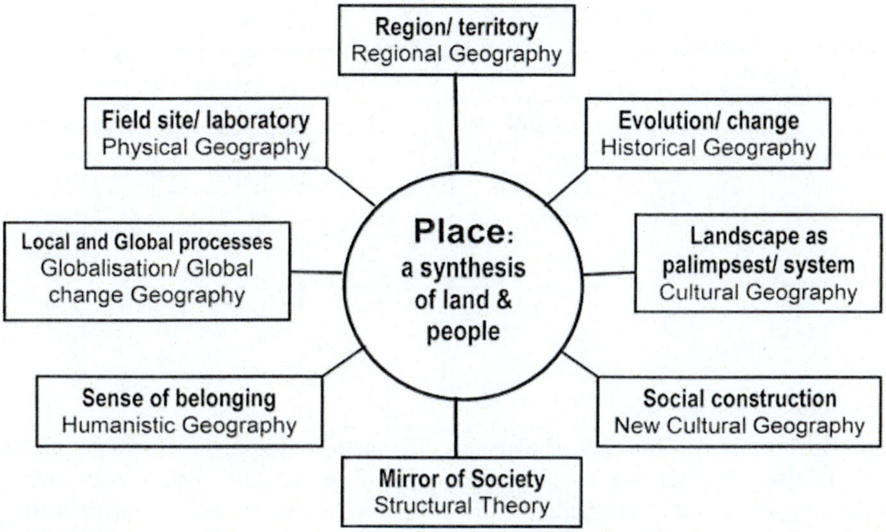

Fig. 1.1 Place as core of education of sustainable development (after Singh 2021: p. 5, reproduced with permission)

sensibility, local–global interfaces, and field studies with co-sharedness together with participatory observations and experiential expositions" (Singh 2021: p. 5).

The concept of the place consists of relational, emotional, cultural, and other structural forms like economic, political, social, educational, etc., and their associated attributes. This conception attempts to 'retain appreciation and understanding of the importance, of the uniqueness in the form of distinctiveness, and at other end consists of the interdependency that led to generality in the form of relativeness'. These two characteristics further provide frames to visualise the place from either 'inside' or 'outside', 'introvert' or 'extrovert', or even altogether in a complex whole. In Indian thought, it is called "mosaicness" where evolved human consciousness, explained with the notion of 'place attachment', what in geographic tradition Wright has introduced the idea of '*geopiety*'—"emotional piety aroused by awareness of terrestrial diversity of the kind of which a geography is also a form of awareness" (see Cosgrove 2000: p. 308, also Tuan, 1976). This is further elaborated by Yi-Fu Tuan (1977), who describes another type of place attachment where a reverential and compassionate connection is developed between a person and their 'terrestrial home' (*cosmos*). In Asian culture, this is narrated as the notion of 'sacredscapes', which consists of several sub-sets of sacredscapes overlapping in vertical and horizontal spaces; they "are the result of cultural developments and have varied multidimensional levels of significance" (Singh 2011: p. 10). In a way, we are lacking a lower-level realisation of the importance of place attachment, which further accelerates the process of negating the moral domain and ethical values in life (see Higgitt 2009). Celebrating 'learning to live together sustainably' under the aegis of the United Nations Decade of Education for Sustainable Development (2005–2014), "understanding the

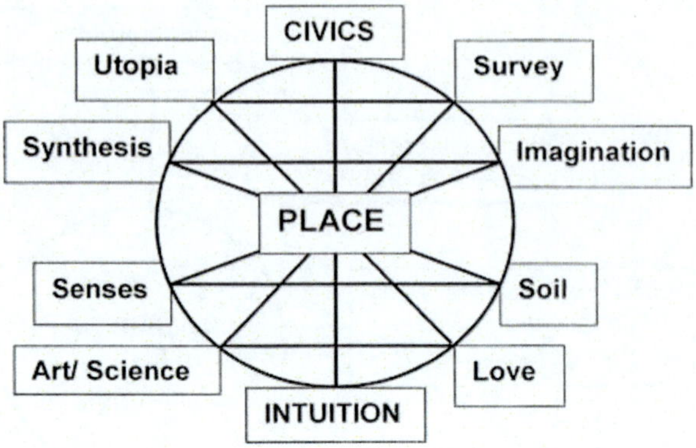

Fig. 1.2 'Civics-Place-Intuition' interfaces after Patrick Geddes (after Singh [2021: p. 6], reproduced with permission)

interdependency and fragility of planetary life support systems, and making better global citizenship that promotes more humane and peaceful life, and develops mass awakening for universal brotherhood, there is a need for the innovative and interdisciplinary courses at the school level that enhance better understanding and reverentially preserving those values of the past that help the humanity in achieving the basic goals of the UNDESD" (Singh 2021: p. 6). It is now realised that global understanding is an essential condition of the twenty-first century, which can be achieved through deeply-rooted education.

The importance of motivating people to make the right choice starting from and getting associated with the "place" was provoked by Patrick Geddes [1854–1932], who termed it as a 'moral issue' (interfacing civics and intuition) and was the concern of cultural conditioning for education, especially in modern terminology as '*ecoliteracy*' (see Fig. 1.2, for details see Singh, 2021). He further provokes this frame of knowledge at contemporary discourses of justice and equity that have interlinking attributes like survey (he means participation and direct observation), imagination, soil (ecological attachment), love, art/science, senses, syntheses, and utopia (Fig. 1.2; see Meller 1990). By emphasising connection and connectedness, placing function alongside aesthetics, and focusing on the stories told about and around the artwork, Fleming (2007) has set an effective standard for making places—the highest art of placemaking; of course, his writing does not clarify as to how a place possesses the instincts of the past and speak to its past, visualise and maintain in the present, and also vision and projects as to how to reach into the future. However, his ideas support the vitality of conscious art of renewal, continuity, and liveability.

1.2 Culture and Global Understanding

Since all the problems are generated from the mind, the solutions will also come from the mind through cultural conscience and revealing. Said Legrand (2021: p. 4), "all the problems we face come from our minds and hearts. There also lie the solutions. As our problems grow more complex and develop unpredictably, we more than ever need to come back to this simple truth". However, this needs a constant practice of awakening the mind on the path of deeply-rooted education and understanding the cultural interconnectedness, and also awakening and rejuvenating the human conscience to actively be part of the universal family as unveiled in the ancient sacred texts of Hinduism, i.e. *Udāracharitanam Vasudhaiv Kutumbakam* (*Mahā Upanishad*, 6.71–72; also, in the *Hitopadesha*, 1.3.71). We should realise global interdependence as the ultimate reality, thus making a sustainable global order through optimal international governance (cf. Lopez-Claros et al., 2020), through a new dynamic world order and global balanced while walking on the path of ethics, values, and virtues (cf. Stückelberger 2020). This could be made operative to move from economic to cultural age—linking cultural conscience and holistic understanding in the light of sustainable development (cf. Schafer 2022). We can learn from indigenous society and their traditional knowledge that culture and cosmos are interconnected in creating conscience from civilisational continuity in reviving the harmonious path we should reconcile our minds and hearts through spiritual awakening (cf. Legrand 2021: p. 259). Ultimately, "if nothing in all the universe is frail as man, nothing likewise is so divine as he!" (Aurobindo 1979: p. 14).

In this era of cybernetics, the commonly accepted paradigm shift is perceiving and practising "Seeing culture as part of something larger to seeing culture in holistic terms as the whole or total way of life so timely, valuable, and indispensable" (Schafer 2022: p. 257). He further adds (ibid.), "Nothing confirms the power, potential, and potency of the holistic perception of culture better than the realisation that culture and cultures are the change agents that are urgently needed during this difficult period in human history. How often have we heard people and organisations say in recent years that it is necessary to 'change the culture'?". The world certainly needs 'wholesale change that is fundamental, foundational, systematic, and systemic'. One has to remember that, "no place remains unchanged except heaven, hell, and outer space, and none of those is fit for human beings" (Lynch 1976: p. 72). It is also debated that "the culture constitutes the set of intellectual knowledge that, acquired through study, reading, experience, and the influence of the environment and reworked in a subjective, critical, and autonomous way becomes a building block of personality" (Niglio 2022: p. 4).

The UNESCO Declaration on Cultural Policies in Mexico City in 1982 highlights empathetically that it is "culture" that gives man the ability to reflect upon himself; it mentions that "It is the culture that makes us specifically human, rational beings, endowed with a critical judgement and a sense of moral commitment. It is through the culture that man expresses himself, becomes aware of himself, recognises his incompleteness, questions his achievements, seeks untiringly for new meanings, and creates works through which he transcends his limitations" (UNESCO 1982: p. 1). Without culture, we cannot build the future of the countries and a better world, because without culture we cannot foster "ecological citizenship" (Duxbury et al. 2017).

The classical model of sustainable development (Brundtland 1987—WCED; Fig. 1.3) has been re-assessed over time and apprehended under the umbrella of the

Universal Forum of Cultures 2007, emphasising the base of Developing Civility—culture, language, and religions for inter-faith dialogue. Three models for culture and sustainable development have been structured engulfing sustainability (Fig. 1.4): So, Social; En, Environment; Ec, Economic, where the three roles of culture in sustainable development (the three circles representing the three pillars, So, En, and Ec) are envisaged. In this model, culture is *added* as the fourth pillar (a), culture *mediating* between the three pillars (b), and culture as the *foundation* for sustainable development (c). The arrows in the model indicate the ever-changing dynamics of culture and sustainable development (Dessein et al. 2015, p. 29).

In the recent thought process of sustainable research and its implication policy, the use of "culture" is implicitly visible and is of vital importance, where the following three basic roles of culture have been discerned (see Fig. 1.4, modified after Dessein et al., 2015: p. 29):

1. *Culture in sustainable development* vis-à-vis social, environmental, and economic attributes or pillars. This way cultural aspects need to be considered in the development processes alongside the above attributes to fulfil the criteria of sustainability.
2. *Culture for sustainable development.* Culture is considered a mediating force between the three classical pillars of sustainability (society, environment, and

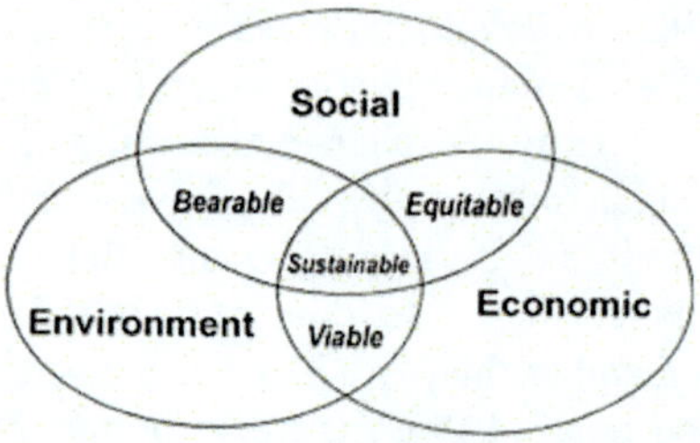

Fig. 1.3 Basic frame of sustainability (after Brundtland Report [1987])

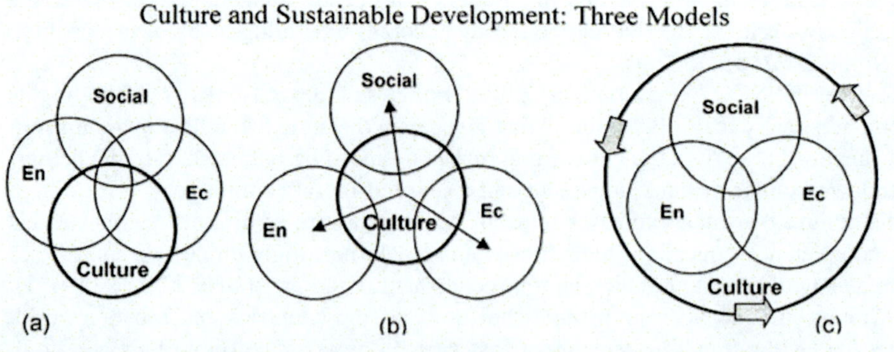

Fig. 1.4 Culture and sustainable development: three models—(a) culture *in* sustainable development, (b) culture *for* sustainable development, and (c) culture *as* sustainable development (after Dessein et al. [2015, p. 29])

economy). In other words, it is accepted that culture processes, facilitates, and translates sustainable development, and therefore, cultural aspects should always be present in sustainability assessments, policies, and planning.

3. *Culture as sustainable development.* Implying that culture is an overarching concept, it contains and influences social, environmental, and economic actions within sustainable development. Sustainability needs to be embedded in the culture, and cultural transition is needed on our way to a more sustainable society.

For investigating the multifaceted roles of culture in sustainable development, the above three basic role models of culture provide analytical "lenses", helping clear understanding and making action plans. In each of the above frames, in regional context, culture may have different meanings, manifestations, and connotations *in*, *for*, or *as* sustainable development; however, it would always be vital energy. Furthermore, they also point to different positions of culture in society and to different strategic implications: Culture as a fourth pillar or the mediating bridge is linked with a better pondering of culture in any decision-making related to sustainable development, while culture as a new paradigm of sustainability suggests a cultural transition towards a society where sustainability is embedded in culture and makes a platform for better future.

In a recent study by Singh et al. (2020), within the above perspective, the two - holy-heritage cities of India, Ayodhya and Varanasi, are selected for illustrating their inclusive heritage development strategies and interlinking SDGs, especially Target 11.4, i.e. (i) Vision: strengthen efforts to protect and safeguard the world's cultural and natural heritage; and (ii) Project Indicator: share of the national (or municipal) budget which is dedicated to the preservation, protection, and conservation of national cultural heritage including world heritage sites, taking in view the local conditions, policies, and orientation.

A recently formed mission of global awakening, viz. RWYC—Reconnecting With Your Culture—is a visionary mission of awakening the youth to prepare for the global understanding and environmental conscience, human services, and preserving our cultural heritage in the cosmic frame of Nature-Culture interfaces through the quality and deeply-rooted education and dialogues, i.e. a march re-appraising interconnectedness between locality and universality—holiness to wholeness (cf. Niglio 2021). This is a call from eternity (humanity) to meet sublimity (divinity); through RWYC, we pray:

Let's believe CULTURE will save our common heritage.
Let every day we work together, also realise together.
Let's follow on, proceed on, and march on this sacred path.
Let's join hands in helping the universal community to realise.
Let's have dreams with action, vision, and be into action.
Let our emotions be awakened through our inner light.
Let this path awaken humanity to meet with sublime divinity.
Let our CULTURE be a vital force to feel God in all life forms.

The RWYC project is also planned to apply a deeply-rooted educational method to introduce among schools, colleges, and universities, the study of history (e.g. the Big History) and cultural heritage—two fundamental themes to achieve an ethically humane world. "The project is based on the UN-issued 2030 Agenda, a program to generate sustainable development. For this initiative, a particular emphasis is placed on point number 4 on "Quality education", supplemented by the principles of inclusion and equity. This item of the 2030 Agenda aims to achieve important and equitable quality education, promoting the opportunity for lifelong learning for all" (cf. Niglio 2021: p. 3). In creating the environmental design, the value of creating a *place* ('placemaking') gives the sense to think locality as a memorable and specific character to which the dweller or users may attach himself and his thoughts and feelings (Banerjee and Southworth 1991: p. 633); this in Indian metaphysics is narrated as the march from the 'realisation' (*anubhava*) to the revelation (*anubhuti*).

1.3 Cultural Landscape: Enveloping Heritage

According to UNESCO, 'cultural landscape' enfolds a variety (distinct and similar) of manifestations of the reciprocated network between humankind and the surrounding natural environment. Cultural landscapes—ranging from cultivated terraces on lofty mountains, and gardens, to sacred places—symbolise the creative genius, social development, and imaginative and spiritual vitality of human beings. Cultural landscapes represent the "combined works of nature and humankind"; thus, they are part of our collective identity, expressing a long and intimate relationship between peoples and their natural environment (see Singh 2010: p. 25); cultural landscape is a '*Design with Nature for Humankind*'. This way 'cultural landscapes' represent visionary symbiosis and interpretive synthesis. As the established notion of 'cultural landscapes' is inspired by the belief that the preservation and protection of globally, nationally, regionally, and locally significant cultural landscapes, both designed and vernacular, are critical to sustaining the continuum of land use and history across generations. Therefore, protection of these cultural landscape resources offers inspirational values and an appreciation of past and present ingenuity, accomplishments, hardships, and hopes, as well as insight into future land use, design solutions, and maintenance of heritagescapes.

Since Schlüter's first (1906) formal use of the term, and Sauer's effective promotion of the idea, the concept of 'cultural landscapes' has been variously used, applied, debated, developed, and refined within academia (see Martin 2005: pp. 175–176), and when, in 1992, the UNESCO World Heritage Committee elected to convene a meeting of the 'specialists' to advise and assist redraft the Committee's Operational Guidelines to include '*cultural landscapes*' as an option for heritage listing properties that were neither purely natural nor purely cultural in form (i.e. 'mixed' heritage) (cf. Fowler 2003). They are today's concern of conservation and revival of aesthetical values aiming to serve global integrity and feeling of interrelatedness (WHC 2003).

It was Carl O. Sauer [1889–1975], a human geographer at Berkeley School, who was probably the most influential in promoting and developing the idea of *cultural landscapes* (for a critique see Mitchell 2000: 27–28). Sauer was determined to stress the agency of culture as a force in shaping the visible features of the Earth's surface in delimited areas (Singh 2013). Within his definition, the physical environment retains a basic platform, as the medium with and through which human cultures act that resultantly forms a 'cultural landscape' (Sauer 1925/1963: pp. 337, 343; see Fig. 1.5): "The cultural landscape is fashioned from a natural landscape by a cultural group. Culture is the agent, the natural area the medium, the cultural landscape is the result". Cultural landscapes are an imprint possessing continuity of human history and representing a closely woven net of the interrelationships between people and their natural environment and are thus fundamental to the identity of the people. In northeast India, these are well represented in the cultural traditions, folklores, and nomenclatures of rainfall characteristics and vegetal association (see Taher 1990). In addition, they tell us the story of a people and how they have transformed the natural landscape into the humanised landscape (see Bandarin and van Oers 2012: 10). Thus, this dialogue is based on the method of interpreting heritage holistically, and it argues that there is a mutual dependency and reciprocity between landscape and culture (Bridgewater and Bridgewater 2004).

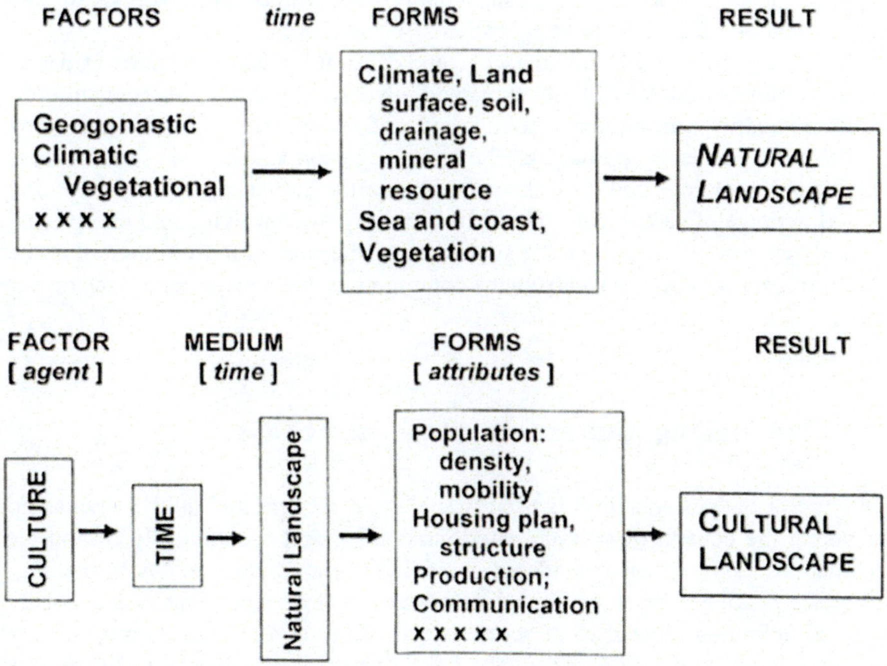

Fig. 1.5 Interfacing natural and cultural landscapes (after Sauer 1925/1963: pp. 337, 343)

Schein (1997), while seeking to retain the identification of cultural landscape with the 'tangible and visible scene', draws upon Massey's (1991) idea of places as 'moments' in a continuing networked process of social relations that stretch across space: 'Landscapes are always in the process of "becoming", no longer reified or concretised—inert and there—but continually under scrutiny, at once manipulable, always subject to change, and everywhere implicated in the ongoing formulation of social life' (Schein 1997: p. 662). That's how "with the introduction of UNESCO's framework of intangible cultural heritage, the notion of cultural landscape has been changed into a more comprehensive way" (see Singh 2010: p. 29).

Shackley describes a cultural landscape as "an integrated complex of cultural and natural resources, whose values derive from their physical quality, as well as from associated human endeavours and traditions" (2001: p. 139). In the Indian context, the use of literature testing the spatial, landscape, and lifeways issues is a rich resource but rarely tapped yet (cf. Singh 2004). It is rightly provoked that "the concept of the cultural landscape and its conservation methodology is a quite effective and appropriate approach to heritage management in South Asia, due to the region's multi-layered and complex eco-cultural heritage. …. [Such] strategies, however, should be based on the understanding that cultural landscapes are evolving and dynamic living systems whose significance is rooted in the interaction of people with their land (*since the ancient past*), and that moderating the rate and nature of change in those interactions is the sustainable way towards the conservation and management of cultural landscapes" (Silva 2017: p. 268).

The concepts of the cultural landscape and cultural heritage are the two sides of the same frame—however, vividly explained in varying contexts, different cultures, disparate situations, disciplinary boundaries, and so on. After all, "understanding cultural heritage means approaching the knowledge and valuation of communities, their history, traditions and then, the symbolic, territorial, ancestral, landscape and cultural heritage of each nation. Thus, the new way of conceiving and interpreting Cultural Heritage encompasses the social sectors because it is from there that the process of creativity begins and from the communities, we must start research on our heritage" (Niglio 2021: p.1).

1.4 Placemaking, Cultural Heritage, and SDGs

The concept of 'placemaking' is used in wide perspectives and quite a diversified manner in the context of socially, environmentally, and economically responsive explanations. The term is considered sometimes contradictory to 'place making' and 'place-making'; therefore, there does not exist a clear, consistent, or universally accepted definition aligned to these terms (cf. Ellery et al. 2021); however, there appears an asset of 'understandings' of a field of practice, which is "used to describe the process of creating the material and social spaces of place so that they are desirable for the public to visit and spend time in" (Courage 2020: p. 623). Based on critical appraisal and review of 120 research papers published during 2016–2020, three "foundational" principles regarding what 'placemaking' is theoretically were evident (cf. Ellery et al. 2021), as given below:

1. *Placemaking as "sense of place"*. This creates an attachment or connection between a person and the place in which they live and work. Sense of place is accepted to be a vital point in our understanding of what 'placemaking', as a process, creates in terms of an outcome within an individual and the cultural group.
2. *Placemaking as a process*. This usually occurs somewhere along a continuum ranging from "change that is imposed upon an individual, or society" through to "change that is created by the individual". The individual's feelings towards the change in place may be positive or negative, depending on the lived experience and perceptions they possess over time in history. The use of a community-led placemaking approach is also more likely to result in the sense of place being both stronger and developed faster, as this individual and community connection to the place now occurs during the change process itself (Ellery and Ellery 2019).
3. *Placemaking as a source of perception*. This has the potential to influence a community at three different levels, viz. (i) individual perception level—learned about a place through the smells, sounds, and sights experienced while immersed in the place itself, and this is where each individual's sense of place is developed; (ii) social connectedness that comes from being part of a neighbourhood community, resulting to co-sharedness; and (iii) physical infrastructure level where most placemaking research, practices, and projects focus because it represents the point at which the role of planning and design becomes significant for professionals.

These three key principles form a foundation for 'placemaking' as a frame, a process, a path, and a strategy in overall sustainable community development. Together, they form a theoretical foundation for placemaking. Here, we have taken these issues in common conscience and consensus, we accept 'placemaking' as a process of making foundational, functional landscapes where Nature-Culture plays an integral, reciprocated, and directly interfacing interaction.

Kevin Lynch [1918–1984], while discussing the implications of placemaking, together with city design, referring to creating new forms of style and conveying meanings and feelings with critical expositions, has proposed six possibilities that could sustain such development, judgement, and decision, and delight too; they are discussed below (Banerjee and Southworth 1991: pp. 501–507):

i. *Basic perceived structure*: that deals with the character of the key points, landmarks, and the connections of their role in successive stages while making social-cultural meanings and identities.
ii. *Experiencing and revealing the journey*: the experientiality of a place that would be a pleasure and peace, and understanding the inherent, manifested, or mythologised meanings.
iii. *Conscious art of renewal*: the art of refabrication, tinkering, and redoing, which may be full of plurality, ambiguity, and contradictions (see Fleming 2007).

iv. *Designing the event*: celebrations can be continued from the past and also be inverted, together with other such performances, e.g. sound and light shows, fairs and festivals, carnivals, dances, etc.
v. *Continuity of routine*: repetition of journeys and performances makes the place constantly alive through interlinking production, installation, and maintenance, and absorbing new imposition.
vi. *Convergence of a system*: engagement with the reciprocity of counter influences between Nature and Culture, and resultantly making a complex system and network of interconnections. This is an overall and holistic frame of knowing and understanding, and also paves the path for the future frame.

The SDG Target 11.4 is the result of a concerted effort from many sectors, as compared to its predecessor the Millennium Development Goals, and strongly emphasises an explicit heritage target. Additionally, it calls for making the human settlements (urban, rural, peri-urban) inclusive and as a process and means for safeguarding the world's cultural and natural heritage. The Target is labelled under the "Urban Sustainable Development Goals", i.e. a set of sustainable development targets related to heritage cities and settlements. The basic frame in revised form taking into view the prevailing condition in holy-heritage cities of Asia (e.g. Ayodhya and Varanasi see Singh et al. 2020) is recently chalked out. Conserving heritage and its associated attributes revives a sense of identity and can inspire new smart and sustainable pilgrimage city and sacred town development patterns, together with an emphasis on the valorisation of the assets of the poor. Moreover, in the other context of SDGs of social sustainability, the heritage city is mostly dependent on social groups of local living peoples, pilgrims, and *sadhus* (Hindu religious ascetics), which result in social beliefs in the form of intangible cultural heritage. According to ancient history, as illustrated in the case study of Ayodhya and Varanasi—the holy-heritage cities of India, various social groups through the religious faiths of Hinduism, Buddhism, Jainism, and Islam meet together in making multiple visitation sites of heritage importance and nexus of harmonious life, resulting to the emergence of sublime cultural landscapes (see Singh et al. 2020).

Cultural tourism plays a great role in socio-economic changes and the promotion of sustainability, especially in holy-heritage cities. As one of the largest industries, tourism is associated with many of the prime sectors of the world's economy. For example, the economic sustainability of Ayodhya and Varanasi (India) is regulated by the annual visit of pilgrimage tourists, recorded ca 1.9 million, in each of the places, in 2019 (see Singh et al. 2020). Taking this view, the Indian government is promoting pilgrimage tourism in the purview of SDGs. Pilgrimage is defined as "a journey resulting from religious causes, externally to a holy site, and internally for spiritual purposes and internal understanding" (Collins-Kreiner 2010: p. 153); thus, pilgrimage tourism will promote, sustain, and maintain economic and cultural sustainability in holy-heritage cities, which bear various interrelated phenomena to which people are affiliated with the belief system, faith, and spiritual merits. A big question still stands—is as to how can we preserve and enhance the cultural and artistic heritage of religious interest, increasingly under threat and so many conflicts in progress, and also use them as a vital force in placemaking and peace formation! Maybe an active platform for interdisciplinary dialogue between experts of the religious and secular community would give a rational and better path to proceed! (Niglio 2017).

Culture and cultural heritage have had an impact on international ethics, global understanding, and laws, which also include the maintenance of peace and security; here, there has been a synergic effort by the United Nations and UNESCO towards the objective of linking respect for cultural heritage and protection of cultural pluralism to the preservation of peace and security at the global level. This can easily be activated and regulated through open dialogue between the humanities and jurisprudence that will be able to better protect and value sovereignty and promote cultural heritage for the development of a better world (cf. Niglio and Joong Lee 2021). The 'cultural heritage', across all the boundedness, constitutes the corpus of fundamental knowledge that is fundamental for a deeper understanding of the origins of human life on Earth. As a pathway in search and exposition of transcultural value, it "promotes the enhancement of diversity and increasingly fruitful dialogue within the different geographies. So today more than ever transcultural heritage, because of its high ethical value worldwide, is a fundamental reference for objectively rethinking the continuation of man's existence on Earth" (Niglio and Joong Lee 2021: p. 10).

The ongoing mission and march promoting shared knowledge and initiatives for enhancing cultural-religious heritage in the world through inter- and intra-cultural dialogues will further energise to realise and reveal the role of religious heritage in making global world order. This is in corroboration with the upcoming ICOMOS Scientific Symposium "*Religious Heritage: Celebrating and Conserving the Places of Religious and Ritual Significance in a Global World and in Changing Climates*"—scheduled during 29–30 October 2022 in Bangkok, Thailand (Niglio 2022: p. 4); of course, the present volume through its paving path to placemaking would also enhance this mission. Additionally, one may also note the role of RWYC (Reconnecting With Your Culture), as already mentioned in the sequence.

Interactivity between the various manifestations in which places reveal themselves, therefore, acknowledges the fundamental creative and cultural quality of the human relationships with place, illustrated with complex relationships. Knapp and Ashmore (1999: p. 16, pp. 20–21) observed that "… just as landscape maps memory and declares identity, … it offers a key to interpreting society … the land itself, as socially constituted, and plays a fundamental role in the ordering of cultural relations … landscape is neither exclusively natural nor cultural; it is a mediation between the two and an integral part….. habitus [*humane world*], the routine social practices within which people experience the world around them". The emergent field of placemaking has noticed a paradigmatic shift in urban design, planning, and policy to engage the community voice, and overall understanding of cultural landscapes and the infrastructure involved therein. There appear a good number of theoretical constructs in analysing placemaking's potential to nurture a holistic community engagement, social justice, human-centric urban environments, art-based representation, the socio-economic structure, image-making, the strategy of sustainable development, and so on (see Courage et al. 2021).

Based on the five cities from different settings, viz. Bengaluru (India), Cape Town (South Africa), Gothenburg (Sweden), Greater Manchester (UK), and Kisumu (Kenya), a recent study remarks that “If the urban SDG is to prove useful as a tool as intended for encouraging local and national authorities alike to make positive investments in the various components of urban sustainability transitions, then it must be widely relevant, acceptable, and practicable” (Simon et al. 2016: p. 60). This is also valid in the case of Indian heritage cities, where one always faces the problem of linking locality and universality. Additionally, the focus of this task has been the challenge of determining how to benchmark and measure performance according to the SMART criteria (i.e. specific, measurable, assignable, realistic, and time-specific), based on specialist knowledge, the thematic literature, and practical experience of the site (see Birch 2015: p. 228; Singh 2022a: p. 6), while taking into account demand, pressure response, and multifactor versus single factor, and also considering two metrics: the traditional dimensions of sustainability (equity, economics, and environment) and later against forming the Liveability Principles of Partnership for Sustainable Communities (PSC), while making a bridge between the age-old traditions and high-tech smart city plans (see Birch 2015: p. 230).

In the purview of understanding, interpreting, and exposing various niches of placemaking, heritage ecology is accepted as a way of deeper knowing and exposition of spirit and power of place (cf. Singh 1995). Heritage ecology helps to explore the meaningful relationship with resultant habitats and inherent meanings—to be taken as core in global understanding. Heritage ecology is also a way to pilgrimage—a way of relating to the land (Earth/Nature spirit) and the people (human psyche), similar to that of marching from realisation to revelation (Singh 2022b: p. 80). The imbued issues are under the process of evolution, existence, and eventuality, on the spatiotemporal scale of ‘succession-sustenance-sustainability’—the foundational platform and evolving pattern, transformation, and scenario in placemaking. Placemaking is the pathways of *succession* (evolution and existence in the ancient past), passing through *sustenance* (continuance, maintenance, organic growth, and uses in the present), and progression towards *sustainability* (planning for conservation, preservation, governance, cultural enveloping, and perspectives as in SDGs) makes the ‘holistic’ (‘cosmic’) frame of understanding and creation of action plan in making the harmonious world of historic-heritage cities: ‘*Sublime, Blissful, Beautiful*’ (“*Satyam-Shivam-Sundaram*” in Indian metaphysics) (cf. Singh and Niglio 2022: p. 10).

The past is to inspire; the future is in dreams; the present is to act; let us understand and practise heritage ecology to foster the preservation and regeneration of historic cities and their integral part of cultural-religious heritage. These issues are critically appraised in the broader domain of geography as the way and the path of awakening. Think *cosmically*, see *globally*, behave *regionally*, and act *locally* but *insightfully* (Singh and Rana, 2020: p. 85). This is an appeal to walk on the path of transformative environmental consciousness towards placemaking. Placemaking belongs to everyone: its message and mission are bigger than any one person or organisation—we are trying to substantiate it by paving it constantly and consistently through care and conscience.

References

Aurobindo S (1979) The supreme discovery. Aurobindo Ashram, Pondicherry.

Bandarin F, van Oers R (2012) The historic urban landscape: managing heritage in an urban century, 2nd edn. Wiley-Blackwell, Chichester.

Banerjee T, Southworth M (eds) (1991) City sense and city design: writings and projects of Kevin Lynch. MIT Press, Cambridge MA

Birch EL (2015) Measuring U.S. sustainable development. Cityscape J Pol Dev Res. 17(1):223–232

Bridgewater P, Bridgewater C (2004) Is there a future for cultural landscapes? In: Jongman RHG (ed) The new dimensions of the European landscapes. Wageningen UR Frontis Series, Vol 4. Springer, Dordrecht, pp 193–200

Brundtland, GH (1987) Report of the world commission on environment and development: our common future. United Nations General Assembly, document A/42/427. Oxford University Press, Oxford

Collins-Kreiner N (2010) The geography of pilgrimage and tourism: transformations and implications for applied geography. Appl Geogr 30:153–164

Cosgrove D (2000) Geopiety. In: Johnson RJ et al (eds) The dictionary of human geography. Blackwell Publishing, Oxford, pp 308–309

Courage C (2020) The art of placemaking: a typology of art practices in placemaking. In: Edensor T, Kalandides T, Kothari U (eds) The Routledge handbook of place. Routledge, London & New York, pp 75–87, 623–633

Courage C et al (eds) (2021) The Routledge handbook of placemaking. Routledge, London

Dessein J, Soini K, Fairclough G, Horlings L (2015) Culture in, for and as sustainable development: conclusions from the COST Action IS1007 investigating cultural sustainability. University of Jyväskylä, Jyväskylä. http://www.culturalsustainability.eu/conclusions.pdf. Accessed 21 Apr 2021

Duxbury N, Kangas A, De Beukelaer C (2017) Cultural policies for sustainable development: four strategic paths. Int J Cult Pol. 23(2):214–230. https://doi.org/10.1080/10286632.2017.1280789

Ellery PJ, Ellery J (2019) Strengthening community sense of place through placemaking. Urban Plan. 4(2):237–248. https://doi.org/10.17645/up.v4i2.2004. Accessed 15 Dec 2021

Ellery PJ, Ellery J, Borkowsky M (2021) Toward a theoretical understanding of placemaking. Int J Commun Well-Being 4:55–76

Fleming RL (2007) The art of placemaking: interpreting community through public art and urban design. Merrell, London

Fowler PJ. (2003) World heritage cultural landscapes 1992–2002 (World Heritage Papers 6). UNESCO, Paris

Hall CM, Lew AA (2009) Understanding and managing tourism impacts: an integrated approach. Routledge, London

Higgitt D (2009) Approaching the UN decade of education for sustainable development: Prospects and challenges. In: Chalkley B, Haigh M, Higgitt D (eds) Education for development: papers in honour of the United Nations Decade of Education for Sustainable Development, 2005–2015. Routledge, London & New York, pp 1–13

Knapp AB, Ashmore W (1999) Archaeological landscapes; constructed, conceptualised, ideational. In: Ashmore W, Knapp AB (eds) Archaeology of landscape: contemporary perspectives. Blackwell, London, pp 1–30

Legrand T (2021) Politics of being: wisdom and science for a new development paradigm. Ocean of Wisdom Press, Somerville MA

Lew AA (2021) Landscapes, peoplescapes, and mindscapes in island tourism. p 109–128. https://projects.upei.ca/unescochair/files/2021/04/Annual-Report-on-Global-Islands-2019-Chapter-4-Landscapes-peoplescapes-and-mindscapes-in-island-tourism-Alan-A.-Lew.pdf. Accessed 6 May 2022

Lopez-Claros A, Dahl AL, Groff M (2020) Global governance and the emergence of global institutions for the 21st century. Cambridge University Press, Cambridge

Lukermann FE (1964) Geography as a formal intellectual discipline and the way it can contribute to human knowledge. Can Geogr 8(4):167–172
Lynch K (1976) Managing the sense of a region. MIT Press, Cambridge
Martin DG (2003) "Place-framing" as place-making: constituting a neighborhood for organizing and activism. Ann Assoc Am Geogr 93(3):730–750
Martin GJ (2005) All possible worlds, a history of geographical ideas, 4th edn. John Wiley & Sons, New York
Massey D (1991) A global sense of place. Marxism Today, Jun, pp 24–29
Massey D (2005) For space. Sage, London
Meller H (1990) Patrick Geddes, social evolutionist and city planner. Routledge, London
Mitchell D (2000) Cultural geography, a critical introduction. Blackwell Publishing, Oxford
Niglio O (2017) Knowing, preserving and enhancing. The cultural-religious heritage. AlmaTourism J Tour Cult Territ Dev 15:152–156. https://doi.org/10.6092/issn.2036-5195/7158
Niglio O (2021) Towards a humanist education: understanding cultural heritage to redesign the future. Acad Lett. Article 3223. https://doi.org/10.20935/AL3223. Accessed 15 Dec 2021
Niglio O (2022) Introduction: structure of the research: intercultural dialogue, places of religion and rituals. In: Olimpia N (ed.) Regenerating cultural religious heritage: intercultural dialogue on places of religion and rituals. Springer Nature Pte Ltd, Singapore, pp 1–10, Chapter 1
Niglio O, Joong Lee EY, eds (2021) Transcultural diplomacy and international law in heritage conservation: a dialogue between ethics, law, and culture. Springer Nature Singapore Pte Ltd, Singapore
Relph E (1976) Place and placelessness. Pion Limited, London, UK
Sauer CO (1925/1963) The morphology of landscape University of California Publications in Geography, 22: p. 19–53; reprinted in his (1963) Land and life, A selection from the writings of Carl O. Sauer. In: Leighly J (ed.) University of California Press, Berkeley, pp 315–350
Schafer DP (2022) The world as culture: cultivation of the soul to the cosmic whole. Rock's Mills Press, Oakville, ON, Canada
Schein RH (1997) The place of landscape: a conceptual framework in the American scene. Ann Assoc Am Geogr. 87(4):660–680
Shackley M (2001) Managing sacred sites: service provision and visitor experience. Continuum, London
Silva KD (2017) Prospects for managing South Asian cultural landscapes. In: Silva K, Sinha A (eds) Cultural landscapes of South Asia: studies in heritage conservation and management. Routledge, Abingdon, pp 258–269
Simon D, et al. (2016) Developing and testing the urban sustainable development goal's targets and indicators: a five-city study. Environ Urban. 28(1):49–63
Singh RPB (1995) Heritage ecology & caring for the earth: a search for preserving harmony and ethical values. Nat Geogr J India. 41(2):191–218
Singh RPB (2004) Cultural landscapes and the lifeworld. Literary images of Banaras (Pilgrimage & Cosmology Series: 6). Indica Books, Varanasi
Singh RPB (ed) (2010) Heritagescapes and cultural landscapes (Planet earth & cultural understanding series, Pub. 6). Subhi Publications, New Delhi
Singh RPB (ed) (2011) Sacredscapes and pilgrimage systems (Planet earth & cultural understanding series, Pub. 7). Subhi Publications, New Delhi
Singh RPB (2013) Indian cultural landscape vis-à-vis ecological cosmology: a vision for the 21st century. Ann Nat Assoc Geogr India. 33(2):36–62
Singh RPB (2021) Place-based ecoliteracy and education for sustainable development: vision from India. In: Dutta B, Chaudhuri P (eds) Global education 2050: enhancing global citizenship & online partnerships (Book Project: XI). MTC Global Trust, Research & Publishing, Bengaluru, pp 1–29
Singh RPB (2022a) Communicating heritagescapes: paving the path for heritage informatics. Inform Stud Heritage. 9(2):3–8

Singh RPB (2022b) Heritage value of religious sites of Hinduism: Asian linkages and reuse for cultural regeneration. In: Olimpia N (ed) Regenerating cultural religious heritage: intercultural dialogue on places of religion and rituals. Springer Nature Pte. Ltd, Singapore, p 63–78.

Singh RPB, Niglio O (2022) Historic cities & cultural-religious heritage: preservation & regeneration. Int J Arch Eng. 9(1):5–14

Singh RPB, Rana PS (2020) Faith and place: Hindu sacred landscapes of India. In: Edensor T, Kalandides A, Kothari U (eds) The Routledge handbook of place. Routledge, London & New York, pp 75–87

Singh RPB, Kumar S, Rana PS (2020) UN SDGs and context of heritage cities in India: a study of Ayodhya and Varanasi. In: Sahdev S, Singh RB, Kumar M (eds) Geoecology of landscape dynamics. Springer Nature, Singapore, pp 121–140

Stückelberger C (2020) Globalance ethics: Handbook for a balanced world post-covid. Globethics.net, Geneva

Taher M (1990) The region as a cultural entity: a humanistic perspective. North Eastern Geographer 22(1–2):55–62

Tuan Y-F (1976) Geopiety: a theme in man's attachment to nature and to place. In: Lowenthal D, Bowden MJ (eds) Geographies of the mind. Essay in historical geography. Oxford University Press, New York, pp 11–39

Tuan Y-F (1977) Space and place: the perspective of experience. University of Minnesota Press, Minneapolis.

Tuan Y-F (1979) Space and place. In: Gale S, Olsson G (eds) Philosophy in geography. Reidel, Dordrecht, pp 387–428

UNESCO (1982) Mexico City declaration on cultural policies. World Conference on Cultural Policies, Mexico City, 26 July–6 August. UNESCO, Paris. https://culturalrights.net/descargas/drets_culturals401.pdf. Accessed 21 Apr 2021

WHC (2003) Cultural landscapes: the challenges of conservation. (World Heritage Papers 7). UNESCO World Heritage Centre, Paris

Chapter 2
Vision and Exposition of Placemaking Under ACLA: Homage and Memorial Tribute to Sung-Kyun Kim

Rana P. B. Singh and Olimpia Niglio

Abstract The interrelatedness of placemaking and cultural landscape in the Asian realm has been explained vividly in the frame of interdisciplinary approach in the writings of (late) Professor Sung-Kyun Kim. Following the Saurian tradition, his studies started in the 1980s at the University of Pennsylvania, USA, and continued longitudinally throughout his life, mostly explaining the case of the folk village of Hahoe and its environs, and urban historic and cultural landscapes of Seoul based on participatory observations, experiential expositions, intimate interaction, metaphysical understanding of archetype and cosmic frame as culturally developed within the system of Pung-su, dwellers' perceptions and visions, and also regularly visit and personally living therein. He has continuously made a successful attempt in making the bridge between locality and universality. All these are narrated in his writings, field notes, designs, exhibitions, eco-parks, botanical gardens, designing streetscapes and eco-suited residences, and park and plaza designs. As a founding and patron, he promoted international dialogues through symposia organised under the aegis of the ACLA (Asian Cultural Landscape Association, headquarters at Seoul National University).

Keywords Cultural landscape · Streetscape · Hahoe · Asian vision · Experiences · Sustainability

R. P. B. Singh (✉)
Cultural Landscapes & Heritage Studies, Banaras Hindu University, Varanasi, India
e-mail: ranapbs@gmail.com
URL: https://bhu-in.academia.edu/RanaPBSINGH/Papers

O. Niglio
Architectural Restoration, University of Pavia, 27100 Pavia, PV, Italy
e-mail: olimpia.niglio@unipv.it

R. P. B. Singh et al. (eds.), *Placemaking and Cultural Landscapes*,
Advances in Geographical and Environmental Sciences,
https://doi.org/10.1007/978-981-19-6274-5_2

2.1 Initiation: Platform and Career

On the evening of 9th September 2011 in the lobby of Hotel Sarovar Portico [Bhavan's College Rd, Khanpur, Ahmedabad, GJ, India], Rana P.B. Singh had first met Prof. Sung-Kyun Kim, who was a special invitee in the 7th National Conference of the ISOLA (Indian Society of Landscape Architecture) at Ahmedabad: 10–12 September 2011, who was the Chief Guest to deliver the inaugural address. Rana was there as a special invitee to talk about the 'Buddhist Cultural Landscape'. As a followed-up programme of CEPT Alumni Students' (Landscape Architecture) Celebration, they sat together and had chat and interaction on the multidisciplinary approaches to the cultural landscape of Asia, in the wider perspective of Saurian tradition (after Carl Sauer [1889–1975], University of California, Berkeley USA). Induration of the three days, they came close to each other and co-shared several ideas about what Kim has envisioned for the similarity, generality, and distinctiveness of the cultural landscapes of Asia. To have the first experience of this perspective, Rana had invited him to visit Varanasi after ISOLA Conference, that's how he came to Varanasi on 17–19 September 2011 (stay: Hotel Hindustan). During this period, Rana guided him in and around the sacred groves and garden, emphasising the religious and healing importance and their contemporary uses of sacred trees and plants. During this period, he took lunch and dinner at Rana's residence in the green-lush old residential quarter in the Banaras Hindu University (New F-7 Jodhpur Colony, Varanasi). Kim was overwhelmed with Indian hospitality, and impressed by the sacred trees and herbal plants in my garden and their uses in a traditional Indian family.

With the above impression, interaction, and mutual understanding, Kim had invited Rana as the keynote speaker at the IFLA—Asia–Pacific Region Congress of Landscape Architecture: 5–7 December 2011, at Seoul National University, Seoul, Korea. Rana had reached there on 4th December and stayed till 9th December 2011, and under his guidance visited several monasteries, sacred sites, sacred groves, and the sacrosanct environment—listening to cultural stories, mythologies, and their present context. This way they had started their joint venture and collaborative programmes for the coming future. On 6th December in the concluding meeting of IFLA-APC, it was realised that Asian culture has been at the margin, and superimposed with Western approaches and conceptual frames! Consequently, after rigorous discourses, professors Rana P.B. Singh, and Ismail bin Said (Malaysia) under Prof. Kim's leadership formed a non-profit informal academic institution, called ACLA—Asian Cultural Landscape Association, and conceived the IFLA-APA meeting as the initial one 1st ACLA Symposium, focussed on "*Sustainable Rural Landscape & Planning in Asia Pacific Region*": 7th December 2011, which was attended by over a hundred participants from different countries of Asia Pacific, and the USA. In this meeting, Rana P.B. Singh had delivered the keynote address on 'Rural Cultural Landscapes of Asia: Vision of Sustainability and Man-Nature Interrelatedness'. Prof. Kim had purveyed and provoked that 'Asia possesses the most dynamic environment of regional identity and diversity in terms of economic development, cultural continuity and change. In landscape architecture, Asian countries record a long history and have

Fig. 2.1 Professor Sung-Kyun Kim (1956–2020) (*Photo credit* Rana P. B. Singh)

developed distinctive and diverse landscape cultural attributes.' This way started a series of ACLA Symposia, under the patronage and leadership of Sung-Kyun Kim (Fig. 2.1).

2.2 Sung-Kyun Kim: Career and Leadership

Prof. Sung-Kyun Kim (Fig. 2.1) was born on 18th February 1956 in Mungyeong (North Gyeongsang Province, Korea), and passed away in Seoul on 19th June 2020 at 4:00 PM in his home due to '*ileus*', gastrointestinal atony—referring to a severe intestinal obstruction that caused to cut off blood supply to the intestines and cause tissue death, resulting to stop peristalsis and prevents the passage of food particles, gas, and liquids through the digestive tract. Having had serious abdominal pain since June 8, he was in an emergency room of a hospital in Seoul on June 11. Afterwards, he stayed at his office in SNU on June 13th but came back home on June 14th, where he passed away on 19th June 2020.

On 17th June 2020 morning, Rana had a last discussion through Fb Messenger, concerning plans of ACLA, integration of ACLA and APELA, organising the 10th ACLA 2020 (26–27 September, B.H.U. Varanasi, India), and future meetings of ACLA-APELA in 11–12 June 2021 (NUAA Nanjing, China) and in 2022 (Sapporo, Japan). Also, we had discussed the initiation of a book series taking in view the proceedings and papers presented in all the eleven ACLA-APELA Symposia. He was excited and assured, as ever in the past, to collaborate and take lead in all the planned symposia, but his destiny stopped him.

Prof. Kim was awarded "academic degrees of BS (LA), Seoul National University (1981), Master of Landscape Architecture, University of Pennsylvania, USA (1984), Master of Arts, University of Pennsylvania, USA (1987), and Ph.D. in City & Regional Planning, University of Pennsylvania, USA (1988). He had started his academic career as Lecturer & Teaching Assistant at University of Pennsylvania, USA (1983–1988), and had served in various capacities, including an Assistant & Associate Professor, University of Seoul (1988–1994), Visiting Professor at University of Washington, USA (1997–1998), Chairman of the Department of Landscape Architecture, Seoul National University (2001–2003), Visiting Professor at University of British Columbia, Canada (2003–2004), Chairman of the Department of Landscape Architecture & Rural System Engineering, Seoul National University (2007–2009), Director of Urban Greening Institute, Seoul National University (2009–2014), Visiting Professor at Universiti Teknologi Malaysia, Kuala Lumpur (2010–2010, 2012), Visiting Professor at Yanbian University, China (2013–2017), Professor at Seoul National University (1994–2015), and finally University Distinguished Professor (2015–2019 June 2020, till his last breath)" (Singh 2020: p. 1).

The academic and institutional positions hold by Prof. Kim "included Director of International Affair, Korean Institute of Landscape Architecture (KILA, 1993–2002), Chief Editor of the Journal of Geographic Information System (GIS) Association of Korea (2000–2002), Director of International Affair, Korean Institute of Traditional Landscape Architecture (1998–2003), Chief Editor of the Journal of Korean Society of Rural Planning (2000–2004), Vice President of the Korean Institute of Traditional Landscape Architecture (2002–2004), Vice President (International Affair) of Korean Institute of Landscape Architecture (KILA, 2007–2008), Vice President of Korean Society of Rural Planning (2008–2011), Vice President of Korean Institute of Traditional Landscape Architecture (2008–2010), President of Korean Society of Rural Planning (KSRP, 2011–2012), President of Asian Cultural Landscape Association (ACLA, 2012–2017), President of Korean Institute of Landscape Architecture (KILA), (2015–2017), and President of Asia–Pacific Environment and Landscape Architecture (APELA, 2015–2020) Forum; and Member—Steering Committee, International Geographical Union, IGU, Commission C16.25 'Landscape Analysis and Landscape Planning' (2016–2020). Prof. Kim's death was a disastrous and a great loss to the whole ACLA community" (Singh 2020: p. 1).

2.3 Sung-Kyun Kim: Patronizing the ACLA and Guidance

Of course, ACLA (Asian Cultural Landscape Association) was informally formed on the 7th December 2011—the day was inaugurated the 1st ACLA symposium; however, it was registered under the jurisdiction of the Government of the Republic of Korea on the 3rd December 2012 (headquarters at SNU Seoul, Korea). Over time, the rules and regulations were formed, the Executive Committee was formalised and regular plans were scheduled and structured.

The logo of ACLA (Fig. 2.2) has been designed by Prof. Kim, which in itself is a testimony of metaphysical exposition of the aims and objectives of this foundation, what he has envisioned. The two sides 'A' symbolises the beginning and returning ('A') through the cyclic notion of continuity 'C', and attached 'L' denotes the locomotive function of longevity. Similarly, the letter symbolism A-C-L-A constitutes distinct metaphoric illustrations, viz. A—*Aspiration* (from affectation to articulation), C—*Connectedness* (from cognizance to cohesiveness), L—*Liveliness* (from learning to liberation), and A—*Association* (from accompaniment to accomplishment). In the light of inherent messages and cosmic vision, he paved the path that the ACLA would take lead in marching for deeper understanding and making bridges between Analysis and synthesis, Culture and nature, Laconic and luminous, and Allegory and allusive (ACLA 2020: p. 3). This would justify the logo symbolism and further help to promote the human march from known to unknown, visible to invisible, simplicity to complexity, finite to infinite, uniqueness to commonality, generality to rationality, periphery-core to core-periphery, fragmentation to consolidation, experience to exposition, realisation to revelation, thought to awaken, mind to soul, transformability to sustainability, …, and so on (see ACLA Newsletter 2020).

Under the guidance and leadership, the role of ACLA in dealing with and promoting its objects is crystallised by Kim and his close associates, as noted below (highlighted in ACLA Newsletter 2019):

- **Asia's unique cultural heritage** lies in its historically rich and culturally diverse settlements and habitat systems forming a complex and highly developed fabric of human habitation. It is through this cultural habitat that the Asian culture flourishes, continuously undergoing a process of change, interacting with new elements, assimilating new ideas, and thoughts, as well as the aspirations and creativity of its people, and the enveloping nature. The habitations are thus

Fig. 2.2 The Logo of ACLA, prepared by Sung-Kyun Kim (*Photo credit* Rana P. B. Singh)

constantly renewing themselves while maintaining their unique and diverse culture and heritage.

- **The uniqueness and the diversity** of Asia's cultural landscapes and heritages can be seen reflected in the morphology, building typologies, activity patterns, social structure, religious beliefs, pilgrimages, sacred sites and gardens, and varying associated traditions that exist in the villages, towns, and varying forms of settlements.
- With the turn of the century **rapid urbanisation** and more recently **globalisation**, the pace of change in all aspects of life has accelerated. The habitat systems and eco-environments of towns and villages are exposed to strong extraneous, fast-growing agents disturbing the delicate balance among the physical, social, cultural, and ecological habitat environments.
- **To have a deeper understanding and cross-cultural integration and to strengthen the capacity of Asian cultural landscapes** to cope with these fast changes, the Asian Cultural Landscape Association (ACLA) is formed to take the initiatives for making Asian Cultural Landscape sustainable, eco-friendly, and envision Green Pilgrimages, the revival of the spirit of sacred sites, and would also seek cooperation from international bodies like UNESCO WHC, UNDEP, IUCN, ICOMOS, Green Pilgrim Cities Initiative (GPCI), Alliance of Religions and Conservation (ARC), Sacred Sites Initiative, 'Bhumi Project', World Wildlife Fund (WWF), IFLA, IGU Commission on Cultural Approach, IGU Commission on Landscape Analysis & Landscape Planning (both for the period 2012–2016), and other such institutions.

The 2nd ACLA with the collaboration of the IGU C12.07 and IGU C12.25 organised the International Symposium on "*Meanings & Aesthetics in Asian Cultural Landscape*" at SNU Seoul, Rep. Korea: 12–14 October 2013 under the aegis of Urban Greening Institute, SNU, and was patronised by Sung-Kyun Kim, and was attended by 50 persons. In his inaugural address, Kim had emphasised the essence and contexts of 'Commonality, Distinctiveness and Transcendentality', with illustrations from Asian countries.

The 3rd ACLA—Asian Cultural Landscape Association International Symposium was on "*Waterfront Asian cultural Landscape*": 07–09 October 2014, at SNU Seoul, Rep. Korea, under the aegis of Urban Greening Institute, SNU. In his address, Sung-Kyun Kim emphasised the need for creating a sustainable waterfront landscape that might be the most important aspect of Asian cultural landscapes and encouraged all the participants to exchange their ideas and share expertise for the cultural landscape at the symposium.

The 4th ACLA International Symposium on "*Agricultural Landscapes of Asia: Learning, Preserving, and Redefining*": 11–13 Sept. 2015, held in the Hotel Puri Ayu, Bali- Indonesia, under the aegis of the School of Landscape Architecture, Udayana University, Denpasar-Bali, Indonesia. A good mass of researchers and senior students of Landscape Architecture actively participated, including 06 moderators for each of the sessions and some observers and recorders.

Kim had taken lead and organised the 1st APELA, Asia–Pacific Environment Landscape Architecture *Forum*, under the aegis of the Korean Institute of Landscape Architecture (KILA) and Seoul National University (SNU), on "*Present and Future of Asia–Pacific Landscape Architecture*": 23–25th November 2015, at Prugio Valley (Yeongdong-daero 337, Gangnam-gu), Seoul, Rep. Korea. This was attended by around seventy participants from 17 countries. The keynote address was delivered by Ismail bin Said (Malaysia), who highlighted the gaps and appraised the potential grounds for cooperation among various Asian countries. Sung-Kyun Kim has provoked about the strategy that APELA will enable to promote exchange and share experience and ideas, building network and collaboration with landscape professionals in our region, and ultimately provide a sustainable and harmonious frame for the whole globe.

The 2nd APELA, Asia–Pacific Environment Landscape Architecture *Forum*, was organised by The Korean Institute of Landscape Architecture (KILA) and Seoul National University (SNU), in continuation of the preceding theme, i.e. "*Present and Future of Asia–Pacific Landscape Architecture*": 17–19th October 2016, at International Conference Hall, SNU Asia Centre, Seoul, Rep. Korea, again under Kim's leadership; he successfully guided to provide a brainstorming discussion platform from various sectors of landscape architecture, including practitioners, researchers, professors, students, community leaders, representatives of national associations, and public official those involved in landscape development and its empowering environment in Asia–Pacific Region. Around sixty papers were presented by scholars and LA practitioners from 11 countries.

The 5th ACLA International Symposium on "*Sacred Sites, Cultural Landscapes, and Harmonising the World of Asia*": 2–5 December 2016 was organised by Suparp Tajai and held at the Faculty of Humanities and Social Science, Lampang Rajabhat University, Thailand, where four keynote addresses were delivered by Sung-Kyun Kim, Rana P.B. Singh (Vice-President- ACLA), Susan Aquino-Ong (ICOMOS Philippines), and Bharat Dahiya (Chulalongkorn University, Thailand). This was sponsored by the IUCN Commission on 'Cultural and Spiritual Values of Protected Area'. In total, 26 papers were presented by participants from 8 countries. In the EC-ACLA meeting, two Joint Secretaries were nominated and approved: Ms. Anita Syafitri Arif (Bali, Indonesia) and Mr. Sarvesh Kumar (India), and they continued to assist; however, on 3rd October 2019, Anita Arif has resigned.

The 6th ACLA International Symposium on '*Urban Cultural Landscape and Urban Regeneration*': 22–24 July 2017, at Laboratory of Urban & Landscape Design, FEFU Far Eastern Federal University, Vladivostok, Russia with the collaboration of IGU Commission—C16.25 'Landscape Analysis and Landscape Planning' was organised by Prof. Alessio Russo. The keynote addresses were delivered by Sung-Kyun Kim and Rana P.B. Singh. In total, 24 papers representing 12 countries were presented there.

Under a joint venture together with ACLA and APELA an International Workshop on the '*Cultural Landscape as National Identity: illustrating Asia*': 17–18 November 2017 was organised at SNU Seoul by Sung-Kyun Kim and Rana P.B. Singh. Here, the national identities are defined by 'portrait, poetics, and perspectives of the images created through the interfacing and reciprocal interaction within nature-culture trajectory', and invited scholars were asked to prepare a national-level report emphasizing these themes. Eleven scholars have presented 8 country-level reports.

The 7th ACLA International Symposium on "*Pilgrimage Cities & Cultural Landscapes of Asia and prospects for Sustainable Tourism*": 23–26 October 2018, was organised at Dr. R.M.L. Avadh University, Ayodhya (U.P., India), with the collaboration of IGU Commission C16.07 "Cultural Approach in Geography", IGU Commission C16.25 "Landscape Analysis and Landscape Planning", and Ayodhya Sodh Sansthan (Department of Culture, Govt. of U.P., India), Ayodhya. In total, 220 participants attended, including from Korea, China, the UK, Thailand, Indonesia, Philippines, Italy, and India. About 62 papers were presented covering the following themes: (i) Sacred Landscape and Sustainable Tourism, (ii) Ritual Landscape and Cosmogram, (iii) Heritage Cities and making of Harmonious World, and (iv) Holy-Heritage Cities and Heritage Planning. The inaugural address was delivered by IGU Secretary-General (*late*) Prof. R.B. Singh (1955–2021; Delhi University), and keynote addresses were by Sung-Kyun Kim and Rana P.B. Singh.

The ACLA-IFLA: CLC International Workshop on "*Cultural Landscapes vis-à-vis Sacred Places: Exposing National Identity*": 18–19 February 2019 at Asia Center, SNU Seoul National University, Seoul, organised as a joint venture of Sung-Kyun Kim and Rana P.B. Singh. Rana Singh delivered the keynote address and presented the case of India while comparing other countries in Asia. Twenty papers were presented on the focal sub-themes. Sung-Kyun Kim, Rana P.B. Singh, and Pravin S. Rana together performed field studies in the Gimhae region (20–22 February 2019) under the aegis of the Indo-Korean Project on "Comparative Study of Cultural Links and Ritual Landscapes between Gaya (Korea) and India" (team, PI-Korea: Sung-Kyun Kim, PI-India: Rana P.B. Singh, Coordinator: Pravin S. Rana).

The 8th ACLA 2019 Conference on "*Perception and Aesthetics of Cultural Landscape in Asia*": 9th–10th October 2019, held at U.T.M. Johor Bahru, Malaysia, was organised by ACLA EC Member Prof. Ismail bin Said and Prof. Joharudin b. Hj. Samion. Sung-Kyun Kim delivered the keynote speech and Rana Singh's special address. In total, 36 papers were presented in the three paralleled sessions by participants from Malaysia, Indonesia, Korea, Italy, China, Thailand, India, Philippines, Singapore, and Japan.

The 9th ACLA International Symposium: 17–19 October 2019, DDU University, Gorakhpur (India) on "*Interfacing Environment, Sustainability, Law, & Cultural Landscapes of Asia*". In total, 225 participants attended, including from Korea, Japan, Chile, Indonesia, Nepal, and India. About 56 papers were presented covering the following themes: (i) Environment, Development, and Landscape, (ii) Sustainability and UN-SDGs, (iii) Cultural Landscape and Cosmogram, (iv) Heritage Cities and making of Harmonious World, and (v) Places of Religion: Sacrality, Spatiality, and Sustainability. The inaugural address was delivered by IGU President Prof. Yukio Himiyama, and keynote addresses were by Sung-Kyun Kim and Rana P.B. Singh (Fig. 2.3).

The 10th ACLA an International Symposium (under the aegis of IFLA–CL) on "*Historic Cultural Landscapes: Succession, Sustenance and Sustainability*": 18–20

Fig. 2.3 Sung-Kyun Kim and Rana P. B. Singh, at Geeta Mandir, Gorakhpur, 18 October 2018 (*Photo credit* Rana P. B. Singh)

November 2019, was held at Asia Research Centre, SNU, Seoul with the collaboration of the IGU Commission—C16.25 'Landscape Analysis and Landscape Planning', under the co-chairmanship of Sung-Kyun Kim and Rana Singh; both have delivered keynote speeches. This had broadly examined the status, role, and vision of Historic Cultural Landscape, HCL, taking together Historic Urban Landscape, HUL, and overall heritagescapes and placemaking in harmonizing the world as professed by Prof. Kim. Altogether 35 papers were presented by scholars from Malaysia, Korea, India, Thailand, Philippines, Chile, Finland, Czechoslovakia, Bahrain, Japan, and China.

With his concurrence and patronage, the 11th ACLA International Symposium was planned and held on '*Placemaking, Cultural Landscapes & Spiritual Tourism*': 26–27 September 2020 at Banaras Hindu University, Varanasi, India, where he was scheduled to deliver the special address, but in the meantime, Kim passed away. However, as a token of tribute the 'Sung-Kyun Kim Memorial Lecture' was delivered by Rana Singh, appraising Kim's contributions to the study of cultural landscapes.

After returning from the 9th ACLA International Symposium at Gorakhpur (India), he came to Varanasi, intending to have to experience the oldest Fossil Park at Salkhan (Sonbhadra), 102 km from Varanasi city and spread over an area of 25 hectares, which is reported as world oldest series of fossils. The age fossils found here are approximately 1,500 million years old and date back to the Mesoproterozoic period according to the geological time scale, and consist of algal stromatolites. Kim and Rana together made a full-day trip on 31 October 2018 (see Fig. 2.4). Kim was deeply interested to see the interconnectedness between the vegetal cover, the fossils, and the surrounding environs, altogether making a spiritual and majestic landscape. Together with his family members of Rana, around six hours of time was used as

Fig. 2.4 Rana P. B. Singh and Sung-Kyun Kim, in the Salkhan, Sonbhadra, 31 October 2018 (*Photo credit* Rana P.B. Singh)

understanding the awe and wonder of nature, and the social event of enjoying a picnic. The fossils appear as rings on the boulders scattered around the area and are regarded as one of the most important destinations to know the Earth's geological and biological past. This fossil park presents evidence regarding the emergence and the initial stages of the evolution of life on the Earth, thus conceived as a geological heritage (see Sharma 2006, and Shukla & Sharma 2016). The Salkhan Fossil Park is three times larger than the Yellowstone Fossil Park of the United States and is quite older than the latter one. Kim had prepared his experiential expositions in his diary with a lot of sketches, photographs, etc.

2.4 Sung-Kyun Kim: Contributions to Placemaking and Cultural Landscape

Professor Sung-Kyun Kim was an applied cultural landscape architect and was mostly involved in creating and promoting students through making a balance between lab work and field study. In this spirit, he has written several books in Korean. Such books (in Korean) include the subject and themes like Computer application techniques for Environmental planning and design (Kim 1977), Urban streetscape design (Kim 1993a), Computer-based applied landscape design (Kim 1993b) Use of GIS in conservation and management (Kim 1995), Ecological aesthetic analysis on urban forests (Kim 1996a), Oriental landscape architectural history (Kim 1996b), Landscape architecture of Seoul (Kim 1997a), Art and landscape (Kim 2001), Walking in the world city forests (Kim 2009a), History of oriental landscape architectural culture (Kim 2009b), and Analysis of environment-friendliness of waterproof and root barrier materials based on fish toxicity test (Kim 2010a, b).

His pioneering monograph (in English), based on his Ph.D. dissertation (1988), is the first publication under the banner of ACLA Press; it deals with poetics, aesthetics, visual, and experiential expositions of Kim's sentiment, attachment, and deep understanding of the spirit of place (Kim 2016). In the closing ceremony of the 2nd APELA on 18th October 2016, his classic book, *Winding River Village: the study of Hahoe*, was released and all the participants were honoured to receive the author's inscribed copies. On the next day, the 19th of October 2016, under the guidance of Sung-Kyun Kim all the participants paid a visit to Hahoe village, an enlisted folk village on the UNESCO World Heritage List (2010). This village was founded in the 14th–fifteenth centuries, and together with another nearby village Yangdong is seen as the two most representative historic clan villages in the Republic of Korea. Their layout and location—sheltered by forested mountains and facing out onto a river and open agricultural fields—reflect the distinctive aristocratic Confucian culture of the early part of the Joseon Dynasty (1392–1910). The villages were located to provide both beautiful physical scenarios and spiritual manifestations of nature (see Kim and Singh 2023).

In the foreword to his book, Singh (2016: pp. xv-xvi) narrates that, "This unique study elucidates how the intricate interplay of the natural topography, architecture, history, sociology, culture, folklore, and even indigenous and traditional science to be taken together to understand the "wholeness" and "mosaicness" perspectives of Asian rural cultural landscape, especially searching interlinks and commonality among Korean, Japanese, Chinese, and other cultural traditions of East Asia. From an uncommon synthesis of all the aspects of the human habitat, this book derives new directions of architecture and dwelling that should deserve to be followed universally".

Prof. Sung-Kyun Kim's conclusion is a signal post for further understanding of village landscapes, i.e. "The villagers love their history. Their history is embedded in every corner of the landscape. Thus, their landscape has been preserved well despite radical changes in the world" (Kim 2016, p. 260). He has further exposed that "The artistic, spiritual, religious, technical, economic, and political modes are collectively integrated into local meaning. The mental organization of the landscape is analogous to its social and technical organization. Nature is metaphorically converted into the culture, that is, meaning, and culture into nature; without such conversion, one would not be able to speak of landscape or one's culture" (Kim 2016, p. 260). Kim has successfully tried to validate the exposition and intimacy of landscapes that Dubos (1972, p. 152) has narrated: "No landscape, however grandiose or fertile, can express its full potential richness until it has been given its myth by the love, works, and arts of man".

This book, written from insiders' understanding, continues the frames of experiential exposition and reverential sensitivity to village landscapes and will set an empirical narration that I hope will be used by fellow spirit-seekers and researchers in the field of Asian Cultural Landscapes. As a memorial tribute to Kim, a 'Hymn to the Landscape of Hahoe' (Singh 2016, pp. xv–xvii) is cited below:

O! Great Land of Hahoe, My head fills with joy;

The scenes of your grandeur, For all of us to enjoy.

To witness the beauty, Of all that You bring;

Empower me greatly, To dance and to sing.

Dearest serene land for all, Who never will sleep;

For everyone, you nourish, And tender your sheep.

Of all you take care, Let everybody go;

The greatest of spirits, I love you so.

To dwell in your environs, Of divine beauty and joy;

The blessing of Hahoe, For all of us to enjoy.

O! intimate friend 'KIM', You've shown us all this;

The wonder and magic, The Landscape of Bliss.

With the support and inspiration of Prof. Kim, Rana Singh paid many visits to Hahoe (during 2012–2019) and felt himself "*being*" there (Hahoe) through belonging (existentiality), "*becoming*" there through adjustment (gracefulness), setting me "*behind*" there to be backing-base (basement), and also "*beholding*" himself with contemplation to become beholden (contemporaneous: living and being at the same time).

In his recent re-visit study of regenerating historic urban landscape with pedestrian-oriented streetscape design, illustrated with Deoksugung-gil Street in Seoul (Kim 1997b), Kim has concluded that "the concept of differential separation introduced to protect pedestrians from the risk of such a car played a major role in protecting pedestrians from cars on the main road, but most of the roads were also the cause of becoming vehicle-first roads. As a result, pedestrians were pushed to the narrow sidewalks, and many of the traditional functions of the street, which had served as a place of living in the past, have gradually disappeared" (Kim 2020, p. 390). His last study (published posthumously) describes Pung-su as a determinant force for evolving cultural landscapes and placemaking, and exemplified with village Hahoe—the village longitudinally studied by him since 1988. This study narrates the inherent *genius loci* and the visuality of cultural landscapes in this village and implicitly explains the visuality and the manifested meanings there that make the whole territory in the cosmogonic frame of sublimity (Kim and Singh 2023).

2.5 Sung-Kyun Kim: Vision and Action

On 4th December 2017, Sung-Kyun Kim was elected as the Chair of the International Federation of Landscape Architects (IFLA) Working Group on 'Cultural Landscapes', where he led to focus on creating cultural landscape programmes in developing countries and encourage understanding and use of Cultural Landscape as a base for Sustainable Development Goals within '*Local action Global vision*'. After joining, he declared the motto of his lifelong passion:

> Vision without action is only a dream,
>
> Action without vision only passes the time.
>
> But, vision with action can transform the world.

In his proposal and plan of action under the umbrella of IFLA CL, he has explained his vision and plans. Since 1992 he had participated in IFLA meets, served as a Korean delegate of IFLA for more than ten years, also as the Treasurer of IFLA Eastern Region from 1999 to 2002, and as the chair of IFLA APR Cultural Landscape Committee from 2007 to 2012. As a core member, Kim successfully organised IFLA World Congress 1992 Seoul-Kyeongju, IFLA ER Congress 1999 Yangyang, and IFLA APR Congress 2009 Incheon. Supported with ACLA, after joining the Chair of IFLA CLC WG, he has emphasised the five basic issues:

i. **Establishment**: to establish a distinct identity of Cultural Landscape in IFLA. It is realised that in the ICOMOS-IFLA Committee on Cultural Landscapes, the role and participation of IFLA have been weak because most of the activities organised by ICOMOS are focusing on the preservation of the Cultural Landscape. There is a strong pace to create a new environment for a sustainable future, emphasizing understanding, cross-cultural comparison, and regional identification, together with increasing participation and regional representation.
ii. **Extension**: to extend and expand the current list of Cultural Landscape sites and areas and create a Social Network System (SNS) to share cultural landscape data, including pictures, drawings, archival documents, ongoing planning strategies, etc.
iii. **Expansion**: to increase networking among Cultural Landscape researchers and increase the opportunities for practical academic interchange at a low cost by operating a system at rational and viable costs and location in the universities and the institutions.
iv. **Creation**: to create cultural landscape programmes in the developing countries to develop educational curricula for courses, and to encourage students to participate in the cultural landscape activities in the frame of 'locality to universality'.

v. **Awakening**: finally, to awaken and encourage understanding and use of cultural landscape as a base for sustainable development goals within 'Local action Global vision', and to establish the research groups for it.

Further, he had added that this is the time for action—using the phrase for IFLA Cultural Landscape: *"Actions speak louder than words."* But in a strange situation of complexity and politics together with regional hegemony in the IFLA committee, he had to resign his formal affiliation as president of ACLA on 4th June 2018; thus, with the concurrence of the EC ACLA members, he handed over charge in full capacity to Rana P.B. Singh (then Vice President ACLA) for the full term (2018–2023). After his passing away, in a difficult and challenging situation Rana Singh with the active support of Olimpia Niglio (Vice President ACLA), and other members, took lead to promote the path of ACLA, as paved by (*late*) 'Master' Prof. Sung-Kyun Kim.

Of course, Prof. Kim had widely travelled from all corners of the world, ranging from Chile and Argentina to South Africa and Australia, Malaysia to Russia and Germany, the USA to many countries in Europe, Iran, etc. He was known for taking initiatives to open studies of cultural landscapes and landscape architecture in many Asian countries; he visited India three times (2011, 2018, 2019), delivered special lectures, and inaugural addresses (ISOLA International Symposium at CEPT Ahmedabad, 7th ACLA 2018 Ayodhya, 9th ACLA Gorakhpur, respectively), and performed field studies in and around Ayodhya, Gorakhpur, and the surrounding Buddhist sites, and Gaya-Bodh Gaya and environs. In all these meets, Rana Singh was with him as coordinator and collaborator.

He left us on the way—the cosmic path of ACLA that he paved, who supervised, guided, and helped Rana personally since 9th September 2011, when he first met him in CEPT Ahmedabad (Gujarat, India); since then every year during 2011–2019 (ten times), Rana visited SNU Seoul, Korea on his special invitation for collaborative projects and delivering special/ inaugural lectures on the cultural landscapes of Asia, and performing field studies to promote Indo-Korean Cultural Interfaces (cross-cultural study of Gimhae-Korea and Ayodhya-India, based on their project, 2016–2020; see Kim and Singh 2023; Singh and Kumar 2023).

Prof. Kim had started ACLA-APELA (SNU Seoul, Korea) Zoom online Lecture Series, and inaugurated it with his slide-show Webinar on the topic "Regeneration of Historic Urban Landscape with Pedestrian-oriented Streetscape Design in Seoul: The Case of Deoksugung-gil in Seoul, Korea" for students and faculty of School of Planning & Architecture, Bhopal (India), held on 20 May 2020. This was based on his latest innovative research paper (Kim 2016, 2020). He had also initiated a followed-up Webinar online lectures in 2020: 15 July (Rana P.B. Singh) and 16 September (Olimpia Niglio), but due to his sad demise, these were cancelled. Also, Olimpia Niglio was awarded a collaborative project fellowship under the NRF Government of Korea to work with Sung-Kyun Kim in late 2020, but it failed to materialise due to Kim's death.

Prof. Kim had framed an Indo-Korean Interfacing Research Project on 'Comparative study of Cultural Landscapes of Ayodhya (India) and Gimhae (Korea, especially sites related to Queen HEO Hwang-ok and King Kim SURO)', and he led the team (Rana Singh and Sarvesh Kumar from India) to visit Gimhae for two days: 19–20 Nov. 2017, and thus was prepared the proposal. Later in 2018, this project took the formal approval by the NRF—Government of Korea ['General Research Collaboration Support Project', 20 Dec. 2018–2019 Dec. 2019, Project No.: 500-20180233], in which Sung-Kyun Kim served as Principal Investigator (Korea), and Rana P.B. Singh as Co-Investigator (India). Unfortunately, the report (in a form of a book) could not get into final shape due to his untimely death; of course, in March 2020, the final report in Korean was completed. However, two research papers are recently published (Kim and Singh 2023; Singh and Kumar 2023; both in this volume).

References

ACLA (2020) Newsletter 10.1, June. https://www.academia.edu/85780675/_470_20_1_ACLA_Newsletter_10_1_19_June_2020_HiLit_ED_New_And_late_Sung_Kyun_Kim_OBITUARY.

Dubos RJ (1972) A god within. Charles Scribner's Sons, New York.

Kim S-K (1977) Computer application techniques for environmental planning and design (Translated and compiled). Sungandang, Seoul [in Korean].

Kim S-K (1993a) Urban streetscape design (Translated and compiled). Kukje, Seoul [in Korean].

Kim S-K (1993b) Computer applied landscape design (Translated and compiled). Kukje, Seoul [in Korean].

Kim S-K (1995) The use of GIS for conservation and management of mountain landscape around city. J GIS Assoc Korea. 3(2):123–134.

Kim S-K (1996a) An ecological aesthetic analysis on urban forests. J Korean Inst Landsc Archit. 24(1):97–108.

Kim S-K (1996b) Oriental landscape architectural history. Munwundang, Seoul [in Korean].

Kim S-K (1997a) Landscape architecture of Seoul. Iche, Seoul [in Korean].

Kim S-K (1997b) A pedestrian oriented streetscape design of "Deoksugung-Gil", Seoul: adapting the shared street concept for traditional streetscape improvement. J Korean Inst Landsc Archit. 25(3):143–153.

Kim S-K (2001) Art and landscape. Panayotis & Effie Michelis Foundation, Athens.

Kim S-K (2009a) Walking in the world city forests. Iche, Seoul [in Korean].

Kim S-K (2009b) History of oriental landscape architectural culture. Daega, Seoul [in Korean].

Kim S-K (2010a) Analyses on environment-friendliness of water-proof materials based on fish toxicity test. J Korean Soc Environ Restor Technol. 13(1):57–68.

Kim S-K (2010b) Analysis on environment-friendliness of root barrier materials based on fish toxicity test. J Korean Soc Environ Restor Technol. 13(4):42–51.

Kim S-K (2016) Winding river village: poetics of a Korean landscape. ACLA Press, Seoul (xviii + 319pp, 2 tables, 183 B-W figures and illustrations; ISBN: 979-11-959082-0-2).

Kim S-K (2020) Regeneration of historic urban landscape with pedestrian-oriented streetscape design: the case of Deoksugung-gil street in Seoul, Korea. In: Kapila DS (ed.) The Routledge handbook on historic urban landscapes of the Asia-Pacific. Routledge, London, pp 377–391. https://doi.org/10.4324/9780429486470_20.

Kim S-K, Singh RPB (2023) Pung-su: Evolving cultural landscapes and placemaking in Korea (chapter 3, in the present book). In: Singh RPB, Olimpia N, Rana, PS (eds) Placemaking and cultural landscapes. Springer Nature, Singapore.

Sharma M (2006) Palaeobiology of mesoproterozoic Salkhan Limestone, Semri Group, Rohtas, Bihar, India: systematics and significance. J Earth Syst Sci 115(1):67–98.

Shukla B, Sharma M (2016) A new assemblage of large-sized microfossils from the Salkhan Limestone (>1600 Ma), Semri Group, Vindhyan Supergroup, India. J Palaeontol Soc India 61(2):287–299.

Singh RPB (2016) Foreword. In: Kim S-K, Winding river village: poetics of a Korean landscape. ACLA Press, Seoul, pp xv–xvii.

Singh RPB (2020) Obituary: Prof. Sung-Kyun Kim (1956–2020): Legacy, vision and exposition of the journey. Web-Pub. https://www.academia.edu/43413832/_619.20_Singh_Rana_P.B._2020_OBITUARY_Prof._Sung-Kyun_KIM_1956–2020_Legacy_Vision_and_Exposition_of_the_Journey.

Singh RPB, Kumar S (2023) Ayodhya, India: placemaking and transformation of historic urban landscape (chapter 8, in the present book). In: Singh RPB, Olimpia N, Rana, PS (eds) Placemaking and cultural landscapes. Springer Nature, Singapore.

Chapter 3
Pung-su: Evolving Cultural Landscapes and Placemaking in Korea

Sung-Kyun Kim and Rana P. B. Singh

Abstract In Korea, for choosing a site and settlement, the *Pung-su* principles, which interconnect the spirit of the natural landscape and human sensitivity, are the common practice in making, maintaining, and manifesting the cultural landscape as archetypal integrity of habitat. Like any other traditional Korean village, Hahoe has also grown up as a natural consequence of spatio-temporal transformation within the Pung-su principle, which in Asian culture is broadly known as the spontaneity of human habitat. The imbued and manifested meanings of the landscape are conveyed, from one person to another and also from one generation to another, in the frame of Pung-su, which is closely identical to *Feng-shui* in Chinese landscape, and in both cases, it means "wind and water." These natural elements are mostly responsible for regulating the cultural notions and traditions in Korean landscapes. With mutual support and interfacing reciprocally, villagers felt themselves "*being*" here through belonging (existentiality), "*becoming*" there through adjustment (gracefulness), setting them "*behind*" there to be backing-base (basement), and also "*beholding*" them with contemplation to become beholden (contemporaneous: living and being at the same time). Illustrated with a study of village Hahoe (inscribed in UNESCO WHL) through several experiential visits (during 2011–2019) to understand and experience the inherent *genius loci* and the visuality of cultural landscapes in this village territory, an attempt is made to understand the visuality and the manifested meanings there in that make the whole territory in the cosmogonic frame of sublimity.

Keywords Pung-su · Hahoe · Spontaneity · Cultural notions · Village territory · Experiential expositions · Placemaking · Geomancy · Feng-shui

---

Sung-Kyun Kim (1956–2020): Deceased

S.-K. Kim (Deceased)
Landscape Architecture, Seoul National University, Seoul, Korea

R. P. B. Singh (✉)
Cultural Landscapes and Heritage Studies at Banaras Hindu University, Varanasi, India
e-mail: ranapbs@gmail.com

R. P. B. Singh et al. (eds.), *Placemaking and Cultural Landscapes*,
Advances in Geographical and Environmental Sciences,
https://doi.org/10.1007/978-981-19-6274-5_3

3.1 Culture and Pung-su: Nature-Human Interconnectedness

Culture is a symbolic meaning system—a pattern of meanings embodied in symbols, which is a vital force for placemaking. It is widely accepted that culture refers to "a historically transmitted pattern of meanings embodied in symbols, a system of inherited conceptions expressed in symbolic forms by means of which men communicate, perpetuate, and develop their knowledge about and attitudes towards life" (Geertz 1973: p. 89). Moreover, a symbol "is used for any object, act, event, quality, or relation which serves as a vehicle for a conception—the conception is the symbol's 'meaning'" (Geertz 1973: p. 91). In tradition and practice, *Pung-su* refers to a shared meaning system encoded in the landscape; thus, Pung-su is both an archetypal set of symbols and a way to combine these symbols into the ritual language that evolved in the past and maintained and continued over time. Symbolic meanings are not simply intrinsic qualities of physical space or landscape. They are imposed upon the landscape by culture, resulting in the formation of the cultural landscape. Cultural meanings stand for the past experiences of a cultural group and each member of the group. This meaning is culturally determined and taught. This way the villagers perceive the hill in the shape of a reclining ox or a flower; the child in Hahoe learns to see the hill as a reclining ox or a flower. Of course, this meaning might be differently projected in a different cultural group; the same hill, for example, might be a tiger or a gourd. Myth, arts, and folk tales illustrate life experiences and also teach inherent and imposed meanings therein.

To explain the idea of attaching or manifesting meanings in religious terms, people cannot escape from the stream of nature-based and culture-based experiences, and also, they cannot cease to manipulate these experiences into patterns that are meaningful to them. People seem to accept that human life persists under the simultaneous influence of supernatural and natural forces. They also know from experience that the body and the mind are directly affected by the forces of nature. Thus, they order their experiences and delimit them through myths, rituals, and religious systems that explain how the world was formed, how they came to exist within it, how they should order our lives, and what they might expect from the earthly experiences. Says Eliade (1957: p. 30), "The sacred reveals the absolute reality and at the same time makes orientation possible; hence it founds the world in the sense that it fixes the limits and establishes the order of the world." We continually impose order on our experience to survive. This order, or "cosmos" in Eliade's terminology, satisfactorily answers the questions people ask about their deep experience; it is a fundamental requirement of human well-beingness. Order emerges continuously from interrelationships inherent in the system's structure, constantly creating a fundamental pattern that survives as long as the whole survives (Dubos 1972: pp. 9–10). Being able to understand an experience, even to predict future experiences, is essential to survival.

It is essential because unpredictable phenomena and threats are sources of great anxiety. In their experience of landscape, people's anxieties are related not only to natural factors, which are unpredictable and uncontrollable but also to social-cultural

Fig. 3.1 Hahoe village, panoramic view (*Source* Courtesy of Sung-Kyun Kim)

factors which lie beyond their control. Especially in an agricultural society like the studied Hahoe village (Fig. 3.1), there are fears of storms and floods, too much rainfall or too little rainfall, and infertile land. The actions of people and groups outside one's circle are as uncontrollable and unpredictable as natural factors. One of the motivations for the Pung-su is to order experience and alleviate the anxiety of unpredictability. Geomancy is characterized as "a unique and comprehensive system of conceptualizing the physical environment which regulates human ecology by influencing man to select auspicious environments and to build harmonious structures (i.e., graves, houses, and cities) on them" (Yoon 2017: p. 1). In Korean folk narratives related to geomancy (Pung-su in Korean, Feng-shui in Chinese), the issue of Nature-Human interrelatedness is a dominant theme and is historically represented in landscape and settlement (see Yoon 2021). The *Pung-su* principle generates the landscape of a clan village, like Hahoe, where the mountain's location and shape in setting the site are essential factors due to the earth's ethereal energy of *gi* (*qi* in Chinese). Thus, in the metaphysical frame, all houses and structures become scattered on the slope that takes a view of mountains and forms a village landscape interconnected with mountains and rivers (Lee-Niinioja 2022: p. 32). It is remarked that Korean geomancers relied more on field observation than their Chinese counterparts, in part because many of them were not able to read geomantic textbooks written in classical Chinese (Yoon 2017: p. 65).

In the choice of the site for a home or a grave, people are directly involved with both natural and social unpredictability. In choosing a home site, people are concerned with natural conditions, such as floods in summer, wind in winter, water for drinking and agriculture, land to cultivate, and so on. They are also concerned with social factors because siting does not involve simply locating a physical convenient place but inserting something into a neighborhood. It is the basis for establishing relations with neighbors, such as a muting competition. Grave siting is also concerned with unpredictable natural and human factors, although it affects people indirectly. When people choose a bad home site, they suffer as long as they live there. When they choose the wrong gravesite, they think they must suffer for generations. Thus, in the siting of a grave as well as a home, people face the anxiety of unpredictability. In this highly unpredictable world, people have yet another anxiety, the difficulty of making decisions. For people siting a house or a grave, there are several alternatives. The criteria for making choices are uncertain, and yet, the decision itself is highly consequential. Pung-su is a pathway to solve these anxieties concerning choices to be made about the landscape.

People realize their dependence on natural forces and then note the constant enforcement of natural forces. They see that the four seasons run their course, they know the direction of prevailing winds, and they observe what kind of site is regularly flooded. They also observe the regularity in the sphere of human life, rise and fall, birth and death, happiness and misfortune. They then see the relationship between these factors and possible sites—the Pung-su idea is notably more about flexible relationships than fixed entities. This way variety and distinct meanings are manifested.

There is a universal principle pervading all things, whether in the realm of physical nature or the sphere of human life. This universal principle expresses itself in the *Yin* and the *Yang*, or the negative and the positive forces, and the Five Elements (air, water, fire, earth, and metal/or sky/space), which account for the organic continuity; however, they are neither directly tested, nor disproved. Pung-su brings the metaphysics of Yin and Yang and the Five Elements to common use. Pung-su symbols formulate a basic congruence between the experience of life and a metaphysical understanding, and in so doing sustain each with the borrowed authority of the other.

People evolve ways of measuring regularities at a local level as well as at a universal level, gauging good or bad fortune in Pung-su, good fortune being wealth, good crops, many sons, and so on. When people observe that time and space are regularly beneficial, then time and space become symbols of good fortune, and their opposites become symbols of bad fortune. People also often regard their landscapes in the same light as their social and economic conditions. Some symbolic images, such as Azure Dragon, White Tiger, Black Tortoise, or Red Bird, are used as symbolic manifestations of types of social character or events. The natural and social influences are the interaction between the symbols. Pung-su is a system of these symbols. Pung-su is a way of integrating the human experience of the past, giving meanings to such experience, and facing the present and future with those meanings. Also, Pung-su is a way of conceiving the invisible and perceiving reality and dealing with it (Feuchtwang 1974).

When people manifest these symbolic meanings to landscapes, the landscape becomes more controllable. When they symbolize the landscapes, the action is consequential. When a mountain is referred to as an animal form, a symbolic manifestation of the natural and social character of the place or people, the fate of the people, living or dead, who reside on or around it, is related to the animal's character, and further activities depend on that form's symbolic meanings. When people believe symbolism projected onto the landscape to be the manifestations of the cosmological forces, they get a sense of control over natural forces and the environment. For instance, if the landscape symbolizes a "flying phoenix," people living in that place will try to keep the phoenix, thereby making a good well for its drinking water along with many small hills, like phoenix eggs, so that the phoenix might enjoy the good drinking water and love the eggs and thus not fly away (see Yoon 1976).

To solve anxiety, people try to regularize the making of decisions in an irregular and uncertain field of choice. For example, to resolve such anxieties of decision or choice, they adhere to principles of divination. Pung-su has powers of divination although they are largely based on practical knowledge based on understanding

Nature-Man interrelatedness. The rules of Pung-su routinize or ritualize the choice of site. The metaphysical and archetypal rules become criteria for the selection of sites when practical criteria have been exhausted. In fact, through the ritualized or routinized Pung-su, the burden of the decision of where to site a house or grave is taken away from the individual. As Geertz (1973: pp. 87–125) points out, man's anxiety is allayed when symbols are employed in rituals. Whenever the practical criteria for action and decision have been exhausted, the action is ritualized by reference to the Pung-su principles, from which people borrow authority.

It is argued by Durkheim and Mauss (1963: p. 77) that the classifications of space and the countless correlations (as expressed in the geomantic compass) in terms of Chinese symbolism were essential attributes of divination; moreover, it is notable that the knowledge of diviners did not form isolated groups of things, but rather groups bound to each other to form a single whole, resulting to the unified knowledge. Simultaneously being a standard system of metaphysics and a form of divination, the elements of Feng-shui were "transposable" into Chinese religion, and over time, it reached Korea (Freedman 1969: p. 331). This approach is also supported by Feuchtwang, who conceptualizes Feng-shui as a way to perceive or conceive reality, but also as a way of dealing with this reality, which is an activist approach to this world (Brunn 2011: p. 17). Feng-shui operates like a cosmological model, applied to reality to serve specific interests. Having previously accounted for Feng-shui symbolism, which is a discourse with language as the analytical model, and combining this with the natural cosmological model composition contained in Feng-shui, Feuchtwang (1974: pp. 249–250) sees these two closely linked in a simple scheme of a model, justifying "its application or exposition progressively more embodied—first in the physical environment and then in social fortunes."

By resolving the anxieties caused by an unpredictable and uncontrollable world, Pung-su increases people's confidence. By living in an auspicious place, locating a building or a grave on a proper site, and facing it in the auspicious direction following the rules of Pung-su, fears and doubts of uncontrollable natural and social factors are minimized. Furthermore, people can then assume the future to be predictable and good for themselves and their descendants. The invisible factors of chance are thought to be under control (Yang 1967: p. 265). They will be comfortable in the place because they face the future with confidence (see Freedman 1969: p. 14). Through ritualized experience, they get an easy and comfortable present life and face a confident future. Space and time are interlocked through Pung-su; psyche and landscapes are blended through Pung-su. Ultimately, what a good Pung-su means is that an individual will be completely at ease in a place (Freedman 1969: p. 14). An old lady dweller told, "A good gravesite is a place in the mourners' favour rather than what the geomancer said."

3.2 Pung-su in Korea

Pung-su is concerned with the harmony of forces and their relation to such groups as the family and clan (through house site) and the village community (through adjacent buildings, roads, etc.). In Pung-su, "a building is not simply something that sits upon the ground to serve as a convenient site for human activity. It is an intervention in the universe, and that universe is composed of the physical environment and men and the relationships among men. … When a man puts up a building, he inserts something into the landscape and between him and his neighbors" (Freedman 1969: p. 7). Thus, the inhabitants tried to get them interconnected with cosmic forces of landscape, making an archetypal representation of interlinking human and divine. The act of building disturbs an existing balance of nature and society. The new balance must be carefully adjusted and accounted for to prevent conflicts.

Of course, the basic frame of Pung-su in Korea came from ancient China; however, Koreans have developed their own system befitting their own environment and landscapes. It is important to understand early Chinese philosophy related to the theory of Pung-su, which emphasizes that all things and events of the world are products of the two elements, *Eum* (*Yin*) and *Yang*. Chou Tun-I (CE 1017–1073), one of the founders of Neo-Confucianism, in his book, *An Explanation of the Diagram of the Great Ultimate* (*T'ai-Chi T'u-shuo*) summarized the doctrine of *Yin-Yang* and the Five Elements (Kim 2013: p. 29):

> The Great Ultimate through movement generates yang. When its activity reaches its limit, it becomes tranquil. Through tranquility, the Great Ultimate generates yin. When tranquility reaches its limit, activity begins again. So, movement and tranquility alternate and become the root of each other, giving rise to the distinction of yin and yang, and the two modes are thus established. By the transformation of yang and its union with yin, the Five Agents of Water, Fire, Wood, Metal, and Earth arise. When these five material forces (*ch'i [ki]*) are distributed in harmonious order, the four seasons run their course.

The Five Agents constitute one system of yin and yang, and yin and yang constitute one Great Ultimate. The Great Ultimate is fundamentally the non-ultimate. The Five Agents arise, each with its specific nature (Kim 2013: p. 29). When the reality of the Ultimate of Non-being and the essence of yin, yang, and the Five Agents [of cosmic organism] come into mysterious union, integration ensues and creative force generates through the interaction between *Ch'ien* (Heaven), constituting the male element, and *k'un* (Earth), the female element. This interaction engenders and transforms myriad things. The myriad things produce and reproduce, resulting in an unending transformation.

To determine the orientation of the family house, their ancestors observed the shadows and examined the sunny and dark sides, the "yang" and "yin" of the country, to see how the constituent principles of the world were distributed. They also observed the direction in which the waters flowed. They must know the religious value, or Pung-su, of the site. Their decision was also governed by practical considerations (supplies of food, fuel, and water) and cultural factors; first among them was the mythical and historical value of the site.

How can one locate an auspicious house site and dwell there in harmony with the natural and cultural environment and thus extract benefit from the location? An old resident expressed that a favorable house location, the direction a house faces, the spatial organization of a house, and the people who live there are all important. The direction a house will face is probably more important than any other element in house geomancy because opportunities for choosing a house site concerning its surrounding environment are very limited and most houses are built within or around settlements. In terms of Pung-su, the "siting" (*jwa*) and "facing" (*hyang*) directions of a house are the most critical factors in determining its auspiciousness. This is, because the directions of important house elements, especially the main gate and the main room, should not conflict with the directions of the siting directions, but should be in harmony with them. Following the classical tradition, one of the distinct houses is located on a hilltop in the center of the ridge from Flower Mountain; thus, the house could get the vital spirit of the mountain (Kim 2013: p. 64).

In most sacred traditions, mountains are the dwelling places of the divinities and also preferred sites of spiritual journeying. The texts dealing with these aspects describe the sacred traditions that attach to mountains in every continent of the world and explore through discussions with travelers, climbers, and adventurers, the transforming effect of mountains on their own lives in the modern world. The experienced visitors narrate the feelings while mentioning the most deeply felt personal journeys through new insights and turning points—finding freedom of worship, celebration, meditation, isolation, and solace; experiences of returning to the source and of spiritual transformation through the deeply-experienced sense of loneliness and wilderness, darkness, and suffering; discovering God's earthly paradise; being overwhelmed by a sense of one's liberating universal wisdom, but also knowing fear in the encounter with the sacred (see Cooper 1997: cover flap, also pp. 312–314).

Certain places in the landscape have a "breath of life" (*ki*, or *Ch'i* in Chinese), and such places are more favorable than others for a town, village, house, or grave. It is noted that the difference "between Feng-shui (or, Pung-su) of the dead (*yin*) and of the living (*yang*) is obvious. Dwelling divination manipulates inherent forces—grave divination transcends this-worldly affairs and reaches out to the world beyond. It forecasts the fate of men as well as of society and attempts to manipulate both history and the future" (Brunn 2011: p. 260). The auspicious place can be acquired only by applying Pung-su principles to the examination of the local landscape. When it is acquired and occupied, people who live on the site or the descendants of ancestors buried on the site can be influenced by the auspiciousness of the place. In other words, one arranges oneself with the forces of nature in the physical world to obtain maximum benefits—wealth and social position, for example. This idea is called Pung-su in Korean, and Feng-shui in Chinese—both of which mean "wind and water." In ancient Indian literature, these two cosmic elements are known as *Prana*, i.e., "*Pra*"—Air—and "*Na*"—water. Pung-su is also considered a metaphysical mapping of the landscapes, which is in a way plasticized as the theoretic construct of wind and water. There is no tradition of geomantic mapping for most of the world's sacred places, so the Hahoe geomancy maps may offer a rare opportunity to help achieve biocentric and ecophilosophical insights about humankind's place in Nature and its relation to the universe of things. Western scholars often describe Feng-shui as a "pseudo-science." It is remarked that "Geomancy maps can be useful and satisfying as intuition-building models of the Neo-Confucian cosmos." A map can be treated as a "morphological" view of the universe. Its visual appearance is a kind of "reality test" of the comprehensiveness of Neo-Confucian cosmology. Thus, it may be possible for outsiders to participate in the Neo-Confucian reality by attempting the following exercise in "therapeutic cartography" (Nemeth 1994: p. 94).

The basic principle of activating Pung-su is the storing of wind, and the second is the acquiring of water. The breath of life (*ki*) is dispersed by wind and blocked by a watercourse. This means that an auspicious place is mainly determined by the location of surrounding mountains which affect its exposure to wind and by the orientation to watercourses. That is why this art of siting houses and graves is called "wind and water" (see Fig. 3.2).

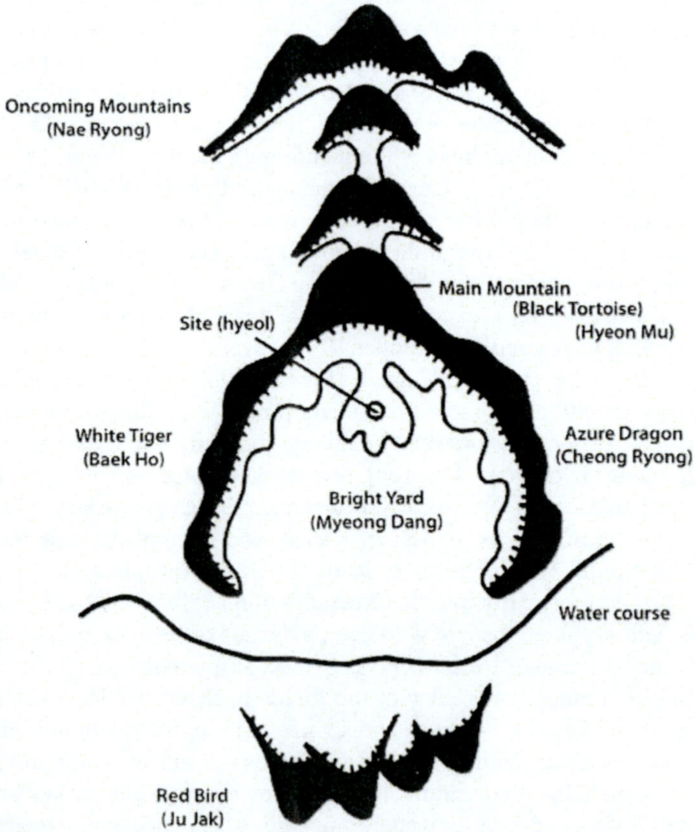

Fig. 3.2 Four mountains and a watercourse for a sacred site in Pung-su (*Source* Courtesy of Sung-Kyun Kim)

The hills and ridges which carry the breath of life are called *dragons*. This may be because, like a dragon, the ridge writhes, runs, turns, and suddenly changes direction. The dragon in oriental society stands for a vast, infinitely changeable, imperial, natural power, which is generally considered good and auspicious. If the ridge has the breath of life, it is called "dragon." If the ridge does not have it, it is called a "barren mountain." Villagers call this Flower Mountain "Nine Dragons" because it has nine ridges. The main dragons ramify into lower orders—trunks (*gan*) and branches (*ji*)—which affect progressively smaller areas and fewer people, and are less and less potent (March 1968: p. 257; De Groot 1897: p. 1009). The breath of life (*ki*) travels on the ground and moves following the terrestrial configurations. Where the *ki* is in transit, in the middle of a ridge, a dragon (*yong*) is of no use. The *ki* must stop and coalesce at a stream junction, forming a situation (*guk*) with local mountains (*sa*; literally, "sands") and waters, within which a site (*hyeol*) may be found. The terrain around a site "should have the formation of an armchair; those [the hills] in the rear should fence off evil spirits brought in by the cold bitter northerly wind; those on the left and right should flank it like embracing arms, and the front should be opened or unobstructed such that view, airiness, and sunlight can be brought in by the yang spirit" (Lung 1980: p. 85). These features are represented by the four animals: the azure dragon on the left side of the site as you face the open view, the white tiger on the right, the red bird in the front, and the black tortoise on the back (Fig. 3.2).

3.3 The Attributes Framing Pung-su and Promoting Placemaking

Theorizing the attributes of Pung-su and selection of sites for village settlements, a noted Korean geographer at the beginning of the eighteenth century, Yi Jung-Hwan (1690–1756), explains six landscape features for examining and selecting a preferable vital site for settlement in his classic book, *Taekni-ji* (1751 in Chinese, translated in Korean in 1971), which is often quoted by the villagers (Yi 1971: pp. 162–165). He explained that, first, observe *Sugu* or the mouth of a watercourse; then *Yase* or the lay of the fields; then *Sanhyeong* or the shapes of the mountains; then *Tosaek* or the color of the soil; then *Suri* or the availability of water flow; and finally observe *Chosan* and *Chosu* or the facing mountain and waters. If one desires to build a house or wants to pass it on to his descendants, one should select the location by observing the geomancy [of the place]. The above six factors are the essence [of the geomantic principles] (Yi 1971: pp. 162–165). They are described in short as below.

Watercourses. If *Sugu* (the mouth of the watercourse) is loosely organized and just spacious, even though the place has a lot of farmlands and enough houses on it, prosperity cannot be extended to the next generation. Those who live there will naturally be scattered and will disappear. Therefore, when one searches for and observes a house site, one should look for a stream whose water discharge cannot be observed and a field enclosed by mountains. While it is easy to find such a watercourse in a mountainous area, it is not easy to find it in a plain. Thus, in the plain, there should be a stream flowing toward the house site. Whether it is a high mountain or a low land (Yin hill), if water flows nearby toward the place, it is auspicious.

The Features of a Field. Generally, man is engendered (is born or is viable and productive) through receiving Yang force. Since the sky is the Yang light, a place surrounded by high mountains where only a small part of the sky can be seen is not a good place to live. For this reason, a broad field is a beautiful place [a good place to live]. The light of the sun, moon, and stars will always shine brightly there, and the various weather conditions, wind, rain, heat, and cold will be mild and moderate. In such a place, many great men would be born and there would be few diseases. [The place] which should be avoided most [in the selection of a place to live] is an area, which has a late sunrise and an early sunset due to the obstruction of high mountains in four directions.

If the spiritual light of the Big Dipper is not seen [in a place] at night, it always has a small Yin force. When the Yin force takes the ascendancy, it causes many ghosts to come in, mountain fog, and miasmas in the mornings and evenings, and thus, people become ill easily. For these reasons, living in a narrow valley is worse than living in an open field. Low mountains surrounding a big field should not be called mountains. They are also called fields, because those areas are not cut off from the light of the sky, and the power of water flows distantly. In high mountains, if there is an open field, it is also a good place to live.

The Forms of Mountains. Generally, the best forms of mountains are, as geomancers say, a feature of high projection for an Ancestral Mountain and a beautiful, neat, clean, and soft feature for a Main Mountain. ... The features that one should most avoid are an Oncoming Dragon [mountain range] which is weak and dull and lacks vital energy, and a mountain shape which is broken or crooked.

The Color of the Soil. Generally, in rural settlements, the soil should be sandy soil that is firm in structure and fine in texture not only at the bottom of the water but also on the edge of the water. Then, there will be cool and clean wells. These constitute the most ideal conditions for a people's livelihood. If the soil is red clay or black sand and gravelly soil or fine yellow clay, it is considered "dead," and water from such soil will be unpleasant (pestilential) without exception. Such a place is not suitable for human life.

The Availability of Water. Generally, man cannot live in a place with no water available. A mountain should have streams; then, it can engender the mysterious transforming power of auspiciousness. The outlets and inlets of water flow should be in accord with the principles [of geomancy]. Such a place is auspicious. ... The village (house) site is different from the gravesite. Because water controls wealth and prosperity, there are many wealthy families and famous villages close to the big water. Even in the mountain area, the place where the mountain streams meet is a good site to live on for a long time, from generation to generation.

Facing Mountains and Facing Waters. If a Facing Mountain is a rugged and ugly stone mountain, a tilted lonely hill [a shape caused by landslides], or a spying or thief mountain [suggested by the summit of a mountain is partly visible in the back of a mountain range], it is not a good place to live. If the mountain profile does not have a rugged and hateful appearance, it is an auspicious mountain. Facing Waters means Outer [distant] Waters. A small stream or river should flow toward the site [in the direction of the dragons (mountain ranges), to combine Yin and Yang forces]. But it is not auspicious for the big river to flow toward the site. In this place, even though one prospers at first, he perishes eventually. Thus, this kind of place should be avoided. … The oncoming water should flow slowly and with many turns from a long distance; it should never flow in a straight line.

3.4 Hahoe (Hahwe): Village on the Winding River

Hahoe (Hahwe), one of the most tradition-bound villages in Korea [36.539054° North and 128.517981° East], and considered the most representative historic village in Korea, is located in Gyeongsangbuk-do Province, southeastern part of the Korean peninsula, and belongs to Andong City. The village is situated at the west end of a peninsula-shaped landform surrounded by a bow-like curve of the river. The river dike, built by the government at the end of the 1970s, winds around the village and follows the curve of the river. On the west dike, several old zelkova and hackberry trees grow. Fields are located between the residential area and the mountains. The river flows from the east, winds around the village, and then turns to the west. Mountains surround and bound Hahoe (Figs. 3.3 and 3.4). The genesis of this type of settlement dates from the late Goryeo dynasty (918–1392). They become typical of Korean villages from the later part of the Joseon dynasty (1392–1910). They are now on the decline with the rapid urbanization and industrialization of Korea. Fortunately, the central, provincial, city, and local governments are taking collaborative responsibility for safeguarding them (Galla 2012).

In 1980, it was designated as a folk village by the Korean government, one of only five such villages in Korea that preserve their old traditions relatively well, physically and socially. Recently, it has been the focus of national interest due to a movement toward the recovery of "cultural roots" and historic preservation in Korea. The landscape of Hahoe seems to have valuable potential for finding vernacular ideas for future environmental design and planning. Surrounded by the Nakdonggang River, the village is home to descendants of the Ryu clan of Pungsan, which still makes up 70 percent of the villagers. The village became even more famous after Queen Elizabeth of England visited on April 21, 1999, and President George H. Bush of the USA in 2005. Population and households of Hahoe village (2008) recorded 125 households, inhabited by 238 people (115 men, 123 women). In 1987, there were 107 households, inhabited by 350 people (174 men and 176 women) (Kim 2013: p. 8). The population growth since 2008 has been stagnant, in a way it declined because many of the villagers now migrated to the city; however, they maintain their occupancy and use their houses for tourist resorts. Most of the village people engage in agriculture. The village population is predominated by *Ryu* families, consisting of around 66 percent of the households, and control over about 85 percent of the village

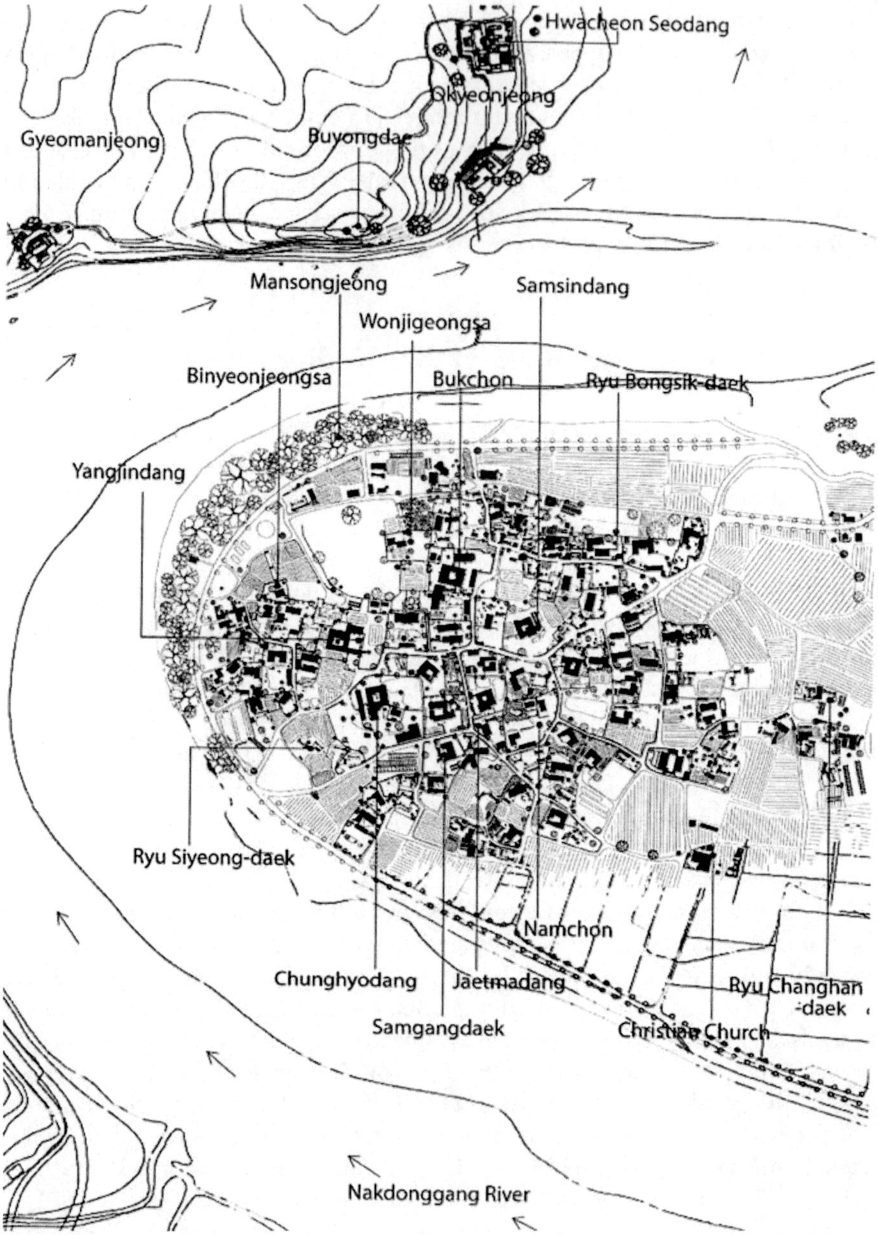

Fig. 3.3 Hahoe village: Space perspective (*Source* Courtesy of Sung-Kyun Kim)

Fig. 3.4 Scenic view of village Hahoe (*Source* Courtesy of Sung-Kyun Kim)

land. Most of the *Ryu* families do not farm their land but rent it to others. The owners and tenants share their products—on a generally half-and-half basis.

In Hahoe village, four kinds of landscape attributes define its distinctiveness: the village settlement, the field, the mountain, and the river (see Fig. 3.5). These four landscape attributes are not only divisions of visual form; they are also the symbolic divisions of the villagers. These landscapes together form the indispensable conditions for an ideal village—they are interrelated, interconnected, and intermingled too. Historically, dwellers conceive the four responsible factors for the selection of the favorite sites for settlements, viz. the geographical features, economic conditions (fields and water), the characteristics of the people, and the beautiful natural surroundings (mountains and water) for recreation (Yi 1971: p. 161).

The two villages of Hahoe and Yangdong were already enlisted under the National Folklore Cultural Heritage (designated on January 14, 1984); however, together they have been inscribed in the UNESCO World Heritage List on July 31, 2010. These villages have preserved beautiful landscape scenarios of heritage values, both the tangible and intangible attributes. These villages still possessed well-preserved houses, and Confucian rituals, which helped to promote research resources for Confucian academies, and scholars of vernacular architecture. Their exemplary characteristics as clan villages are evidenced by the well-kept Korean traditions of ancestral rites, village worship, and communal games, and their maintenance of a harmonious system of farming.

The following criteria of WHL (see UNESCO 2005, 2021) have been approved for nominating the village in the list:

Fig. 3.5 Hahoe village, a typical folk house in the natural setting (*Source* Courtesy of Sung-Kyun Kim)

III To bear a unique or at least exceptional testimony to a cultural tradition or to a civilization that is living or which has disappeared;

IV To be an outstanding example of a type of building, architectural or technological ensemble, or landscape that illustrates (a) significant stage(s) in human history;

V To be an outstanding example of a traditional human settlement, land use, or sea use that is representative of a culture (or cultures), or human interaction with the environment especially when it has become vulnerable to the impact of irreversible change;

VI To be directly or tangibly associated with events or living traditions, with ideas, or with beliefs, with artistic and literary works of outstanding universal significance. (The Committee considers that this criterion should preferably be used in conjunction with other criteria, as cited.)

The nominated property of these folklore heritage villages includes four areas: (i) the residential area with houses, (ii) the spiritual area, (iii) the village setting, and (iv) the natural landscape of rivers and mountains.

The boundary of the nominated property in Hahoe is demarcated by linking the visible panoramic points of the remote mountain peaks surrounding the village landscape. This boundary matches with the "Cultural Heritage Protection Area" authorized by the central government. The property boundary of Byeongsanseowon is managed by Confucian Academy following the estate bylaws; however, the village has been protected under the National Heritage Protection Act since 1984. The buffer

Fig. 3.6 Bird's eye view of Byeongsanseowon (*Source* Courtesy of Sung-Kyun Kim)

zone boundaries are determined in correspondence to follow the surrounding mountain peaks and the riverscapes. The criteria of delineation include the four mountain sceneries viewed from the village heights and as well as the river bed viewed from Byeongsanseowon Confucian Academy; these altogether constitute the famous 16 scenic sites of Hahoe Village (for example, Fig. 3.6). The heritage zone of the Hahoe consists of two parts, i.e., nominated properties (501.2 ha) and the surrounding buffer zone (566.1 ha); thus, the total area reaches 1067.3 ha. For Hahoe village, the boundary of the Cultural Heritage Protection Area covers the shared buffer zone and, in some instances, even extends to nearby surroundings.

The physical characteristics of the village have remained intact in their natural settings, site layouts and scenic points, neighborhood landscapes, and individual buildings in terms of their style and building materials. The intangible heritage of the village, such as the *yangban* (nobility) tradition, community cultures, and folk activities, is also well preserved, conserved, and continued even today as a symbolic icon of the cultural history.

Existing since the fourteenth-fifteenth centuries, village Hahoe together with Yangdong are projected as the two most representative historic clan villages in the Republic of Korea. Their layout and environmental setting—wrapped by forested mountains and inclined to face toward the river course and all-around scattered agricultural fields—all together remind the distinctive aristocratic Confucian culture of the early part of the Joseon Dynasty (1392–1910). These settlements include residences of the head families, represented by a majority of timber-framed houses of other clan members, also pavilions, study halls, Confucian academies for learning, and clusters of one-story mud-walled, thatched-roofed houses, formerly used for visitors. The landscapes of mountains, trees, and water around the village are framed in ways visible from pavilions and retreats; these aesthetic structures and scenes have been vividly described by poets of the seventeenth and eighteenth centuries.

The village has been located to provide both physical and spiritual nourishment from their surrounding landscapes, based on the spiritual-metaphysical basis of Pung-su. Pung-su has been one of the most important elements in regulating cultural behavior as it relates to the Korean landscape and is illustrated vividly in Village Hahoe. It has been transmitted from one generation to another in this village. The metaphysical and archetypal purview of the village setting is described by Lee-Niinioja (2022: p. 32), "The spiritual areas are divided into practicing places of Confucianism and Shamanism. The Confucian areas include study halls and pavilions for the literati's studies, recreation, and private Confucian academies for worshipping individual sages and teaching. The areas of Shamanism contain the shrine for the village spirit, the shrine for supplications of childbirth (*seonangdang*), and the big old tree in the village worshipped as a village spirit (*samsindang*)."

In the village ritual, the mountains and water link heaven and earth, establishing the vertical order, while in Pung-su, they establish the horizontal order, with four mountains in the back and front, on the right and left, and water in the front (Fig. 3.7). These natural elements are mostly responsible for regulating the cultural symbolism and traditions in Korean landscapes. Through these reciprocal and interfacing systems, villagers felt themselves "*being*" here through belonging (existentiality), "*becoming*" there through adjustment (gracefulness), setting them "*behind*" there to be backing-base (basement), and also "*beholding*" themselves with contemplation to become beholden (contemporaneous: living and being at the same time). In Hahoe, mountains and waters are never far from the people's minds. Whether it stemmed from Pung-su or not, the veneration of mountains and water is very ancient. The villagers have lived close to their land, and it is one of their most treasured possessions. Their close ties with the land have led them to revere nature, and natural powers have become fantastic concepts or peculiar symbols. From the earliest times, villagers have been content to perceive simply and devoutly the godlike quality of mountains and water, of trees and sky. They perceive divinity wherever nature is mightiest. The highest mountain is held to be the protector of the country it dominates. It keeps the surrounding land in its power, defending it from floods, diseases, and other evils. In each smaller province, there is a smaller sacred mountain. And every village has its sacred mountain behind, which is connected to the superior mountain physically or symbolically, and it has a modest shrine to the local divinity, from whom the villagers ask blessings.

There remains a deep sense of aliveness in Hahoe villagers, a feeling and an intimate attachment to the landscape; the mountains and water are more firmly rooted, more creative, and more vigorous than anywhere else in Korea. It may be because the village is isolated, thus less westernized, but closer in terms of space and more intimate in terms of interrelatedness. It also may be because of villagers' attachment to their auspicious land. The relationship with the landscape can be said to have manifested itself in their religion, philosophy, painting, and poetry and in constant devotion to the landscape itself (cf. Tuan 1975).

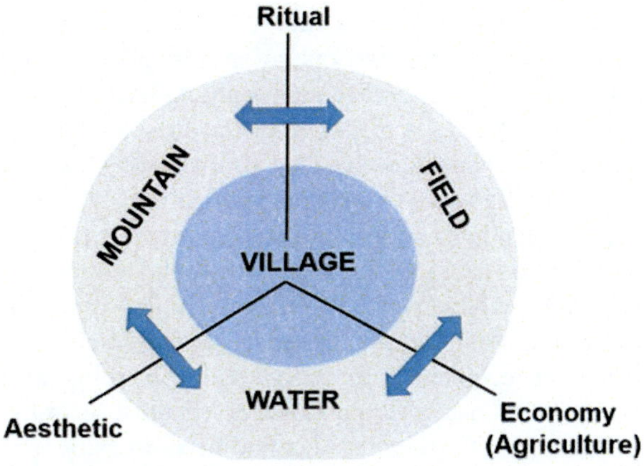

Fig. 3.7 Interfacing mountains, water, and fields in the Pung-su system (*Source* Courtesy of Sung-Kyun Kim)

The village, above and beyond its buildings and streets, is full of significance, and the scenic and sensuous landscapes, rather than being just physical and geological features, and an amalgam of recording of history in which the rocks and trees are experienced as ancestors and spirits. Hahoe possesses a landscape personality where "time is absorbed into place, and place into mind" (Glassie 1982: p. 664). The past is a determinant of the present, and the present is the key to understanding the past and a frame for visioning the future; thus, past is a hard reality and organic wholesome that exists in the basement. The village, as a "place," is "an organized world of meaning" (Tuan 1977: p. 179) and also is "the present expressions of past experiences and events and hopes for the future," thus making a transformative channel of placemaking (Relph 1976: p. 33).

Through time, particularly through the persistence of tradition, the relationship between inhabitant villagers and landscapes has been a very powerful one in which each reinforces the identity of the other, and in which the landscape is very much an expression of communally held beliefs and values systems and interpersonal involvements, interaction, and interdependency. Since the ancient past, the traditions continue and maintained where the landscapes of Hahoe function collectively as a powerful and symbolic living system among villagers (Galla 2012).

3.5 Epilogue: Concluding Remarks

Most East Asian countries, like Korea, China, and Japan, always possess a tradition of appreciating the inherent role of nature in life. This realm emphasizes their landscape more on the scenic and aesthetic sense of naturalism, such as water, mountain, river, and forest, and has recorded a long history of established and well-recordable description and manifestative interpretation of the interfaces between landscape and cultural traditions. Moreover, their traditional cultural identities are obvious when there exists a mono-ethnic society in their countries (Kim 2011: p. 46).

The Korean Pung-su is similar in many contexts to Chinese geomancy, which "in all its conceptual and operational broadness, more than anything else denotes an archetypical mode of thought and explanation, arising spontaneously out of the human intellect as a response to the inescapable pondering of the innate human condition. It considers fate and chance events and explores possible correlations in the combined psycho-social-material world" (Brunn 2011: p. 257). Moreover, "it may even be seen to serve as a matrix for rewriting and reconstructing reality. Variations on the same theme may be found in India, in large parts of Southeast Asia, in the old Confucian world including China, Korea, Vietnam, and Japan, as well as in medieval times in Europe, where it is now returning in the curious version of Chinese Feng-shui reworked in the USA. It is something we share in one form or another as a countermeasure to political, intellectual, or religious attempts at bringing the world to 'one', the relentless economic compulsions and political forces attempting to monopolize rationality and restrict our freedom of reasoning" (Brunn 2011: p. 257).

Of course, essentially, over time Korean geomancy derived from Chinese geomancy, however in the process of adaptation and transformation, the Korean system has developed its distinctive form, which is unique in Korean landscape and serves as the most fundamental metaphysical frame for placemaking and placeness (see Yoon 2017: p. 110), as illustrated in Hahoe village, where deeper life and richer life knowledge of life are preserved, maintained, and enjoyed with pride and grace by the villagers.

A recent study on the prospects of the rural landscape in Korea also corroborates the case of Hahoe; the study concludes that "the successful management of traditional rural landscapes can create sustainable local and regional development and models of sustainable development while drawing on traditional practices of sustainable use of resources. Through such sustainable development, management of these cultural landscapes can play a great role in people's lives and secure more supporters, and ultimately contribute to a sustainable future" (Jung and Ryu 2015: pp. 11, 235).

The Provincial Developmental Plan of the Korean government has focused also on Hahoe village, which represents an integral part of the Confucian cultural region. Under this project, the US $40.1 million has already been invested in the construction of an entrance path and gate to the village, and tourist infrastructure, including parking lots and a traditional shopping mall for visitors. The local government provides a subsidy of over 40 percent of the income generated from entry fees to the heritage territory, which on average comes to about US $3 million per year. These funds are used for direct investment and are managed by the Village Conservation Council. In addition, a similar proportion of revenue comes from the parking charges and rental fees from the new store complex at the entrance point to the village; its management is looked after by the Conservation body of Andong city. In addition to financial support for the historic villages, the central government also provides an additional subsidy, which is directly paid to families living in and managing the ancient houses (Galla 2012).

A comparative study of Hahoe Village with Shirakawa-mura (Japan; see Singh 2019), as they largely share common characteristics, concludes, "such as being rural farm villages in East Asia and being managed by village preservation societies composed of the local residents. After WHS designation, Hahoe Village inherited Shirakawa-mura's management technologies for handling the changes after WHS designation through international exchange. This sharing system can help in preserving heritage values of WHS, so the international exchange can be one of many significant advantages of WHS designation for sustainable management of historic villages" (Kim 2016: p. 13).

A study noted that "in the case of Hahoe village, the brand impact of World Heritage tended to be limited only to domestic tourists. For improving the brand impact that constitutes sustaining the number of tourists of the village, planners need to consider both increasing re-visitation of domestic tourists and improving visitation of foreign tourists" (Kim 2019: p. 9). It is noted that "the tourism development plans, involving both hardware improvement (such as constructing a new visitor centre, extending a parking lot, and reorganizing visitor traffic with a shuttle bus and ticketing system), and also software improvement, including increasing visitors' possibilities of traditional experiences like *Seonyujwibulnori* (the traditional fireworks) and *Hahoebyulsinguttalnori* (the traditional mask dance), were consistently established and utilized for delivering a better tourism experience and satisfaction to visitors" (Kim 2019: p. 8). Hahoe has already proved its validity on the UNESCO WHL criteria of (v) and (vi). It is recently rightly remarked, "Without a doubt, Hahoe's tangibility and intangibility at the centre and periphery of the village are ready to provide its distinct culture in heritage making through the *fabric*, *stories*, and *culture* in each epoch since its establishment. This endeavour extends to the preservation and regeneration of Hahoe Village by any means among the community members to collectively remember the prosperous legacy of the *yangban* culture" (Lee-Niinioja 2022: p. 40). In terms of tangibility and intangibility of Hahoe village, the conceptual frame of "collective memory" as developed by Halbwachs (1925) is worth mentioning here; says Lee-Niinioja (2022: p. 39), "Symbols, architecture, and literature are references for binding people to past generations and influencing their memory. Every collective memory relies on specific groups described by space and time; the group builds the memory, and the individuals do the work of remembering." This is fully exemplified in Hahoe village.

The tradition of keeping a balance between old traditions and modernity, conservation of the landscapes, and the heritage of Hahoe village and environs offers valuable insight into local community engagement, especially the value of traditional village officials. It is rightly remarked that "Village community culture requires a neighbourhood approach that brings all elements together. The conservation and management of Hahoe village have been made possible only through the active participation of the residents. Top-down funding could easily lead to an external perspective in conservation, which not only marginalizes the heritage values of the residents themselves but also leads to a lack of awareness among residents on dealing with conservation in the face of rapid change" (Galla 2012: p. 239). Let the heritage village may maintain its cultural ensembles that grown in the historical past, and are valuable as a resource today, especially concerning heritage and cultural tourism.

References

Brunn O (2011) Fengshui in China: geomantic divination between state orthodoxy and popular religion. Man & nature in Asia series, no. 8, 2nd edn. NIAS, Nordic Institute of Asian studies, Copenhagen

Cooper A (1997) Sacred mountain: ancient wisdom and modern meaning. Rudolf Steiner Press, London

De Groot JJM (1897) The religious system of China, vol. 3, book 1, part III, ch. XII. Librairie et Imprimerie, Leide, pp 935–1056

Dubos R (1972) A god within. Charles Scribner's Sons, New York

Durkheim E, Mauss M (1963) Primitive classifications [1903] (trans: Needham R). Cohen and West, London

Eliade M (1957) The sacred and the profane: the nature of religion. Harcourt, Brace & World, New York

Feuchtwang S (1974) An anthropological analysis of Chinese geomancy. Vitagna, Vientiane

Freedman M (1969) 'Geomancy'. Presidential address 1968. Proceedings of the Royal Anthropological Institute of Great Britain and Ireland, 1968–70. RIAGBI, London, pp 5–15

Galla A (2012) Village on the winding river: historic villages of Korea: Hahoe and Yangdong. In: Amreswar G (ed) World heritage: benefits beyond borders. Cambridge University Press, Cambridge, pp 231–241. Online publication. https://doi.org/10.1017/CBO9781139567657.025. Accessed 13 Nov 2021

Geertz C (1973) Interpretation of cultures. Basic Book, New York

Glassie HH (1982) Passing the time in Ballymenone. University of Pennsylvania Press, Philadelphia, PA

Halbwachs M (1925) Collective memory. University of Chicago Press, Chicago

Jung HJ, Ryu J-H (2015) Sustaining a Korean traditional rural landscape in the context of cultural landscape. Sustainability 7:11213–11239. https://doi.org/10.3390/su70811213

Kim S (2016) World heritage site designation impacts on a historic village: a case study on residents' perceptions of Hahoe village (Korea). Sustainability 8(258):1–16. https://doi.org/10.3390/su8030258

Kim S (2019) Tourism impacts continuity of world heritage list inscription and sustainable management of Hahoe village, Korea: a case study of changes in tourist perceptions. Sustainability 11(2573):1–11. https://doi.org/10.3390/su11092573

Kim S-K (2011) Identity of Asian traditional cultural landscapes. In Maheshwari D (ed) Cultural landscapes [Proceedings of 7th ISOLA national conference on "cultural landscapes", 10–11 September]. ISOLA, Ahmedabad, pp 30–47

Kim S-K (2013) Winding river village: poetics of a Korean landscape. ACLA (Asian Cultural Landscape Association) Press, Seoul

Lee-Niinioja HS (2022) Andong Hahoe village: The legacy of Joseon's yangban culture and its heritage making. Int J Archit Eng. 9(2):31–42

Lung D (1980) Feng Shui, an intrinsic way to environmental design with illustration of Kat Hing Wai in the new territories of Hong Kong. J Hong Kong Branch R Asiat Soc. 20:81–92

March AL (1968) An appreciation of Chinese geomancy. J Asian Stud 27(2):253–267

Nemeth DJ (1994) Enlightened Underdevelopment. Nat Geograph J India 40:87–100

Relph E (1976) Place and placelessness. Pion, London

Singh RPB (2019) Continuing culture and meeting modernity: the world heritage villages of Shirakawa-Gō and Gokayama, Japan. In: Bharne V, Sandmeier T (eds) Routledge companion of global heritage conservation. Routledge, London, pp 128–151

Tuan Y-F (1975) Geopiety: a theme in man's attachment to nature and to place. In: Lowenthal D, Bowden MJ (eds) Geographies of the mind: essays in historical geosophy in honour of John Kirtland Wright. Oxford University Press, New York, pp 11–39

Tuan Y-F (1977) Space and place: the perspective of experience. University of Minnesota Press, Minneapolis

UNESCO (2005) The criteria for selection. https://whc.unesco.org/en/criteria/. Accessed 13 Nov 2021

UNESCO (2021) The operational guidelines for the implementation of the world heritage convention. Operational Guidelines WHC.21/01, 31 July 2021. https://whc.unesco.org/en/guidelines/. Accessed 13 Nov 2021

Yang CK (1967) Religion in Chinese society. University of California Press, Berkeley

Yi C-H (1971) Taekniji. Ulyumunhwasa, Seoul (trans: Yi I-S from the Chinese] (In Korean)

Yoon H-K (1976) Geomantic relationships between culture and nature in Korea. Taipei: The Chinese Association for Folklore [PhD Dissertation], Asian Folklore and Social Life Monographs, vol. 88. University of California Press, Berkeley

Yoon H-K (ed) (2017) Pung'su: a study of geomancy in Korea. SUNY Press, Albany, NY

Yoon H-K (2021) Ideas relating to humanity–nature relationships in Korean folk narratives reflecting geomantic values. Landsc Res 46(6):782–792. https://doi.org/10.1080/01426397.2021.1898570

Sung-Kyun Kim was a distinguished professor of Landscape Architecture at Seoul National University, and Founder and Patron of ACLA—Asian Cultural Landscape Association. During 2012–2019, both the authors collaborated in several studies under the ACLA; they together performed field studies in Hahoe several times.

Chapter 4
Cultural Landscapes: Integrating Culture and Nature to Uplift Global Sustainability Through the Lenses of the UN SDGs 2030 Agenda

Patricia M. O'Donnell

Abstract Cultural landscapes form the shared commons of each community, while rural agriculture, woodlands and waterways bear the imprint of human uses as they support biocultural diversity. Our shared landscapes intertwine and entangle biological and cultural diversity, natural and human habitats, and tangible and intangible heritage. Cultural landscapes are vessels of heritage where livelihoods, spiritual meanings, traditions and practices express the values of landscapes and waterscapes. Human settlements of all sizes and types are vessels of biotic and cultural diversity. Varied twenty-first-century challenges urge us to embrace these interrelated aspects of diversity within cultural landscapes. Working toward the protection, future vitality of cultural landscapes is a deeply sustainable activity. With 75% of the terrestrial surface deeply altered by humanity, in both positive and negative ways, cultural landscapes offer an opportunity to help address the big issues of our times by engaging biodiversity, climate, justice, inclusion, resources, wealth and poverty. As we implement the United Nations Sustainable Development Goals (UN SDGs) 2030 Agenda with its 2020 to 2030 decade of global and local action and the parallel UN 2021–2030 Decade of Ecosystem Restoration, our actions, based in an understanding of entangled and inseparable Nature and Culture, offer a platform for effective cultural landscape undertakings.

Keywords Cultural landscape · Sustainability · Biocultural diversity · Nature-cultures · Heritage · Historic urban landscape

---

P. M. O'Donnell (✉)
ICOMOS IFLA International Scientific Committee on Cultural Landscapes, PO Box 321, 501 Lake Road, Charlotte, VT 05445-0321, USA
e-mail: odonnell@heritagelandscapes.com

R. P. B. Singh et al. (eds.), *Placemaking and Cultural Landscapes*,
Advances in Geographical and Environmental Sciences,
https://doi.org/10.1007/978-981-19-6274-5_4

4.1 Introduction

The broad field of cultural landscapes embraces tangible and intangible cultural assets and ecological and human-shaped natural assets. Cultural landscapes embody the integration of people and place in a forward looking, mutually sustaining relationship that values heritage, habitat, resilience, inclusion and more. The visionary 2015 to 2030 global agenda, articulated in the United Nations Sustainable Development Goals (UN SDGs), offers sound goals and comprehensive targets to transform our world, relying in part on our shared heritage of places, to achieve that transformation.

The often-separate constructs of cultural diversity and biodiversity interrelate in cultural landscapes, with varying degrees of natural and cultural assets and values within each. These interrelationships require more complete study and understanding to achieve conservation and future vitality for cultural landscapes. Biological diversity occurs in natural and human-shaped habitats, from untouched nature to human transformed thriving nature. As vessels of cultural diversity, cultural landscapes enrich places and peoples as an inheritance and a legacy for the unborn thus continuing to shape traditions, beliefs and world views. Biodiversity and cultural diversity are landscape assets, where nature and culture are entangled and inseparable, addressed and integrated as biocultural diversity.

It is important to recognize that cultural landscapes generally function as carbon sinks, aiding in meeting the need for carbon drawdown, the point at which more carbon is inserted than that released worldwide. For example, soils and plants capture and embed carbon contributing to drawdown. That drawdown process can and should accelerate in cultural landscapes in a conscious manner to achieve greater sustainability.

Raised voices all over the world address timely vectors of sustainability. This paper focuses on sustainable cultural landscapes within the contexts of climate change action, biocultural diversity, local place commitment and intergenerational inclusion to address the urgent needs for transformational professional and citizen actions achieving place-based solutions, responding to pressing global issues. Works that sustain place build global understanding by uplifting cultural landscapes toward cultural, environmental, social and economic harmony that aligns to people, planet, peace, prosperity and partnership. To address the plethora of current challenges, sustainable development is foundational. Cultural landscapes present a vehicle for sustainable development that can achieve the visionary UN SDGs goals and a significant number of specific targets. The cases described provide evidence of the process of understanding and effecting protection, interventions and management toward enhanced cultural landscape sustainability. Cultural landscape research, study and practice, in all aspects, are worthy of our highest endeavors to be deeply and broadly sustainable.

4.2 The Concept of Sustainability and the UN SDGs

The future we want integrates culture and nature to achieve the targets of the UN SDGs 2030 Agenda. Fully cognizant that these times hold immense challenges, the United Nations States Parties approved this fifteen-year visionary and inclusive global agenda in September of 2015. For only the second time worldwide, following on the less comprehensive eight UN Millennium Development Goals, nation states agreed to collaborate from 2015 to 2030 on addressing 17 goals and 169 specific measurable targets (UN 2015). The UN SDGs encompass a wide range of common issues: from climate action to life below water; sustainable cities and communities to gender equality; to life on land and peace, justice and strong institutions; and responsible consumption and production to partnerships; and much more. Nearly seven years since the adoption of this transformational agenda, much of global humanity remains ignorant or fails to engage. It is past time to move forward making substantial progress on these goals and targets.

The definition of sustainable development arose in 1987, as the UN World Commission on Environment and Development reported that environmental and social aspects of all nations were linked in development that could persist into a balanced future (UN 1987). Over time contemporary use enlarged sustainability as supported by three pillars: economy; society; and environment, referring to a balance among the three pillars that connotes success in sustaining places and peoples (UN 1992, 2002). With the progress of the Millennium Development Goals 2000, and the UN SDGs 2015, the culture sector has further expanded the sustainability model to include culture, which can be seen as an encompassing umbrella that permeates all three pillars (UN 2000). Another important refinement is captured in shifting from resources to assets in this model exchanging resources that can be expended, degraded and abused with assets that have inherent value as they exist today (see Fig. 4.1). Cultural landscapes permeate the full range of places and human activities and interactions, activating several targets of the UN SDGs, within a range of effectiveness and importance.

4.3 The Biocultural Diversity of Cultural Landscapes

Culture and nature coexist and co-evolve in cultural landscapes. A range of pressing issues impact cultural landscape assets that require actions at the local level applying both place-based knowledge and universal guidance. To think and act holistically for cultural landscapes, the combined works of nature and humanity, it is useful to consider the constructs of biological diversity and cultural diversity. Are these separate or is there an overlap and interrelationships to be understood and acted upon? In Montreal in 2010, a joint conference of UNESCO and the Convention of Biological Diversity set forth a declaration acknowledging the vital importance and links between biological and cultural diversity for the "survival of traditional

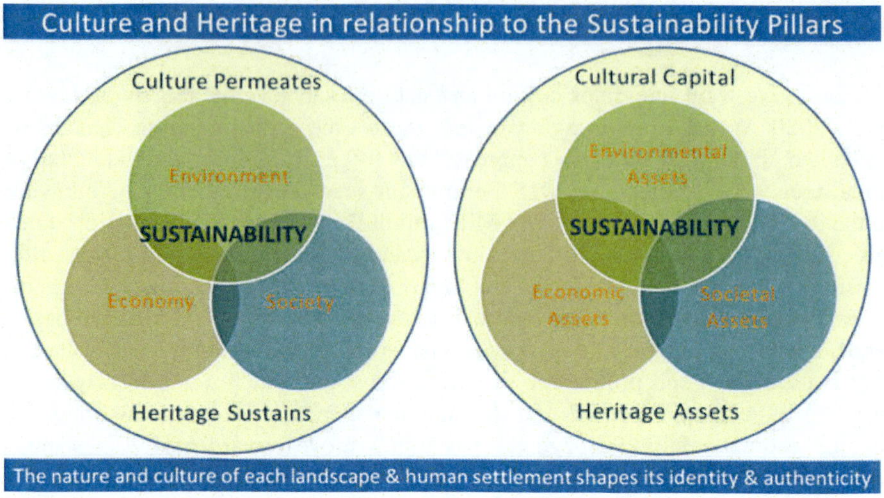

Fig. 4.1 Sustainability is defined as achieving a balance among environment, economy and society values benefits, permeated by culture on the left. While on the right, these are each viewed as Heritage Assets that embody cultural capital (Courtesy O'Donnell/author)

and contemporary societies, both urban and rural" (UN CBD 2010). Further, a 2019 declaration notes that biocultural diversity embraces the interconnected, mutually reinforcing interdependence and coevolution of biological and cultural diversity and cites that biocultural diversity "refers to the diversity of life in all aspects—meaning not only the biological diversity … but also cultural and linguistic diversity" (Canada 2019). The term biocultural diversity has been recently employed highlighting a symbiotic relationship that integrates conservation and spiritual meanings of diversity "between habitats and cultures, and between ecosystems and cultural identity indeed, religious rules and rituals often strengthen this relationship and are characterized by a conservation ethic" (Negi 2010). A model, developed by the author, captures this integration showing overlaps and opportunities (cf. Fig. 4.2). The construct and ramifications of biocultural diversity are integral to increasing traction on appropriate and helpful human responses to sustaining the values and benefits of diverse cultural landscapes for today and the unborn of all species.

Similar to biocultural diversity, the term "naturecultures" has emerged to capture the notion of entanglement and inseparability (Haraway 2003). These landscapes vary in the degrees and quality of natural assets of biodiversity, native species, habitat and ecological systems and cultural imprints of features, meanings and values. An effective sustainable development framework for cultural landscapes understands and protects the diversity of these intertwined natural and cultural assets. Given the pressures of current events and trends, from forced migrations to COVID-19, recent dialogues enlarge to include notions of planetary justice for the earth and all lifeforms, to more deeply connect places and peoples for all. These directions return to resonance with protecting and improving cultural landscapes, the 75% of

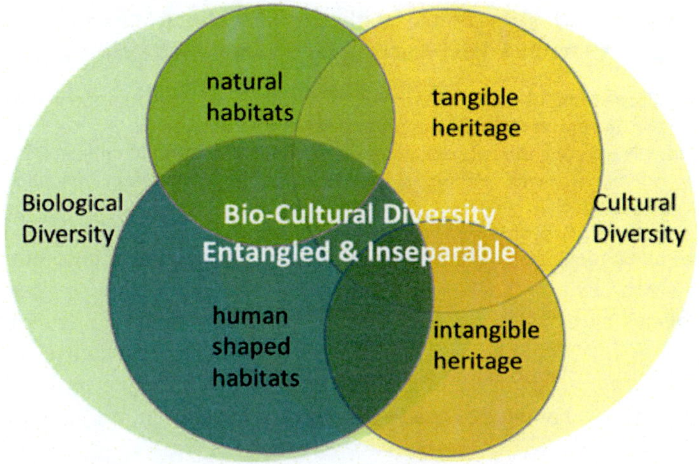

Fig. 4.2 Earth's natural and cultural heritage of biocultural diversity intersects in contemporary life as inseparable assets to be cherished (Courtesy O'Donnell/author)

the deeply altered planetary surface. Comprehensive and inclusive actions conceive cultural landscapes as combined natural and cultural assets as vessels of biocultural diversity.

Exploring these assets of cultural landscapes reveals, in many cultures, an inaccurate and unhelpful separation of cultural and natural assets of landscapes, the results of which plague the planet today mired in a climate emergency prompted by human activities. The notion of human dominance is a primary suspect in shaping patterns of planet-wide extraction and degradation for human gain in the present, void of sustainable development considerations that include the generations to come. An opposing set of beliefs and practices within some cultures, particularly those of Indigenous peoples and local communities dedicated to their place, unites humanity and nature. These traditional, multigenerational practices consider people as part of, not separate from or having dominion over, nature. Relying on place-based knowledge and endemic techniques of management and care, persisting cultural landscapes sustain diverse assets of all species and elements, without depletion over time, focusing on mutual benefits, coexistence and coevolution (Negi 2010). Detailed relevant guidance is found in a group of international declarations and conventions, such as the UN Declaration on the Rights of Indigenous Peoples (UNDRIP) adopted by the UN General Assembly in 2007, among others. The text of this declaration includes the recognition "that respect for indigenous knowledge, cultures and traditional practices contributes to sustainable and equitable development and proper management of the environment" (UN DRIP 2007).

To reinforce the construct of biocultural diversity and interdependence, an important foundation in spiritual teaching and religious beliefs applies to personal commitment to planetary well-being. Many creeds express the importance of the earth and human responsibility to act in alignment with the lands, waters and array of species with whom we share this fragile but resilient planet. In an important ecumenical statement, Roman Catholic Pope Francis notes that there is a (Vatican 2015):

> … wrong understanding of the relationships between human beings and the world … our "dominion" over the universe should be understood more properly in the sense of responsible stewardship… Neglecting to monitor harm done to nature and the environmental impact of our decisions is only the most striking sign of disregard for the message contained in the structures of nature itself.

This statement echoes those from many spiritual sources that guide humanity to protect and serve our planet. Viewed from the cultural landscape perspective, that service to places and the people who value them aligns to sustainable conservation, protection, interventions and management. In order to achieve sustainable cultural landscapes, advocates and experts must apply effective tools of engagement, empowerment, knowledge sharing, planning, regulation, traditional practices and finance.

The outcomes of these nature and culture integrations of biocultural diversity capitalize on the time-proven traditions, beliefs and practices world-wide in the face of the plethora of issues about carbon, climate, justice, governance, well-being, funding and more. All these issues are addressed across the 17 UN SDGs.

4.4 Sustainability and Cultural Landscapes

What is the relationship of sustainable development to cultural landscapes? Based in the ongoing coevolution of the relationships of land and people, the ascribed landscape values and care provided by people underpin sustainability into the future. Cultural landscapes can be considered into three categories: designed, evolved and associative (WHOG 2021). At first glance, cultural landscapes appear to relate best to the environmental pillar of sustainability, while those of society and economy may be less central. This shallow viewpoint lacks the more accurate assessment of the interrelated dynamics of society, economy, environment and culture and other crosscutting aspects that are integrated throughout cultural landscapes.

An important aspect of cultural landscapes, perhaps less recognized, is the relationship of local language to place. In indigenous cultures, language is the gateway to landscape conception and beliefs that contribute to collective identity of people, place and life forms. Language frames and expresses perception, values, philosophy and so much more. The words of particular aspects of the landscape are often unique, arising from localities, but not readily translated. For example, a study of place and place names reveals that the Western Apache employ words to recognize place and embody meanings (Basso 1996). As one native of Rapa Nui, an island in the Pacific Ocean notes "Language is inseparable from our way of being, our thoughts, our feelings, our joys and much more. It is through our language that we show who we are. If our language disappears, the whole socio-cultural foundation of our community of speakers is put at risk" (Haoa 2021). A recent UN update notes that out of "7000 global languages, 6000 are indigenous and 40% of these are threatened with extinction". While fully part of the contemporary global world, Indigenous cultures predominantly see themselves as a part of nature, not separate from or in dominion over. A rising global dialogue recognizes the multigenerational relationships of indigenous and minority communities as land stewards of their places. Played out through cultural landscapes, these stories and practices are critically important lessons of sustainability to be understood and extended in responses to contemporary pressures and crises.

4.5 Budj Bim, Australia, Indigenous Landscape

Useful examples can be drawn from places where the long-term management of Indigenous cultural landscapes by local peoples across generations are found. In these places, the interactions of people and place support the health and well-being of both land and humanity, sustaining a well-managed inheritance into the future for generations to come. The Budj Bim Cultural Landscape of the Gunditjmara Aboriginal people, a World Heritage Site located in southeastern Australia and inscribed in 2019, bears witness to a land and sea-based productive deep-time eel aquaculture that is transmitted to each successive generation. This ancient network is sustained through a dynamic knowledge system addressing a managed waterscape of constructed channels, weirs and dams that trap store and harvest kooyang, a short-finned eel (WHL 2019a, b). In the Budj Bim example, land, sea and people achieve sustainability through placed-based community actions, built on collective multigenerational traditions. Familiarity with the nomination content for Budj Bim indicates that specific aspects of the UN SDGs are addressed in the continuing practices carried out by the Gunditjmara Aboriginal peoples over six millennia. Starting with 2 *zero hunger*, 2.1 seeks to ensure *access to safe, nutritious and sufficient food* for all. Goal 8 *decent work* targets 8.5 *full and productive employment* and 8.6 *youth employment, education and training* may also address the constant efforts to maintain and manage the aquaculture system. A review of the protection and management of Budj Bim indicates that target 9.1 *develop quality, reliable, sustainable and resilient infrastructure* may apply as well as target 11.4 *strengthen efforts to protect and safeguard the world's cultural and natural heritage.* The specific target 14.2 *sustainably manage and protect marine and coastal ecosystems* relates to the ongoing process of working with lava rock and soils to form dry and fresh water channels to foster the growth of the eel. At Budj Bim, the cultural traditions, knowledge, practices and ingenuity of these aboriginal peoples are being passed on to the coming generation. As the Gunditjmara people manage this land and seascape providing a valued food source and food security as well as a product to sell, these targets may be addressed.

This is a bioculturally diverse system, as the nomination notes, "The Budj Bim Cultural Landscape exemplifies the dynamic ecological-cultural relationships evidenced in the Gunditjmara's deliberate manipulation and management of the environment" (WHL 2019a, b). The inscription on the World Heritage List of the Budj

Bim Cultural Landscape raises the appreciation of this deep-time aquaculture system managed by traditional owners for themselves, in harmony with their neighbors and the local authorities, in a manner that aids in future recognition and protection. Taken together, this aquacultural landscape reflects the entangled aspects of culture, society, economy and environment sustained over time by Indigenous people as owners and caretakers of their place.

4.6 Jackson Park, Unites States, Designed Historic Landscape

Drawing on a recent work of heritage landscapes to uplift a degraded nineteenth-century urban park, the collaborative project reversed insensitive changes over time and addressed ecosystem restoration, historic design, local use and resilience together. Jackson Park is a community asset as well as a heritage asset listed as a National Historic Landmark noting significance at the national level in the United States of America.

At Jackson Park in 2014, the low-performing conditions sparked a construction project based on ecological restoration of lands and fresh water bodies in this Olmsted-designed landscape. As that work was planned, the importance of the designed landscape heritage as an Olmsted-designed masterwork of landscape architecture was raised. The integration of the 1892 to 1897 Olmsted firm landscape architecture design and construction for the World Columbian Exposition and for the rebuilding of the park after the demolition of the fair required research, study and analysis of as-constructed landscape character for project application. Clarity of historic landscape character and contributing features formed a basis for integration of the designed landscape and the ecological enrichment for biodiverse habitat.

One clear message from local voices was that the park felt unsafe. Local users could not move through along well-built routes, or see around themselves as walked or ran. Because of that lack of a sense of personal safety, positive uses were limited, while undesirable and illegal ones could fill that vacuum. In response to this need, the shaping of functional circulation added both project construction access and future community uses. Adding to the dual purposes of historic character and habitat improvements, project plans directed the creation park paths, based on the historic path system character, to enable construction access and for park users to move through the landscape effectively and safely and enjoy experiencing the park land and waterscapes. Within that circulation system, areas overlooking the lagoons along these routes were built to replace the missing historical resting areas and adapt locations to current topography and shoreline configuration.

Photographs taken before and after the construction project capture a particular area where a truncated lagoon left a derelict scar (see Figs. 4.3 and 4.4). Excavation, grading, plantings and paths soften new margins of the former lagoon to shape a habitat rich, scenic passage in the parkland. Overall the project used more than 300

Fig. 4.3 Before the project degraded land and waterscape at the Haynes Bridge in Jackson Park (Courtesy O'Donnell/author)

diverse native plant species that were all native within the Chicago 500-mile radius 1400 new native trees 330,000 new native shrubs over 1,000,000 flowering forbs and grasses, across 10 ecosystems types. This greatly improved habitat was designed by a collaborative team of landscape architects, ecologists and civil engineers, to shape appropriate design and to reflect the two Olmsted office design of the 1890s recasting lost and degraded spatial organization, access, views and details.

This planning and implementation project links directly to several of the UN SDGs integrating sustainable development into both project shaping and outcomes. There are several goals touched on in this project including Goals 10 Reduced Inequalities, 11 Sustainable Cities and Communities, 13 Climate Action, 14 Life below Water, 15 Life on Land and 17 Partnerships for the Goals with specific targets met such as 11.7 that provide universal access to safe, inclusive and accessible, green and public and 14.2 sustainably manage and protect marine and coastal ecosystems. By uplifting an urban park sustainably, it becomes more resilient. The work addresses Goal 13 Climate Action: by enhancing the ability of the park to accept and respond to climate shocks of heavy rainstorms and high winds; by stabilizing terrestrial and aquatic soils with native plants; by improving habitat for biodiversity; by responding to climate-related drought with tough native plant selection. Lagoon water quality was improved with native plantings along water margins and on banks.

Goal 15 Life on Land was enhanced through biodiversity improving overall habitat, and for park users the enhanced landscapes offered daily enjoyment, and

Fig. 4.4 After the project work, a vibrant scenic landscape and path to enjoy it (Courtesy O'Donnell/author)

positive contact with nature, in a designed park offering scenic beauty. Inequalities were addressed by improving circulations for physical access and enhancing a sense of safety through visual access, welcoming positive uses by park neighbors, city residents and visitors. Walking, running, boating, fishing and other park use activities support overall health and well-being. Today, this historic park functions better for the community with the 10,000 linear feet of pathways for daily use and plantings predominantly below eye to reinforce a sense of personal safety. A richer ecology is also evident with diverse biomes of healthy land and water plants hosting pollinators, birds, fish, reptiles and more. The project welcomed locals and visitors to a safer, more scenic and biodiverse place, enhancing resilience throughout the public park landscape.

4.7 UNESCO HUL and Indian Urban Heritage

A third group of examples concentrates on the heritage of urban cultural landscapes of community, memory and spiritual value, an important subject of the 2020 symposium that brought forward this book. The UNESCO Recommendation on the Historic Urban Landscape (HUL), adopted as a global soft law for urban heritage guidance in

2011, is a particularly useful tool for addressing historic cities, towns and villages. HUL is an approach to urban heritage that recognizes the complexity of human settlements, places shaped by the nature of the site and the activities of people over time and continuing to evolve. Urban heritage generally fits into the World Heritage cultural landscapes definition of an evolved continuing cultural landscape and may also include designed and associative cultural landscapes and places within historic settlements. Both tangible heritage and intangible heritage reside in urban places. In HUL, the historic urban landscape is defined as: (UNESCO HUL 2011):

8. The historic urban landscape is the urban area understood as the result of a historic layering of cultural and natural values and attributes, extending beyond the notion of "historic centre" or "ensemble" to include the broader urban context and its geographical setting.
9. This wider context includes notably the site's topography, geomorphology, hydrology and natural features, its built environment, both historic and contemporary, its infrastructures above and below ground, its open spaces and gardens, its land use patterns and spatial organization, perceptions and visual relationships, as well as all other elements of the urban structure. It also includes social and cultural practices and values, economic processes and the intangible dimensions of heritage as related to diversity and identity.

The HUL approach aims to conserve and manage the urban environment sustainably, recognizing the dynamic and the multi-functional qualities of settlements. Four tool groups are advanced in HUL, requiring a context of good governance to fully activate. They include: 1. Civic Engagement Tools; 2. Knowledge and Planning Tools; 3. Regulatory Systems; and 4. Financial Tools. As heritage professionals, we must recognize and activate these tool groups to uplift the heritage of cities.

Urban conservation initiatives may draw on all or some of these tool groups. For example, a project to recognize and uplift urban heritage may begin with a community-based identification process to document local heritage, employing elements to tool groups 1 and 2.

As a nation with a long, rich past, there are a wealth of urban heritage settlements across India pressured by the dense population and an urban poor that gravitates to degraded historic areas. From 2005 to 2014, JnNurm, the Jawaharlal al Nehru National Urban Heritage Renewal Mission, brought a focus to the incredible depth of urban heritage across India. For the historic core of Jaipur, an ancient Rajasthan city founded in 1727, core city heritage mapping in 2005 identified a broad overview of built heritage and a detailed listing, to serve as an early step. The development of a Jaipur Heritage Management Plan, 2007, was foundational, marking a first for contemporary Indian city planning tools. The JnNurm initiative on historic city bazaar revitalization underpinned subsequent renewal actions for the walled city bazaar to improve streets and heritage buildings. The Jaipur bazaar is an exceptional urban heritage of local economic activities and traditions (Fig. 4.5). However, a specific Jantar Mantar World Heritage Site Management Plan added to the knowledge and planning toolkit for both managing and monitoring Jaipur urban heritage.

Fig. 4.5 Jaipur bazaar area is a vessel of local tangible and intangible heritage, uplifting the economy and society through traditions and practices (Courtesy O'Donnell/author)

Training programs to bring skills into the future conserved and reinforced craft skill development to underpin appropriate conservation works on urban built heritage renewal, as well as to provide employment for craftspeople. As another important benchmark, the Jaipur integrated Master Plan 2025 which has integrated the heritage management plan for the core city into that city-wide plan (INTACH 2015).

In terms of the UN SDGs, and external viewpoint could consider aspects of Goal 1 No Poverty, Goal 3 Good Health and Well-Being, Goal 8 Decent Work and Goal 11 Sustainable Cities and Communities all converge in the efforts carried out to uplift Jaipur.

Perhaps partially in response to these efforts, India brought forward a successful inscription the Jaipur City as an exceptional, indigenously planned city located in a level valley continues to express its spatial organization, laid out on a grid system. The World Heritage nomination details the relevant criteria (WHL Jaipur 2019):

> ii- to exhibit an important interchange of human values, over a span of time or within a cultural area of the world, on developments in. . . town-planning, as a planned city departing distinctively from medieval city layouts and in response to Mughal guidance on monumental urban form; iv- to bear a unique or at least exceptional testimony to a cultural tradition or to a civilization which is living; and vi to be directly or tangibly associated with events or living traditions ... with artistic and literary works, for its craft traditions.

Another important example from India comes from Ahmadabad, with a walled city core sited in the fifteenth century and evolved to the present. As an early focus, a Ford Foundation funded study in 1984 provided recognition of Ahmadabad's heritage, as evidence of urban decline, demolition of traditional havelis and intrusions into the city's cohesive historic fabric-sparked concerns. Stabilization and community-based projects resulted. A 1997 walled city plan for revitalization adds important aspect of the urban cultural landscape, the need to provide more pedestrian space for ease of movement, as well as the needs for traditional building restoration. Urban landscape heritage funding has been applied to the Manek Burj wall, the bird feeders and revival of subterranean water tanks. Specific projects have addressed stabilization and renewal of the public, religious and private historic structures. Initiatives to uplift knowledge of the urban heritage include a heritage walk, a freedom walk including sites of Mahatma Gandhi, a new public open space interpreting a local poet, rickshaw driver training, neighborhood signage and street theater performances.

In 2017, it was declared India's first World Heritage City, recognizing heritage values. As the capital of the State of Gujarat, the nomination notes that "a rich architectural heritage from the sultanate period, notably the Bhadra citadel, the walls and gates of the Fort city and numerous mosques and tombs as well as important Hindu and Jain temples of later periods. The urban fabric is made up of densely-packed traditional houses (pols) in gated traditional streets (puras) with characteristic features such as bird feeders, public wells and religious institutions" (WHL 2017). Criterion ii is employed in the inscription to note the settlement planning, historic architecture and a high level of local craft expressions at various scales. Criterion v is also cited as a hierarchy of public community spaces, streets and close packed neighborhoods that create community cohesion. In Ahmadabad, the streets, squares, small charbaghs or four-part quadrilateral gardens, religious buildings and grounds and the river all make contributions to the city's heritage (see Figs. 4.6 and 4.7).

An exemplary undertaking in Delhi, India begins with a recognized monument and enlarges to include a nursery site for employment and an adjacent neighborhood. The expansion of the heritage intervention addressing Humayan's Tomb, a 27-hectare garden tomb complex, World Heritage inscribed in 1993, grows to encompass uplift of the adjacent Nizammudin Basti neighborhood (Fig. 4.8; INTACH 2015). These initiatives are carried out by the Aga Khan Foundation, Aga Khan Development Network and Aga Khan Trust for Culture. The work began at the tomb main site restored over several years, and in 2009, neighborhood mapping identified issues and problems to address in Nizammudin Basti. The neighborhood benefitted from early projects to engage the community toward improving the landscape of streets and pedestrian access and recapturing five acres of anti-socially occupied, exclusionary public parks to return them to inclusive community access and use. Neighborhood programs developed to present the concept of caring for community heritage through street plays, performances, visuals and discussions. A series of neighborhood urban renewal undertakings address health, housing, sanitation, waste management, education and vocational training. This work embodies the firm belief that uplifting places for the people who live there also benefits pilgrims and tourists. In this Delhi example

Fig. 4.6 A traditional street gate or *pura* in Ahmadabad defines areas and ornaments public space displaying traditional stone masonry skills (Courtesy O'Donnell/author)

of urban heritage revitalization, the HUL approach has been thoroughly activated for heritage and community benefits to bring forward a more sustainable city.

A final example draws on the cultural landscapes of integrated "naturecultures" exploring the relationships between valued natural sites and adjacent communities where conservation and control of human activity are paramount to protection of biocultural assets. In China, the Migratory Bird Sanctuaries along the Coast of Yellow Sea-Bohai Gulf of China (Phase I) is World Heritage listed as a site of nature (WHL 2019a, b). Applying only criterion x to contain the most important and significant natural habitats for in situ conservation of biological diversity, including those containing threatened species of outstanding universal value from the point of view of science or conservation, the inscription in 2019 enumerates the wealth of

Fig. 4.7 A small charbagh, quadrilateral garden, at the Sultan Ahmad Shah Mosque, provides a welcome open space and place of respite in Ahmadabad (Courtesy O'Donnell/author)

Fig. 4.8 Restored Humayan's Tomb, Delhi, India, hosts a large school group learning about this historic place (Courtesy O'Donnell/author)

life in these wetlands, mudflats, shorelines and seascapes to include 680 vertebrate species of birds, mammals, fish, amphibians, reptiles and zoobenthos with a significant number of species that are rare or threatened. Present year-round or in migratory patterns along the East Asian-Australasian flyway, these diverse species rely on this earth and sea scape for survival. The lands are adjacent to these important Yellow Sea natural sites and heavily populated with human activity, road systems, villages, towns and agriculture, all potentially polluting and degrading these critical environments. While World Heritage inscription is not a panacea, it is helpful in raising the profile and significance of an inscribed property to a national and international level.

Can the influence of this World Heritage inscription be assessed in relationship to the UN SDGs? As an example, drawing on the inscription narrative, a selection of specific goals that address the conservation and community actions that this property relies on serves as an instructive exercise. The relevant elements of the UN 2030 Agenda goals and targets include the following (UN 2015):

Goal 3. Good Health and Well-Being, particularly target 3.9, By 2030 substantially reduce hazardous chemicals and air, water and soil pollution

Goal 6. Clean Water and Sanitation, with a focus on 6.3 By 2030 improve water quality by reducing pollution, eliminating dumping and minimizing release of hazardous chemicals and materials

Goal 9. Build Resilient Infrastructure, Promote Inclusive and Sustainable Industrialization and Foster Innovation, particularly 9.1 Develop quality, reliable, sustainable and resilient infrastructure, including regional and transborder infrastructure, to support economic development and human well-being,

Goal 11. Make Cities and Human Settlements Inclusive, Safe, Resilient and Sustainable. particularly 11.4 Strengthen efforts to protect and safeguard the world's cultural and natural heritage and 11.a Support positive economic, social and environmental links between urban, peri-urban and rural areas by strengthening national and regional development planning

Goal 12. Responsible Consumption and Production focusing on 12.2 By 2030, achieve the sustainable management and efficient use of natural resources, and 12.5 By 2030, substantially reduce waste generation through prevention, reduction, recycling and reuse, as well as 12.8 By 2030, ensure that people everywhere have the relevant information and awareness for sustainable development and lifestyles in harmony with nature

Goal 13. Action to Combat Climate Change and Its Impacts, to address 13.1 Strengthen resilience and adaptive capacity to climate-related hazards and natural disasters

Goal 14. Life Below Water, Conserve and Sustainably Use the Oceans, Seas and Marine Resources for Sustainable Development particularly focusing on 14.1 By 2025, prevent and significantly reduce marine pollution of all kinds, in particular from land-based activities, including marine debris and nutrient pollution and 14.2 By 2020, sustainably manage and protect marine and coastal ecosystems to avoid significant adverse impacts, including by strengthening their resilience, and take action for their restoration in order to achieve healthy and productive oceans **Goal 15**. Life on Land, Protect, Restore and Promote Sustainable Use of Terrestrial Ecosystems, Sustainably Manage Forests, Combat Desertification, and Halt and Reverse Land Degradation and Halt Biodiversity Loss addressing 15.1 By 2020, ensure the conservation, restoration and sustainable use of terrestrial and inland freshwater ecosystems and their services, and 15.5 Take urgent and significant action to reduce the degradation of natural habitats, halt the loss of biodiversity and, by 2020, protect and prevent the extinction of threatened species.

It is obvious that conservation at the nature-culture interface requires strong regulation and actionable guidance that is supported across society for the protection and betterment of all species. A series of Chinese and local level governance frameworks, guidance, regulation and infrastructure seeks to conserve and effectively manage the Migratory Bird Sanctuaries by clearly defining required human practices and constraints to support the necessary protections. This inscribed heritage nomination notes that there is an urgent need for broader national and multi-national strategies to ensure conservation of tidal areas as this flyway traverses two hemispheres and 22 countries. This recent property inscription advances the potential for a conscious integrated planetary conservation approach where enlightened purposeful cooperation between the countries along the East Asian-Australasian flyway may prompt action on shared concerns for the planet's species. An initiative to collaborate across the flyway addresses UN SDGs Goal 17 Partnerships.

4.8 Conclusion

In conclusion, the time is now to advance our works in cultural landscapes integrating nature and culture for highly sustainable solutions. Pope Francis captures in his appeal what we face, "The urgent challenge to protect our common home includes a concern to bring the whole human family together to seek a sustainable and integral development" (Vatican 2015).

Effective guidance, from recent cultural landscape endeavors, aids in charting the way forward to address the interrelated challenges of our times. Working for the protection and enhancement of our shared heritage of cultural landscapes can increase our ability to engage in inclusive dialogues shaping shared outcomes at local levels that resonate together to form the needed momentum toward sustained carbon drawdown, effective climate response and planetary survival.

References

Basso KH (1996) Wisdom Sits in Place: Landscape and Language Among the Western Apache. University of New Mexico Press

Canada Commission for UNESCO (2019) The North American Declaration on Biocultural Diversity. The Atateken Declaration. May 2019. https://www.cbd.int/portals/culturaldiversity/docs/north-american-regional-declaration-on-biocultural-diversity-en.pdf

Haraway DJ (2003) The Companion Species Manifesto: Dogs, People, and Significant Otherness. Prickly Paradigm Press, Chicago

Haoa MV (2021) UNESCO The Tracker Culture & Public Policy. December. p. 6. https://en.unesco.org/sites/default/files/tracker_2.0_issue_15_eng_0.pdf

INTACH (2015) Compendium of Good Practices Urban Heritage in Indian Cities. https://www.niua.org/pearl/sites/default/files/books/GP-IN4_HERITAGE.pdf

Negi CS (2010) Traditional Culture and Biodiversity Conservation: Examples From Uttarakhand, Central Himalaya. 1 August. https://bioone.org/journals/mountain-research-and-development/volume-30/issue-3/MRD-JOURNAL-D-09-00040.1/Traditional-Culture-and-Biodiversity-Conservation--Examples-From-Uttarakhand-Central/10.1659/MRD-JOURNAL-D-09-00040.1.full

Persic A (2010) Linking Biological and Cultural Diversity: UNESCO 2010 Declaration on Biocultural Diversity. UNESCO Division of Ecological and Earth Sciences, Man and the Biosphere Program, International conference on Biological and Cultural Diversity, Montreal

United Nations Conference on Environment and Development, Rio de Janeiro, Brazil. 3–14 June 1992

United Nations Convention on Bio-Diversity (2010) Declaration on Bio-Cultural Diversity, Montreal, p. 1 "Acknowledging the important progress made in building the knowledge on the links between biological and cultural diversity and to mainstream it into practices to ensure environmental, economic, social and cultural sustainability and human well-being"

UNESCO (2021) The Tracker Culture & Public Policy. December. https://en.unesco.org/sites/default/files/tracker_2.0_issue_15_eng_0.pdf

UNESCO (2011) Recommendation on the Historic Urban Landscape, adopted by the General Conference at its 36th session. Paris, 10 November

United Nations General Assembly (2007) Declaration on the Rights of Indigenous Peoples (UNDRIP), 107th plenary meeting, 13 September

United Nations General Assembly (1987) Report of the World Commission on Environment and Development: Our Common Future. 20 March. Transmitted to the General Assembly as an Annex to document A/42/427—Development and International Co-operation: Environment; Our Common Future.

United Nations (2000) Millennium Declaration (Resolution 55/2), 8 September

United Nations (2015) Transforming Our World: the 2030 Agenda for Sustainable Development. https://sdgs.un.org/publications/transforming-our-world-2030-agenda-sustainable-development-17981

United Nations (2002) World Summit on Sustainable Development. Johannesburg, South Africa, 26 August to September 4

Vatican, R.: Encyclical Letter Laudato Si' of the Holy Father Francis, On Care for Our Common Home. https://www.vatican.va/content/francesco/en/encyclicals/documents/papa-francesco_20150524_enciclica-laudato-si.html. 24 May 2015

World Heritage (2021) Operational Guidelines WHC.21/01, 31 July. https://whc.unesco.org/en/guidelines/

World Heritage List (2019a) Budj Bim Cultural Landscape, Australia. https://whc.unesco.org/en/list/1577

World Heritage List (2017) Historic City of Ahmadabad, India. https://whc.unesco.org/en/list/1551

World Heritage List (2019b) Migratory Bird Sanctuaries along the Coast of Yellow Sea-Bohai Gulf of China (Phase I), China. https://whc.unesco.org/en/list/1606

Chapter 5
Branding the Image of Religious Heritage in India

Samidha Mahesh Pusalkar

Abstract This research attempts to examine the image branding and placemaking of cultural and religious landscapes in India through ongoing tourism initiatives. With reference to the massively successful "Incredible India" campaign and multiple state tourism campaigns, the research is based on the case studies of key tourism initiatives introduced in the country over the years. Based on the concept of 'tourist gaze', multiple visuals depicting cultural and intangible religious heritage are analysed in detail to understand the deeper meanings behind them and to find what new perspectives they wish to portray to the potential tourists. Religious tourism and its marketing generate a huge amount of revenues for state governments, and as a result, branding the places with living religious heritage is seen to be an ongoing trend in India. However, there is a necessity to acknowledge the effects this so-called branding will have on the religious heritage and its core communities. Hence, issues such as representation of identities, possibility of false impressions of destinations and commodifying cultures are analysed in this study to better understand the present working of these initiatives in India.

Keywords Religious heritage · Culture · Tourist gaze · Destination image · Placemaking

5.1 Introduction

Visual communication can be viewed as pervasive and ground-breaking in contemporary society. Advertising through visuals or visual manipulation can likewise be utilized to encourage enrolment in a 'globalized citizenship', where society and social individuality, religion and thoughts of individuals are progressively consistent and portable. Schroeder (2015) describes people being "organized around attention,

S. M. Pusalkar (✉)
Preservation of Architectural Heritage, 36° Cycle, Department of Architecture and Urban Studies, Politecnico di Milano, Milano, Italy
e-mail: samidha.pusalkar@polimi.it

R. P. B. Singh et al. (eds.), *Placemaking and Cultural Landscapes*,
Advances in Geographical and Environmental Sciences,
https://doi.org/10.1007/978-981-19-6274-5_5

in which strategic communication incorporates visual images designed to capture attention, build brand names and create mindshare". This can imply that the travel industry depends intensely on visuals as consumers draw on various imagery to build up their opinions towards a destination.

Travel industry's advertising efforts present pictures of places that are coordinated for target showcases through explicit promoting channels. Thus, their destination advertising is an intentional endeavour to create motivating mentalities and ideal behaviour from travellers. To support this practice, Morgan et al. (2003) express that "Limited attention has focused on how the marketing of destinations can reflect socio-political, economic and cultural change". Even while conducting culturally diverse research, a pivotal factor in making space for exchanges is connected to visual and semantic portrayals. In contradiction, these have also remained main sources of misinterpretations and socio-cultural stereotyping among different societies and cultures at different phases of experience and exchange.

It is widely known that from the beginning of international tourism, travel leaflets and TV ads have employed significant impact in making a "destination image" of the host nation, state or city for tourists. This in a way makes them a guide to help the outsider to become more familiar with the new place. In destination promotion literature, social facts are usually designed as a façade of the host culture. This "interstitial reality" assumes the facility to ultimately shape the destination image. In its own approach, this influences the perceptions and pre-purchase selections of tourists whose collective gaze results in establishing a symbolic unit wherever he/she gazes at. This "gaze" [1] is ultimately structured around culturally specific notions of what is extraordinary and worth viewing (Urry, 1990).

Visuals created to brand this destination image steer the "gaze", and the language used in them turns into an instrument to tempt by changing itself into an amazing social arbiter between the host and visitor societies. As a result, a charming mix rises with "Visuals of the Destination" and "Language of the Visitor" that allures people to visit the destination, working eventually as a mediator among societies. Even in the age of Instagram reels and TikTok videos, these visuals keep on doing likewise and are still viewed as a standout among the most broadly utilized and advantageous sources of information to the visitor around the globe. They satisfy the obligation of connecting places and cultures by forging dialogues. This is evident especially through the adverts that promote a place through its religious heritage. As religion is one of the most common aspects that connects people all over the world, it has now become a tool to brand places and attract tourists.

However, when it comes to branding the destination image through its religious significance, there are many aspects that need to be taken under consideration. This research thus attempts to first analyse and then understand the role of such visual adverts especially in portraying the India's sacredscapes[2] and their transition into a tourist experiencescape.[3] With reference to the massively successful "Incredible India" campaign and state tourism campaigns, the initial research is based on the case studies of selected tourist brochures and television advertisements. Later, the research will also break down how they have taken a stab at controlling and directing the "tourist gaze" through visuals and use of language in interpretations with a push to reflect upon the complexity of the representation of identities and communities.

5.2 Start of Incredible India Campaign

India is a mosaic of societies, traditions, cultures, beautiful natural backdrops and spiritualism, which offer variety of tourism forms to experience here. It is the variety of cultures that is considered as the fundamental element of its travel industry. This multi-cultural and spiritual image has continued throughout; however, it was very hard to persuade the tourists who needed to be shown that there is considerably more to India than just Taj Mahal, huge palaces and Hindu temples. There are legacies left by native and foreign rulers combined with Hindu, Muslim, Mughal, Portuguese, British, Sikh, Buddhist and Jain elements to create the blend of Indo-Islamic, Indo-Gothic and many other architectures and cultural styles. India was and still is like an open museum filled with trails of history for the world to see. The government realized that this treasure had the potential to help and make an innovative and strong tourism branding initiative for India with an aim to create unique experience for tourists.

This brought the need of increasing tourism brands in India. Thus, during the 4th Five Year Plan 1969–1974, a tourism plan was proposed to encourage "diversification by introducing adventure tourism, sports tourism, leisure tourism, wildlife tourism and Buddhist circuits to reach special target markets" (Mittal 2013). Still, even in early 2000s, India was among the few countries that had a lower number of foreign tourists coming while large number of tourists from India travelling internationally. Hence, in order to change the situation, the Ministry of Tourism initiated a campaign to promote India as a tourist destination in 2002. The message "Incredible !ndia" was issued by the government as a slogan. The Ministry of Tourism then started to make a deliberate attempt to put more competence to balance in its efforts to encourage tourism. The Ministry involved the advertising and marketing company Ogilvy & Mather (India) (O&M) to set up a new strategy. The initiative envisioned India as a fascinating tourism destination, showcasing distinctive aspects of Indian culture, history and tradition such as yoga and spirituality.

From 2002 to 2003 onwards, the ministry team used the logo's exclamation mark through illustrations in early promotions. The idea behind switching the letter 'I' in the logo with '!' was to create significant impact and promote India as a high-end tourist attraction (Geary, 2013, p. 45). In later years, promotional strategies started highlighting the religious and spiritual values of India. Consequently, Indian states were also encouraged to independently create their own regional tourism initiatives that could be enveloped in the "mother" brand. After 2006, tradition and adventure took centre stage. Additionally, on the World Tourism Day in 2017, President Ram Nath Kovind collaborated with Ministry of tourism and culture along with Archaeological Survey of India to introduce the "Incredible India 2.0 Campaign", and "Adopt a Heritage"[4] scheme, which concentrated on innovative marketing strategies, with significant focus on social media and newly launched Incredible India Website. Over the decades, the representations created and advertised through these initiatives have helped numerous states and India as a whole to continuously improve their identity and brand image with their targeted demographic, both domestic and foreign. In the last decade, promoting religious and spiritual heritage of India was seen to be focus of many state campaigns as well as the Incredible India initiative.

5.3 An Overview of Visual Representation in Indian Tourism Campaigns

Tourists frequently regard visits to historic and religious sites as platforms for cultural and educational dialogues (Timothy and Olsen 2006). They expect authentic experiences linked to a particular location, not just relaxation in a resort that could be anywhere (Macleod 2006). Interpretation or visual representations of these locations are used to help visitors anticipate these encounters while reducing some of tourism's negative impacts. Furthermore, they may assist visitors in properly appreciating what they are experiencing and, in some cases, also convey acceptable tourism behaviours that reduce impacts on the site and its religious use.

Edwards and Ramamurthy (2017) in their paper state that many photographs used for tourism initiatives in India follow compositional guidelines that promote a sense of peace and timelessness in order to fulfil the conventional tourism standards of India as an unchanging destination of ancient cultures that offers an escape from modernization. Scenes were taken in high resolution, with deep horizontal lines providing a sense of constancy. Visuals offered accents of colour in a landscape rather than identifying interaction. Some of the examples include photographs of massage-giving that emphasize on a state of calm; yoga poses are perpetually kept. Water is frequently portrayed with just a ripple; deserts and beaches are barely influenced by human presence. The photographic style is close to traditional artistic stock photography, which induces symbolic and deracinated fresh photographs with smooth, yet vibrant colours and a clear, static backdrop as seen in Fig. 5.1. These pictures featured beautiful representations of world-famous Indian symbols such as the Taj Mahal. Photos of the majestic mountain range and other geographical parallels were also used to depict the diverse natural beauty of India. Impressive depictions of "escape" and "exploration" depicting the backwaters of Kerala were indeed prevalent. In other words, through the implementation of the Incredible India logo, a vibrant panoramic image of the country was designed around a relatively generic series of touristic visuals (Geary, 2013, p. 42).

The photographs connoted a way of distinction that's moving in specific histories or languages and somewhat offers a general atmosphere prepacked in an exceedingly variety of photography that gives comfort in its well-meaning familiarity and vanity. The framing of India as passively hospitable destination is done by adding images

Fig. 5.1 Incredible India campaign posters (Spiritual Destinations, n.d.)

of a rural Indian woman throughout their campaigns. She may be a silent figure, showing in red or pink saree or in village dress together with her head lined and her face never fully open. She has an exotic expression who walks barefoot in the areas of deserts (Fig. 5.1), courtyards, monuments which fascinates the traveller whose approved and unhampered gaze is left to occupy and consume the landscape.

Furthermore, these photos echo the gender specific and racialized tones, framing the West as feminine, sensitive and possessive, in contrast to a powerful, logical, masculine East. It characterizes India as a respectable and welcoming country, while the confident, modern photographic vogue appeals to international markets by implying co-existence and transparency. The parallel between the capitalist economic model and the Hindutva ideologies is further expressed by the Indian femininity as being in accordance with the Hindutva and its patriarchal tradition.

These images were used to promote and advertise in some worldwide channels, such as CNN, BBC and Discovery along with some travel channels. National Geographic and Condé Nast Traveller used captions like "Only one country can change the way you see the whole world" and "A masterful evocation of the country's spiritual heritage combined with the sensuous images, creates a kind of nirvana that can be surpassed only by a journey to India itself" (Kant 2005).

With appealing visuals, Incredible India turned into an aegis to brand the initiative of the Ministry of Tourism even at the state levels which anticipated fundamental beliefs and values of the nation tending to the diverse Indian cultures. The campaigns were executed with a determined dedication of branding India as the tourist destination for foreigners. However, with its major success, Incredible India began to pose as a threat to individual state tourism industries and in a way forced them to develop on their own with a different identity.

For a more detailed analysis of the placemaking strategies, two of the most successful state campaigns are discussed in the following sections of this chapter to understand their take on India's representation. Kerala state tourism campaign is chosen for the study considering its focus on spirituality and meditation whereas Gujarat tourism campaign is chosen due to its representation of intangible and religious heritage.

5.4 Case Study 1: Journey to God's Own Country

Spiritual tourism is a form of leisure travel that aims to appease the spirit and is therefore emotionally rewarding. It may have a religious aspect because, in addition to visiting sacredscapes and participating in pilgrimage trails or retreats, it can also take place in natural and cultural surroundings and provide relaxing and wellness experiences for the body. Using this, Kerala state initiated a powerful image branding plan with a value proposition of "Gods Own Country" in all its campaigns when the umbrella of Incredible India campaign almost dominated the entire tourism brand wagon. It situated the state as an affordable and fascinating tourist destination, making it an ideal brand for all seasons. It stirred the 'state branding' for contending states to get a different image of their own by almost stealing the thunder from the successful Incredible India campaign. A creative showcasing procedure and a strong team initiative helped Kerala to rise from simply a state in India to a standout among many famous tourist destinations in the world. Throughout this journey, its centre moved from displaying tourist attraction to promoting the spiritual experience. Far-reaching, time-explicit and well-considered showcasing methodologies to make visuals for advertisements have helped Kerala remain in the worldwide spotlight. Its media campaign was driven by a widely inclusive blend of print and TV, out-of-home and new media with investment in national and worldwide travel adverts. Campaigns such as the "Home of Ayurveda" and "Temples of Kerala" showcased Kerala as THE place for ancient and traditional medical sciences and pilgrimage destination.

The visuals advertised (Fig. 5.2) show different pictures with great number of colours in each visual. The blues of the streams and oceans in Kerala, or the skies shape the ideal backdrops. The key thought in the visual mosaic of the pamphlets appears to reaffirm a dormant idea of 'Unity in Diversity', whether it be through various colour schemes or distinctive topics such as the splendid reds exhibiting the "Theyyam" dance form or Kathakali through expressions.

As mentioned before, these adverts also highlighted experiences focused on spirituality and wellness. The pictures demonstrate the visitors resting or sitting in a casual state of mind, sometimes eyes shut, with a general articulation of grinning happiness. The care and warmth of those offering the services are unquestionably detectable through the expressive nature of masseur/masseuse's face. On the other hand, those receiving the service are unexpectedly shown closing the eyes and not having any visual impressions of the scene. This predominates the increasingly significant vibes of peaceful spiritual and body encounters. In about all the photographs regarding this

Fig. 5.2 Posters of Kerala's tourism campaign (Kerala Tourism, n.d.)

matter, it is the "look" of the professional that comes to symbolize the warmth and welcome that is advertised.

As shown in the examples, the visuals were usurped by the scents, sounds and material forms, which brought something far outperforming articulation through pictures and content in the handouts. This advancement in their perspective on "The India Experience" was associated with a kind of "significant purgation". By depicting themes like "spirituality", "rituals", "nature" and the unmissable spirit of "collective conscience" in their campaign, Kerala became an appealing destination.

After Kerala's success, state tourism brands like Madhya Pradesh, Rajasthan, Kashmir and Maharashtra were also coming forward with strong initiatives. During the same time, Gujarat state was also endeavouring to make an individual brand for itself and cut a differential specialty for itself as a tourist place with massive potential.

5.5 Case Study 2: Breathe in a Bit of Gujarat

Gujarat state is well-known for its rich heritage, wildlife sanctuaries, spiritual locations, deserts, beaches and is packed with many unique selling points. The state with its cuisine, festivities and culture always made people aware of its presence on the map. Because of the different sanctuaries and pilgrimages, Gujarat particularly attracted spiritual tourism in the past. Still, the need to change the perception of native and foreign tourists was needed. To address this, government hosted 5th Vibrant Gujarat Summit on 12 January 2009 in Ahmedabad. The focal point of the summit that year focused on 'Globalization of Gujarat Tourism' in the state. The thought for a strategic branding initiative to propagate the state as a tourist destination was developed in this summit. The decision was made to get Amitabh Bachchan, a Bollywood megastar as the brand ambassador and face of the campaign (Sharma, 2015). And with this the campaign known as "Khushboo Gujarat Ki (Breathe in a bit of Gujarat)" was born. To compliment the beautiful visuals, O&M office came up with this tagline as to use while promoting the entire campaign. The situating of "Khushboo Gujarat Ki" was proficiently mixed with different brand components into a story. A conscious effort was made with the use of the word Khushboo (Aroma/Fragrance) to metaphorically depict the experience in Gujarat to work as the guiding principle recommendation creating a strong relationship with the brand.

Gujarat turned into an example on branding a state and changing its image entirely by using a people-focused marketing initiative. The campaign showed unity in diversity of the Gujarat state with Mr. Amitabh Bachchan featuring and promoting its religious and cultural heritage. Shot over 17 destinations in three years, "Khushboo Gujarat Ki" advertisements were televised on all social platforms in English, Hindi and several regional languages.

In September 2011, Pandey's team decided to put emphasis on the detailed depiction of religious heritage, festivals and USPs of the state in the 1st phase of promotion. For this, key destinations were decided with an aim to print brochures and to shoot television advertisements featuring Mr. Bachchan. The ads featured the sanctity of Dwarka, Somnath temple and Dholavira among other destinations. Creative branding of festivities like Rann Utsav showing richness in legacy and folklore, handicrafts of Gujarat, International Kite Festival season and Monsoon Festival at the Saputara hill station, Navaratri with Dandiya and Garba dance forms and Sun temple of Modhera was executed in January 2012. Then in the next phase, the focus was put on the community's culture and architecture with an additional intent to promote Parsi, Sikh & Jain communities and their culture. For this, Saputara, Ambaji temple, Buddhist trails at Uparkot and Adalaj featured in the adverts.

Unlike Kerala where spirituality and meditation focused visuals were making a strong case for tourists to visit, the board went for pilgrimage, tradition, history and performance when it came to Gujarat state (Fig. 5.3).

With the recommendations of the brand ambassador Mr. Bachchan, the travelogue format was chosen for promoting the campaign on TV and YouTube, while print promotions were executed in the advertorial format. In Amitabh Kant's article (2005), Bharat Bala—the film director for TV adverts—explained that the simplicity was applied to both the imagery and to the music track that would complement them. One of the key objectives, according to him, was to make a film that would accurately represent the scenery and the people of India as visitors would encounter them after arriving. They attempted to depict the very same, genuine feelings that tourists would feel once they visited India on their own. This can be seen even through the smallest details in the adverts, for instance, the outfits Mr. Bachchan wore while promoting. In the adverts of some religious sites, he was seen wearing traditional Indian clothes in shades of red or yellow, whereas in some places, to convey the explorer's spirit, he was seen wearing white or Khaki colours.

Fig. 5.3 Visuals and screenshots of Khushboo Gujarat Ki campaign ads (Photo Gallery, n.d.)

Moreover, in the ads showing religious places, the slogan was lengthened to "Breathe in the faith, breathe in the blessings, breathe in a bit of Gujarat". The same was done with the spiritual places such as ad on Buddhist caves where Mr. Bachchan appealed to "Breathe in the fragrance of prayers, breathe in the meditative silence, breathe in a bit of Gujarat". These subtle details made an impact on the people seeing the adverts. In a way, it made them experience the places from their homes through these visuals even if it was just for a minute or two.

These visuals were continuously shown on prime-time TV channels, news channels and even international channels to have maximum impact. Even after so many years of its launch, the campaign still maintains a steady growth of 15% or more.[5]

With the success of both Kerala and Gujarat state tourism brands, many state tourism departments were inspired to promote their religious heritage as well. Bihar state's already existing religious tourism campaign also came into light during same period. Bihar, known as the land of monasteries (Blissful Bihar, n.d.), had developed "Buddhist circuits" under the new criteria of "pilgrimage with pleasure". As stated by Geary (2013), "It was designed to cater to the financially lucrative religious diaspora". It got much-awaited boost in the last decade when branding the religious heritage became a go-to plan for state tourism departments. This new trend became evident when inspired by Tamil Nadu state's religious tourism campaign, currently, Telangana state is also set to get a makeover in branding historic temples. The state government has announced a plan of action to support "Temple Tourism" in a major way by creating a temple grid (Patan 2017). Similarly, according to reports, now,

Uttarakhand state Tourism Minister—Satpal Maharaj—is also eager to build the state's ancient temples and encourage tourism around them (Anon 2020). He even met with the Uttarakhand Tourism Board (UTDB) to discuss this issue, along with the potential of making a temple route for travellers. Seeing as Uttarakhand is home to the four key points for Hindus, the sacred Char Dhams, namely Badrinath, Kedarnath, Gangotri and Yamunotri as well as host of many ancient temples, religious tourism is always a significant revenue generator for the state tourism industry.

5.6 Dynamics Between Religious Heritage and Tourism

With these cases, the shift towards religious tourism branding is now evident. However, to fully grasp the situation in India, understanding the relationship between religious sites and tourism industry is necessary.

Religious tourism is described as a form of tourism that is predominantly driven by religion whether solely or coupled with other motivating factors that have a religious site as a destination; and that may or may not be associated with engaging in celebrations or religious activities. Nolan and Nolan (1989) published the widely accepted classification of religious tourist destinations, based on three categories that overlap among each other: pilgrimage temples, religious tourist attractions and areas where religious festivals are held.

These sacredscapes and their cultural contexts are continuously reshaped by the everchanging socio-cultural and political systems (Singh 2013). Additionally, they easily get moulded by society's understanding and interpretation of them. So, their visual representation and interpretation specifically created for tourism purposes may convey a more commercial view of these places. This is because tourism's main product is experience and tourism organizations portray them with an aim to draw visitors and generate revenue. In terms of tourism marketing, as Mossberg (2007) suggested that they are branded as "experiencescapes". Due to this, these destinations start absorbing many of the cultural values depicted. But as they progress towards becoming globalized tourist places, regions next to religious sites start getting turned into crowded and busy trading areas. As a result of traders' fast thinking in making use of any available means for earning revenue, the lanes leading to many religious heritage sites get turned into congested shopping areas. Same is the case in India where even the communities located near these religious places have moulded themselves into a new culture-focused commerce. Due to this, the authentic experiences associated with the traditional practices, arts and crafts also got commodified.

This situation totally contradicts the central intent of branding these places as 'experiencescapes' as their promotion itself harms the entire experience of them. Moreover, it starts highlighting the shift of their value as a pilgrimage or religious retreat to a 'purchasable pleasure'. The visitors come to the destination with certain expectations of experiences and visual treats that were promised to them via these visual adverts, but this shift hinders the true experience of visitors. In both the case studies, we have seen how the spiritual and religious places are portrayed to brand the destination image of Kerala and Gujarat state, respectively. However, as realistic as these visuals may seem, it is a common knowledge that these adverts tend to be superficial. This contradicts the indexicality[6] of these visuals because the tourism place branding initiatives only show the good side of the places. Even though it may be acceptable for the commercial purposes, the issues of how these adverts and their aftermath affect the community or even religious practices of that place need to be addressed. MacCannell in 1976 made the argument of authenticity in this case (Scott 2014). Tourists searching authentic experiences are similar to secular pilgrims who seek to impart meaning in their lives through these out-of-the-ordinary experiences. Therefore, an inauthentic perception towards a place ultimately translates into a lack of sense of place and lack of awareness in any profound symbolic meaning or significance of sites, resulting in failure of acknowledging their identity (Cresswell 2004). If this is the case, then, are the present techniques of image branding acceptable for the portrayal of cultures, religious places and the legacy of its kin by putting them up for show? What is required now, maybe, is a more profound and sensitive way of portrayal.

5.7 Evaluation-Based Image Branding

With the increasing commercialization and the sand traps of market standards, the fundamental aspects like self-will of the traveller or communities' connection with the religious site wind up buried in idealistic translations of place branding. Moreover, as Go and Govers (2009, p. 51) discuss, it is almost superficial to assert that places can "change their identity by simply changing colour; a new logo, a new marketing campaign and perhaps new management". But branding any place is much more complicated than this. Currently, the dominant view of place branding has a very limited role—firstly defining an identity for the place, then identifying new methods which can define this identity's relevance to a variety of audiences and lastly attempting to convince and attract these audiences to the place that is getting advertised. Same is the case with image branding through religious places. However, it is crucial to understand that this implies that their place identity is somewhat static. Kalandides (2011) argues on this stating that any place is not a static construct and there should be no limitations while conceptualizing places. Places don't have a single permanent identity since they have constant internal conflicts.

To understand his argument better, after referring his works on identification five elements of place identity[7] (Kalandides 2011), the place identity of religious or spiritual places can be understood with similar approach—place image (a non-evaluated characteristic of the place), materiality (tangible heritage aspects associated with the place and its surrounding area), institutions (organizations, administrative bodies which have control over the access, regulations and management of that place), relations (referring to the involvement of other stakeholder groups associated, i.e. craftsmen, religious groups, politicians and commerce-based citizens in the area), and people and their practices (core communities and their traditions, rituals and other intangible aspects associated with the place).

The development of a brand image is, thus, dependent on several factors namely culture and its material representations as well as intangible attributes and traditions that contribute to a community's ideals. With this perspective, place branding enables the integration of planning strategies, knowledge of resources and skills in managing the urban heritage development. Considering the phenomenology (Norberg-Schulz 1980) of historic urban environments, especially with religious and spiritual significance as well as communal values, the place branding can be defined as 'the process of discovering, developing and implementing ideas and policies required to (re)build the distinctive character and sense of place, towards a new local identity carved between tradition and development' (Daldanise 2020). This opens a window of opportunity for generating additional socio-economic prospects by exploring new community-focused policies that can revive shared images, not just for travellers but also for communities.

A new framework focusing on religious heritage enhancement can be prepared which consists of innovative tools, such as Community Impact Evaluation (Lichfield 2005). This evaluation method using the five elements of place identity can help determine which cultural values play an important role in branding the image of that place and how they can be used as a tool. It can further acknowledge the relevant stakeholders and their roles in the entire image-making process and act as a new method for heritage enhancement, while still being 'culturally conscious'. In this perspective, due to this evaluation-based empirical approach, a shift from mere place branding to an inclusive evaluation-based branding would be possible. This strategic process would not only address the issues of representation of identities, possibility of false impressions of destinations and commodifying cultures but also make the entire process of image branding to be a local "tailor-made" strategy for and from communities towards culture-led regeneration (Daldanise 2020). With this strategy in place, every place would have a different image branding process.[8]

In this context, Moilanen and Rainisto's (2009) five stages place branding process "(a) start-up and organization, (b) research stage, (c) forming brand identity, (d) making and enforcing the plan, and (e) implementation and follow-up" can help the evaluation-based strategy further. Additionally, if combined with Kavaratzis's process (2009) of formulating a local vision, the method would be used to open a dialogue with the local citizens, place branding managers and all potential partners. Through this approach, an authentic yet impactful representation of religious and cultural heritage can be done for tourism image branding in India and abroad as well.

5.8 Concluding Remarks

In the most genuine sense, with all the accomplishments of the place branding campaigns, India's heritage woke up through writings and visuals. However, it has also tangled the truth and sowed the seeds of fake impressions. For tourists, trusting the visual forms seemed to be somewhat closer to "a stealthy peep into truth" and perhaps an assurance of "value for money" in terms of the uncertainty caused by the experience. Culture in this way is being looked upon as a commercial commodity with a price tag on it. If not dealt with delicately, this exploitation and commoditization of religious heritage can affect the tourism sector along with the general population with a 'revenue-driven thought process' as its main cause. So, instead of merely reading the extrinsic messages piled together in separate adverts, perhaps more focus should be given on understanding the authenticity of those adverts.

This study opens a dialogue for future research in the fields of image branding, especially those focusing on religious heritage, religious tourism, pilgrimage and its commercialization in India and abroad. It also draws attention to how these heritage places are portrayed and by whom, and further discusses the importance of community evaluation framework and strategies that may guide policymakers in implementing collaborative image branding campaigns through religious and spiritual places.

Notes

1. To examine this, the hypothesis of "Tourist Gaze" propounded by John Urry as a reference will be utilised for a conceptual framework. He discussed 'phenomenon of commodification of peers' and a destination turning into 'an object of consumption'.
2. A sacred site has an existing important value for a religious man. According to Singh (2017), "the sacred landscape combines the absoluteness of space, relativeness of places, and comprehensiveness of landscape, thus altogether result in a 'wholeness' carrying the inherent and imposed spirit of 'holiness', which is to be called sacredscapes".
3. O'Dell (2005) defines experiencescape as "a place where human interactions, pleasure, entertainment and enjoyment can occur to create an experience".
4. The 'Adopt A Heritage' Project's main goal was to hand over control of heritage sites to public and private sector individuals and firms for infrastructure development of public facilities.
5. These statistics are of pre-COVID-19 times.

   For more information, refer: Das, S. (2017). *Gujarat tourism grows 17%, even 7 years after campaign*. Business-standard.com. Retrieved 2 September 2021, from https://www.business-standard.com/article/economy-policy/gujarat-tourism-grows-17-even-7-years-after-campaign-117061900051_1.html.
6. According to Kuhling (2017) "indexicality refers to a photograph as index, meaning a photograph indicates 'what was there' in the moment that a photograph was been taken, like smoke indicates fire". Pinney (2014, p. 452) further explains that "it is this authority of the photographic camera that gives the traveller/photographer the impression of 'knowing' what they have seen. It occurs through documenting places and people in real time/space conjunctions".
7. Kalandides's five elements of place identity: (1) place image, (2) materiality, (3) institutions, (4) relations and (5) people and their practices.

8. Kavaratzis and Hatch (2013) states that "every place is different and there are no 'one-size-fits-all' branding processes".

References

Anon (2020) Uttarakhand tourism to train spotlight on ancient temples [Online]. Times of India Travel. https://timesofindia.indiatimes.com/travel/travel-news/uttarakhand-tourism-to-train-spotlight-on-ancient-temples/as79717151.cms

Aulet S, Vidal D (2018) Tourism and religion: sacred spaces as transmitters of heritage values. Church, Communication and Culture 3(3):237–259. https://doi.org/10.1080/23753234.2018.1542280

Balgobind S (2013) An incredible investigation. Erasmus University. https://thesis.eur.nl/pub/15590/. Accessed 18 April 2021

Bandyopadhyay R (2016) Who owns the past? The politics of religious heritage in contemporary India. Tourism Review 71(3):234–243. https://doi.org/10.1108/TR-09-2016-0035

Blaney A, Shah C (2018) The Aesthetics of Contemporary Indian Photography in an "Incredible India." Photogr Cult 11(1):3–18. https://doi.org/10.1080/17514517.2017.1410322

Cerreta M, Daldanise G, Sposito S (2018) Culture-led regeneration for urban spaces: Monitoring complex values networks in action. Urbani izziv 29(supplement). https://doi.org/10.5379/urbani-izziv-en-2018-29-supplement-001

Cresswell T (2004) Place: A short introduction. Wiley-Blackwell, Oxford

Daldanise G (2020) From place-branding to community-branding: A collaborative decision-making process for cultural heritage enhancement. Sustainability (Switzerland) 12(24):1–23. https://doi.org/10.3390/su122410399

Dowson RR (2020) Eventization of religious heritage: Case studies and practical considerations. In: International journal of religious tourism and pilgrimage. Dublin Institute of Technology, p. 27–36. https://doi.org/10.21427/kygk-g772

Edwards L, Ramamurthy A (2017) (In)credible India? A critical analysis of India's nation branding. Commun Cult Crit 10(2). https://doi.org/10.1111/cccr.12152

Foroudi P et al (2016) A framework of place branding, place image, and place reputation: Antecedents and moderators. Qualitative Market Research 19(2):241–264. https://doi.org/10.1108/QMR-02-2016-0020

Geary D (2013) Incredible India in a global age: The cultural politics of image branding in tourism. Tour Stud 13(1):36–61. https://doi.org/10.1177/1468797612471105

Go F, Govers R (2009) Place branding: Glocal, virtual and physical identities, constructed, imagined and experienced. Palgrave-Macmillan, Basingstoke, England

Gohil N (2015) Role and impact of social media in tourism: A case study on the initiatives of Madhya Pradesh State tourism. Int J Res Econom Soc Sci. http://www.euroasiapub.org

Hakala U, Lemmetyinen A, Kantola SP (2013) Country image as a nation-branding tool. Mark Intell Plan 31(5):538–556. https://doi.org/10.1108/MIP-04-2013-0060

Kalandides A (2011) The problem with spatial identity: revisiting the "sense of place". J Place Manage Dev 4(1). https://doi.org/10.1108/17538331111117142

Kaneva N (2011) Nation branding: Toward an agenda for critical research. Int J Commun. http://ijoc.org

Kant A (2005) The Branding of India. Incre India Mag. April, pp 32–43

Kavaratzis M (2009) Cities and their brands: Lessons from corporate branding. Place Brand Public Dipl 5(1):26–37. https://doi.org/10.1057/pb.2008.3

Kavaratzis M, Hatch MJ (2013) The dynamics of place brands: An identity-based approach to place branding theory. Mark Theory 13(1):69–86. https://doi.org/10.1177/1470593112467268

Kerrigan F, Shivanandan J, Hede AM (2012) Nation branding: A critical appraisal of incredible India. J Macromark. Sage Publications, pp 319–327. https://doi.org/10.1177/0276146712445788

Kuhling C (2017). Rewriting tourist photography [Western Sydney University]. http://hdl.handle.net/1959.7/uws:47965

Kumar U (2017) Analysis of tourism advertising campaign: With special reference to incredible India

Larsen J (2014) The Tourist Gaze 1.0, 2.0, and 3.0', In: Lew A, Hall C, Williams A (eds) The Wiley Blackwell Companion to Tourism, 1st edn. John Wiley & Sons, Ltd., pp 304–313

Levi D, KoCher S (2009) Understanding tourism in religious heritage sites. Focus 6:17–21

Lichfield N (2005) Community impact evaluation. Routledge. https://doi.org/10.4324/9780203991282

Macleod N (2006) Cultural tourism: Aspects of authenticity and commodification. In: Smith M, Robinson M (eds) Cultural tourism in a changing world: Politics, participation and (Re)presentation. Channel View Publications, Clevedon, UK, pp 177–190

Mehta-Karia S (2011) Imagining India: The nation as a brand imagining India: The nation as a brand Master of Arts 2011. University of Toronto

Miles S, Paddison R (2005) Introduction: The rise and rise of culture-led urban regeneration. Urban Stud 42(5–6). https://doi.org/10.1080/00420980500107508

Mittal M (2013) Management of tourism in Punjab an appraisa [Punjabi University]. http://hdl.handle.net/10603/28136

Moilanen T, Rainisto S (2009) How to brand nations, cities and destinations. Palgrave Macmillan UK, London. https://doi.org/10.1057/9780230584594

Morgan NJ, Pritchard A, Piggott R (2003) Destination branding and the role of the stakeholders: The case of New Zealand. J Vac Mark 9(3). https://doi.org/10.1177/135676670300900307

Mossberg L (2007) A marketing approach to the tourist experience. Scand J Hospitality Tourism 7(1). https://doi.org/10.1080/15022250701231915

Nolan M, Nolan S (1989) Christian Pilgrimage in modern Western Europe. University of North Carolina Press, Chapel Hill

Norberg-Schulz C (1980) Genius loci: Towards a phenomenology of architecture. New York, Rizzoli, New York

O'dell T (2005). Experiencescapes: Blurring borders and testing connections. Experiencescapes: Tourism, Cult Econ 3–17

Shyju PJ et al (2020) Local stakeholders' perspectives on religious heritage and tourism development in Varanasi. Int J Tourism Cities 6(3):529–544. https://doi.org/10.1108/IJTC-10-2019-0194

Patan A (2017) Temple grid to promote tourism [Online]. The Hans India. https://www.thehansindia.com/posts/index/Telangana/2017-05-29/Temple-grid-to-promote-tourism/302992

Paul S (2016) "Incredible India": Revisiting the location of cultural heritage in India—A semiological perspective and beyond. Int Res J Interdiscip Multidiscip Stud (IRJIMS) A Peer-Rev Monthly Res J 2(9):86–92

Pike S (2005) Tourism destination branding complexity. Journal of Product and Brand Management 14(4):258–259. https://doi.org/10.1108/10610420510609267

Pinney C (2014). CHAPTER 16. Notes from the Surface of the Image: Photography, Postcolonialism, and Vernacular Modernism. In: Jay M, Ramaswamy S (eds) Empires of vision: A reader, New York, USA, Duke University Press, pp 450–470. https://doi.org/10.1515/9780822378976-022

Problems and Prospects of Accessible Tourism in India (2010). www.iittm.org. Incredible India Annual Report 2017–18 (2017).

Sarma A (2019) Marketing of religious tourism; a case study on Ambubachi Mela Kamakhya Temple, Assam. In: Proceedings of the 8th International Conference on Education, Teaching, Learning & Innovation (ICE19Thailand Conference). Bangkok-Thailand. www.globalbizresearch.org

Schroeder JE (2015) Consumption, visual. In: The Blackwell Encyclopedia of Sociology. Oxford, UK, John Wiley & Sons, Ltd. https://doi.org/10.1002/9781405165518.wbeosc125.pub2

Scott D (2014) The tourist: A new theory of the leisure class. Ann Leisure Res 17(1) https://doi.org/10.1080/11745398.2014.890513

Sharma A (2015) Impact of promotion on destination reputation: A case of Gujarat. Int J Hospitality Tourism Syst 8(1):23–31

Sharma S (no date) The not so 'Incredible India': Understanding how and why India fails as a global tourism destination

Singh RPB (2013) Sacred landscapes, cosmos and shared wisdom: The Asian Vision. In: The 50th World Congress of IFLA (International Federation of Landscape Architects). 2013 Auckland NZ

Singh RPB (2017) Sacredscapes of Banaras (Kashi/Varanasi): Cultural landscape and cosmic geometry: Vol. XIII. https://banaras.academia.edu/RanaPBSINGH/Papers

Srivastava D (2012) Steering the tourist gaze: Travel brochures as cultural mediators between France and India. Synergies Royaume-Uni et Irlande n° 5:291–305

Superbrandsindia (2014) Kerala Tourism Report. Tiffany M. Crotti R (eds) (2017) The Travel & Tourism Competitiveness Paving the way for a more sustainable and inclusive future. Geneva, World Economic Forum

Timothy D, Olsen D (2006) Tourism, religion and spiritual journeys. In: Timothy D, Olsen D (eds), Routledge. https://doi.org/10.4324/9780203001073

Timothy DJ (2002) Sacred journeys: Religious heritage and tourism. Tourism Recreat Res. Taylor and Francis Ltd., pp 3–6. https://doi.org/10.1080/02508281.2002.11081214

Urry J (1990) The Tourist Gaze. Sage Publications London, New Dehli

Van Bühren R, De Ascaniis S, Cantoni L (2018) Special issue on tourism, religious identity and cultural heritage. Church, Commun Cult. Routledge, pp 195–198. https://doi.org/10.1080/23753234.2018.1548904

Vasavada F, Kour G (2016) Heritage tourism: How advertising is branding the intangibles? J Heritage Manage 1(1):22–34. https://doi.org/10.1177/2455929616640688

Yao D et al (2020) From religious belief to intangible cultural heritage tourism: A case study of Mazu belief. Sustainability (Switzerland) 12(10). https://doi.org/10.3390/su12104229

Samidha Mahesh Pusalkar PhD student in Preservation of Architectural Heritage, 36° cycle, Department of Architecture and Urban Studies, Politecnico di Milano; Head Prof. Mariacristina Giambruno, Supervisor Prof. Stefano Francesco Musso; Email: samidha.pusalkar@polimi.it.

Chapter 6
Cultural Landscapes: Essence and Application Perspectives in Georgia

Nodar Elizbarashvili, Sandro Gogoladze, Gela Sandodze, Lado Grigolia, and Teona Qobalia

Abstract The natural diversity of Georgia is unique in the world, especially—in the subtropical and temperate zones. More than seven dozen natural landscapes have been created here, starting in the humid subtropical or semi-fertile wooded forests, ending in wetlands and semi-wetlands landscapes. Georgia, despite its small territory, is distinguished by the diversity of the living world, as well as natural and cultural landscapes. Such diversity makes it difficult to study the geographical-ecological (geo-ecological) features of the country, to fully reveal the cultural modifications of the landscapes. According to the UNESCO nomination, 12 types of cultural landscapes are known in the world. Their total number reaches 120, according to which Europe stands out. Almost half of the cultural landscapes are represented here. Among the countries in the world, France and Italy lead in the number of cultural landscapes (8–8 units), followed by Great Britain, Germany, China (5–5 units), Portugal and Spain (4–4 cultural landscapes). No cultural landscape has been created in Georgia under UNESCO status. However, Georgia is known in the world as the homeland of wine and wheat, a country of ancient culture and traditions, an attractive geographical area with natural and landscape diversity. The article discusses the definition of cultural landscape, its essence, types and their geographical features. There are also three areas for cultural landscape nomination in Georgia, which differ in scale (size).

Keywords Cultural landscape · Determination · Geography · Agro-landscapes · Human activity

N. Elizbarashvili (✉) · S. Gogoladze · G. Sandodze · L. Grigolia · T. Qobalia
Department of Regional Geography and Landscape Planning,
Tbilisi State University, Tbilisi, Georgia
e-mail: nodar.elizbarashvili@tsu.ge

R. P. B. Singh et al. (eds.), *Placemaking and Cultural Landscapes*,
Advances in Geographical and Environmental Sciences,
https://doi.org/10.1007/978-981-19-6274-5_6

6.1 Introduction

The cultural landscape is the result of a long (historical) interdependence of human and nature, part of the national heritage. It reflects well the local culture and identity. It can be related to an important historical event as well as traditional economic and social activities. The cultural landscape on the national level is part of the world heritage, it belongs to everyone. It can be called a mirror of our coexistence with nature. Through it, it is possible to have an idea of the potential and ecological features of the landscape, economic and social values. Preservation of the cultural landscape is a precondition for the preservation of national identity, national dignity. Caring for him and his correct perception will improve the quality of life of future generations and deepen their patriotic mastery. Cultural landscapes are protected by the 1992 World Heritage Convention. According to him, the protection of cultural landscapes can contribute to the realization of the principles of sustainable development, while preserving or improving the natural (biological diversity) and historical-cultural value of the landscape.

Numerous methods are used to study the geographical features of the cultural landscape. Among them are: general geographical, historical–geographical, cartographic, geo-informational, landscape analysis and sintesis, landscape planning, landscape-ecological, social, etc. It can be said that the study of cultural landscape is not fundamentally different from the study of natural landscape, especially when we are interested in their modern state and potential. In reference to the article, a number of scientific sources and geographical methods of field research were used, through which the above-mentioned characteristics of the cultural landscape were studied. On the basis of the landscape analysis and landscape (Beruchashvili, 1995; Beruchashvili and Zhuchkova 1997; Methodology, 1983) synthesis, concrete results are obtained: the theoretical concept of spatial and time approach in landscape research is developed and confirmed, certification of landscapes of Georgia and Caucasus is done, geographic information systems and databases are developed, landscape and ecological characteristics of Georgia are studied, and methodological fundamentals of landscape planning are developed.

The concept of spatial and time analysis and synthesis of natural territorial complexes, developed by Prof. Niko Beruchashvili (1947–2006), has always been an important basis for landscape researches and development of different scientific directions, including: landscape ethology, geophysics of landscape, dynamics of landscape, studying of conditions of cultural landscapes (agro-landscapes), structures and functioning of landscapes, military geography, landscape planning, modeling and forecasting of geographical processes, etc. The current status of landscapes, natural and anthropogenic condition, conflicts, potential and sustainability of areas were evaluated by geo-ecological analysis and synthesis (Elizbarashvili 2005, 2009, 2016, Elizbarashvili et al., 2018) and Methodology of landscape planning (Piloting 2009; Landscape, 2009). A geo-ecological investigation of a landscape is implemented in some stages, with the following stages as the most important ones: landscape-ecological analysis (inventory) of the territory (general geographical, landscape, social-economical, ecological) and estimation. During the general geographical analysis, the geographical location of the study territory, area, borders of the neighboring regions, common physical-geographic, social-economic and historical–geographical features are considered. During the landscape analysis, the scale needed for the investigation purposes, natural potential of the territory, interrelation between the environmental components, basic features of structure and functioning, dynamics and ethology are identified. It must be noted that the characteristics of natural and socio-economical potential of the territory important for the geo-ecological investigation are properties of the relief and geological formation, climate and climatic resources, waters and water resources, plants and herbal resources, bio and landscape diversity, historical and cultural diversity.

6.2 Definition, Essence and Geography of Cultural Landscape

The formation of the cultural landscape as a term is linked to the development of landscape painting in Europe. This process dates back to the sixteenth century, when it became popular in painting to depict individual sections of the environment, earth shapes, or effective landscapes. The spread of the German term "landscape" in Europe belongs to the same period. The term "cultural landscape" appeared in the scientific literature (German school) in the early twentieth century. Thus, the history of the term—cultural landscape—dates back a century, although its active discussion began in 1992, after the Rio Historical Conference. The study of the cultural landscape (so-called landscape science) began in the early twentieth century, attributed to the German geographer Otto Schluter. It considered the geographical environment in two ways: natural or original (without human intervention) and a collection of cultural landscapes. According to him, the main task of geography was to study the process of transforming a natural landscape into a cultural landscape.

The vast majority of researchers in this field agree that:

1. The cultural landscape is developed on its natural foundation,
2. Cultural landscape preserves the main geographical elements (relief, geological structure, climate, waters, soil, accommodation, communications, etc.), and
3. The cultural landscape mainly (essentially) "obeys" the processes taking place in the natural environment (Elizbarashvili 2005, 2016, Elizbarashvili et al. 2015; European 2000).

Originally considered a cultural landscape as a landscape with high aesthetic and functional characteristics. It was the living environment of people or ethnos, in the formation of which spiritual and material values played an active role. In a separate sense, the cultural landscape: the product of the history, material and spiritual culture of the people living in this or that place; It is the result of processes initiated and interconnected by nature and society. Cultural landscape can be: specially created (by an architect or gardener), folk (as a result of daily activities of family or villagers), historical (reflecting a historical event or activity), ethnographic (a place that includes many natural and cultural elements of historical heritage).

There are different approaches (visions) for defining a cultural landscape that take into account its historical, ecological, agrarian, ethnocultural and other values. The main ones are:

- Urban values—takes into account the historical value of the inhabited landscape;
- Ecological values—considers the environmental, recreational and environmental value of the landscape;
- Religious values—refers to the religious value and status of the landscape, for example, a monastery complex, a place of baptism, etc.; and
- Traditional and ethnocultural values—such as historical monument, historical building, archeological monument, place of historical battles, ethnocultural space, historical industrial or industrial area, historical water regulation building, traditional agricultural production, etc.

Thus, the cultural landscape can be: a city or its separate district, a separate historical monument and place, a monastic ensemble, a historical settlement, engineering communications, an archeological monument, canals, vineyards and gardens, parks, natural monuments and related cultural objects, etc.

Such diversity of the cultural landscape is a prerequisite for its different definitions. Because of this, many terms with similar content are known: historical landscape, agrarian landscape, natural-agrarian landscape, historical-natural landscape, geocultural region, natural-cultural landscape, historical-cultural landscape, urban-landscape, place-urban, urban Cultural complex, historical–geographical landscape, etc. (Elizbarashvili 2010; Elizbarashvili et al. 2018; Piloting 2009). However, it is a fact that in any case, the cultural landscape is part of the natural and cultural heritage. It clearly shows the peculiarities of the origin and development of this or that country (territory), the interdependence of man (local community) and nature (https://tclf.org/places/about-cultural-landscapes).

The purpose of creating a cultural landscape is related to a number of factors. They are described in seven keywords: man, ethnos, nature, history, culture, function and high meaning. By combining such words we can get a complete definition of the cultural landscape: the cultural landscape is the best example of the historical coexistence of man and nature and an area of high ethnocultural significance. Cultural landscapes are recognized around the world (by UNESCO) for displaying and preserving ethnoculturally and naturally important 12 values these are:

1. Religious values and religious architecture—such a cultural landscape is created in 14 countries around the world. These include: Italy, Japan, Afghanistan, Kenya, Poland, Portugal, USA, France, China, Canada, Australia, etc.
2. Traditional agriculture—this kind of cultural landscape is created in 15 countries around the world. These include: Andorra, Australia, Austria, Germany, China, Italy, Mexico, South Africa and more.
3. Place of historical significance (settlement, historical building, historical road (crossroads). Such cultural landscapes are created in 18 countries of the world,

including: Argentina, Lithuania, Saudi Arabia, South Africa, Sweden, Syria, Vietnam, Spain, Ukraine, etc.
4. Cultural Landscape—this type of cultural landscape is created in 6 countries of the world and is mainly related to prehistoric wall painting. Such countries include: Canada, China, India, Kazakhstan, etc.
5. Harmony of urban and natural landscape—this kind of cultural landscape is created in 6 countries around the world. These include: Brazil, Germany, Israel and others.
6. Unique landscape architecture and landscape design—this kind of cultural landscape is created in 10 countries around the world. These include: France, Iran, Spain, Liechtenstein, etc.
7. Urban and economic evolution—such a cultural landscape is created in 2 countries of the world (Austria and Portugal).
8. The interdependence of nature and man—such a cultural landscape is created in 7 countries around the world. These include: Norway, Russia, Lithuania, Spain, etc.
9. Traditional form of extraction and use of natural resources—such a cultural landscape is created in 8 countries around the world. These include: Czech Republic, Ethiopia, Japan, Nigeria, Great Britain, etc.
10. Ethnoculture—Polynesia (France).
11. The evolution of nature—such a cultural landscape is created only in Gabon.
12. Botanical Park—Singapore, UK.

Thus, among the values of the cultural landscape, areas related to religious, traditional agriculture, historical sites, landscape architecture, forms of traditional use of natural resources are distinguished in meaning and quantity.

The cultural landscape has been a UNESCO World Heritage Site since 1992 and has been granted this status by dozens of territories. According to the World Heritage Convention, "the cultural landscape is the product of the joint action of man and nature." It, by its status, should contribute not only to the preservation of material and spiritual cultural monuments, but also to biodiversity. The area of such a cultural landscape in the Convention on Cultural Heritage includes:

The territory of the monument, the borders of which pass through the protection and regulation zone;

- Protection zone,
- Regulation zone, and
- A historical zone with special historical, urban, architectural and cultural value.

Different categories of cultural landscape are defined by different organizations. For example, UNESCO has been known to protect many cultural, natural and historical sites around the world since its inception. Founded in 1942, UNESCO is a leading international agency with a rigorous scientific vision for cultural landscapes and their preservation. Its guidance document states that the cultural landscape encompasses a wide variety of manifestations of the interactions between humanity and its natural environment. They should be selected based on their distinctive universal significance as well as their representativeness. It should be a clearly defined geocultural, easily recognizable, orthologically developed landscape with an essential social role (https://www.archronicle.com/types-of-cultural-landscapes; https://whc.unesco.org/en/culturallandscape/#2).

The geography of cultural landscapes recognized by UNESCO varies considerably from continent to continent. Europe is distinguished by their number. Here, 58 sites have cultural landscape status in 19 states. Leading countries are France and Italy (8–8 units each), Great Britain and Germany (5–5), Portugal and Spain (4–4 cultural landscapes).

In second place is Asia. There are 30 cultural landscapes, though in 21 states. Among them are China (5 units), Turkey (4) and Iran (3). On both continents of the Americas, 14 cultural landscapes are found in 10 states. Brazil (3), Cuba and Mexico (2–2 each) are distinguished by their number. 13 cultural landscapes are represented in 9 African countries. Among the states of this continent are South Africa, Nigeria and Senegal. 5 cultural landscapes have been observed in Australia and Okinawa. Australia stands out for their number (3) here.

6.3 Traditional Agricultural (Agro) Landscapes—Peculiarities of Structure and Functioning

According to the occupied area, the largest part of the cultural landscapes are agro modifications of natural landscape or agro-landscapes. The study of their structural and functional features is one of the main tasks and directions of applied geography.

In the geographical literature, agro-landscape is referred to in various terms. Among them: agro-territorial complex, natural-industrial geosystem, natural-anthropogenic territorial complex, agricultural landscape, etc. Despite so many variations, the essence of agro-landscapes is linked to several common postulates. These are:

1. Agro-landscape is formed, developed and functions in accordance with natural processes and public interests;
2. Agro-landscape is in its essence a natural complex that is transformed by man;
3. The structure and functioning of the agro-landscape is mainly related to the natural regime of the geographical environment in which it was formed and which is constantly controlled by man by bringing it in maximum harmony with the natural environment;
4. The more the agro-landscape is in harmony with the natural environment, the more profitable and effective its public importance is;
5. Agro-landscape should simultaneously combine the function of maintaining an agricultural and healthy environment; and

6. Agro-landscapes include all agricultural lands, including meadows and pastures. It is true that in the latter the natural processes are not constantly regulated by man, though they are kept in the desired state.

Assessing the structure and functioning of the agro-landscape according to the concept of spatio-temporal analysis and synthesis is a task of interesting scientific-practical importance. Their structure and functioning are related to the natural and agro-technical processes, as well as the ethological features of cultivated plants. We can compare natural and agrarian landscapes according to the vertical structure. For example, crops of cereals in terms of vertical structure are similar to meadow or steppe vegetation, perennials (orchards)—light forests, etc.

There are impacts on the functioning of agro-landscapes (plowing, sowing, irrigation, fertilization, pest control, harvesting, etc.) that have significant dissonance in its natural state, although it is essential for the functioning of the agro-landscape. Some of them can be analogous to natural conditions. Consider (e.g., rain—analogous to irrigation, harvest—analogous to simplification of vertical structure, etc.).

However, most of the agro-technical measures lead to the formation of specific agrarian conditions or Agrostex. Agrostex is a daily condition of the agricultural land, which is conditioned by seasonal rhythm (natural mode), weather and the nature of agro-technical measures. Agrostex can be divided into two groups:

1. From harvest to the beginning of agro-technical measures, and
2. From the beginning of agro-technical measures (soil cultivation) including harvesting.

The first group includes agrotextures that completely match the natural conditions of the original landscape. They also include pluvial (rain-related) and nival (associated with frost and snow) conditions due to weather variability.

The conditions of the second group are of several types. One part of them coincides with the phenophases of cultivated plants (e.g., emergence or leaf formation—complication of vertical structure). The second part sometimes involves several phenophases at once or, conversely, one phenophase is associated with a much longer period than the duration of Agrostex. The third part of Agrostex s is also found, which is related to the arid or semiarid state. At this time, the agro-landscape will not be able to function normally without human intervention. Although we equate irrigation with rain, this is when substantial changes occur in the structure and functioning of the agro-landscape (distribution of synovial indicators). It is possible to consider the fourth part of Agrostex, which is related to the fight against weeds or plant diseases. Despite their short-lived nature, such action substantially alters the functioning features of the agro-landscape.

The study of the formation and geographical features of agro-landscapes is an interesting scientific task. They represent well the types of traditional farming as well as the forms of harmonious coexistence of society and nature.

6.4 Cultural Landscapes of Georgia (for UNESCO Nomination)

Georgia is located in the South Caucasus, between the Black and Caspian Seas, among the South Caucasus Mountains. Georgia covers a territory of 69,700 km^2 and its population is almost 3.7 million. By area and population, Georgia ranks the 25th in Europe. Georgia is distinguished by highly diverse natural condition, and biological and landscape variety. 20% of the territory of Georgia is intact natural environment and 40% of it is occupied by primeval forest. There are all types of plants and soil in Georgia typical to Europe. The country ranks the second with mammals and birds, the third with reptiles and the fifth with the endemism of their species and quantities (Biology 2000; Geography 2000).

Georgia is distinguished for the highest landscape diversity in Europe. The altitude of the relief in Georgia changes within the limits of 0–5201 m. 54% of its territory is presented by mountains, 33%—by hills and knolls, and the rest of it—by lowland and plain. There are all forms of reliefs known in the world present in Georgia. The geographical location of Georgia, its dissected relief and variety of the climate in the country stipulate the high values of the landscape diversity. There are 13 natural types and 72 genera of landscapes in Georgia, which, if considered in proportion with the territory of the country, is the highest indicator in Europe (Biology 2000).

The highly diverse landscapes of Georgia complicate the study of the geographical-ecological (geo-ecological) properties of the country. Among such properties, horizontal and vertical structure of landscapes, energy and substance transformation in the landscape (functioning), forms and scales of influence on the landscapes, landscape stability and function, and landscape condition and potential have a particular importance.

The natural diversity of Georgia is outstanding in the world. This is particularly true with the subtropical and moderate belts of the country presenting over seventy natural landscapes, from humid subtropical or light semiarid forests through moist and Alpine landscapes. Such a variety is the result of several factors, with the following ones being most important: geographical location (location along the border of the tropical and moderate belts), the Black Sea, which never freezes, altitudinal zoning (up to 5200 ml) and several-thousand-year-long history of the economic use of the territory (Biological 2000; Geography 2000; Elizbarashvili 2013; Elizbarashvili et al. 2018) (Fig. 6.1).

With its landscape variety (related to the area), Georgia is an outstanding country both, in Europe and in the world. Georgia, as a part of the Caucasus, is: (a) among the world's top 25 biologically richest and endangered "hot spots," (b) among the world's top 200 most sensitive and vulnerable eco-regions with particularly rich biodiversity, (c) a habitat of endemic birds, (d) among the world's agro-biodiversity centers and (e) one of the "hot spots" with large herbivorous animals (Geography 2000, 2013; Kharatishvili 1990; Elizbarashvili 2012).

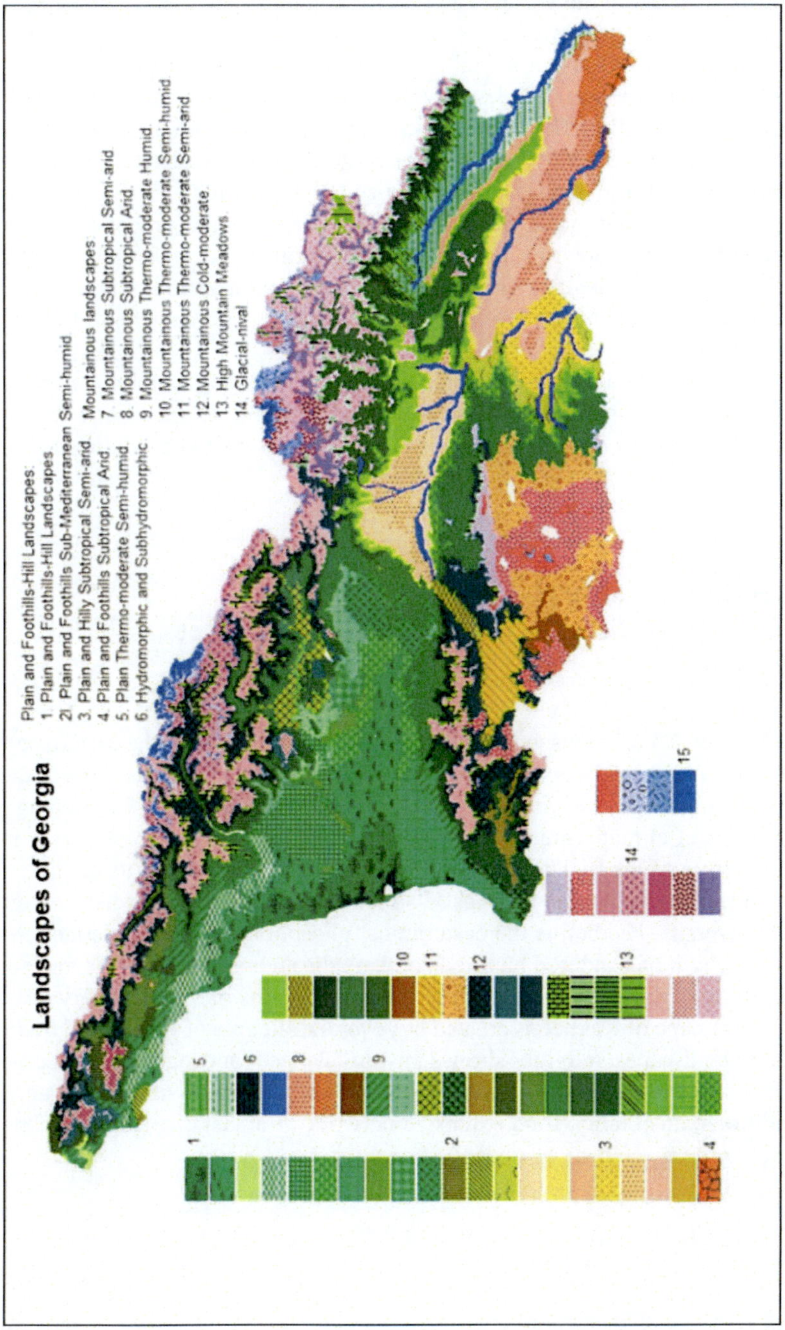

Fig. 6.1 Landscapes of Georgia (Beruchashvili 1986)

Georgia is the homeland of wine and one of the most important centers of viticulture and winemaking in the world. The oldest surviving specimen of grape seed dates back to eight thousand years ago. Numerous works and documents of antiquity describe the unique culture of wine making and consumption in Georgia. This is narrated by the works of Homer's Odessa (VIII century BC), Apollonius Rhodesel's Argonautics (III century BC), Strabo (III–II century BC) and Xenophon (V century BC). It is confirmed that the term "wine" in other languages has spread from Georgian.

The number of cultural landscapes recognized by UNESCO in the world is growing by 5–7 units every year. Such a process is an essential precondition for the development of many countries. Protecting the cultural landscape is an important basis for raising national identity, national dignity. Its popularization, planning and care will improve the quality of life of future generations and contribute to their patriotic upbringing.

In Georgia, we find all forms of cultural landscape, which are even known in the world. At this point, several cultural landscapes of different scales can be distinguished. Among them, in our opinion, the following deserves to be nominated for UNESCO.

6.5 Large-Scale Cultural Landscape—Historical–Geographical Province of Kakheti

The area of Kakheti is 12.2 thousand square kilometers. It is located in the eastern part of Georgia. The unique landscape of Kakheti is created by the high mountains of the Caucasus (4 km high) and the vast plain of Alazani (40 km wide and 110 km long). There are 4 types of climate (humid, semi-humid, semi-arid, arid), which is rare in the world (in terms of area). The diversity of the climate contributes to the diversity of the landscape. Kakheti is the most distinctive part of the Caucasus in terms of landscape diversity. Kakheti is the easternmost phosphorus of the Christian religion, which is why it has endured hundreds of wars throughout history. The ancient monastic complexes are located here, which attracts millions of visitors every year. The cultural landscape of Kakheti is created by several millennia of viticultural traditions, beautiful landscapes, religious shrines, historical places and legends, numerous historical and cultural monuments, archeological sites and cities—museums, agricultural traditions, hospitality, food extensive network of areas, etc. (Figs. 6.2 and 6.3).

Fig. 6.2 Alazani Valley and Caucasus mountains

6.6 Medium-Scale Cultural Landscape—Truso Valley

Truso valley is located in the northern part of Georgia, in the gorge of the river Tergi. Its length is 25 km. The vertical height of the valley exceeds 2 thousand meters, which is the best example of natural zoning for such a short distance. The natural landscapes include floodplain forests as well as subalpine, alpine and nival zones. The valley is distinguished by the geological structure of different periods, with various relief forms, travertines and mineral waters. Many of the historical monuments are defensive structures and mountain rural settlements. The valley is home to the highest settlement in Europe—the village of Resi (2400 m above sea level). The population almost no longer lives here, although the valley is known for its mountain cattle-breeding potential and traditions (Fig. 6.4).

Fig. 6.3 David Gareja monastery

Fig. 6.4 Truso valley

6.7 Medium-Scale Cultural Landscape (#2)—Khevsureti Mountain Region

Here people, in a difficult geographical environment, have lived and worked since the Bronze Age. Khevsureti's neighbors are the peoples of the North Caucasus Mountains—Chechens, Ingush, Dagestanis and others. Historically, Khevsurians have always defended the northern border of Georgia, from where they were attacked.

Most of the villages in Khevsureti were defensive structures. There are many towers and castle-villages, which served a unique defensive purpose. They blend well with mountain relief forms. Khevsureti is distinguished by unique traditions of cattle breeding, hunting, folk medicine, local production and food production.

Khevsureti mountain natural landscapes, effective landscapes, endemic vegetation and wildlife, historical fortress-villages (Shatili, Mutso, Khakhabo, etc.), architectural monuments, agricultural traditions of the mountain population, ecological culture and historical sites are presented in the National Park. The creation of the cultural landscape of Khevsureti will further enhance its special, general human significance (Fig. 6.5).

Fig. 6.5 Khevsureti–Fortress-village Shatili

6.8 Small-Scale Cultural Landscape—Vardzia Canyon (Mtkvari River Valley)

The beauty of Vardzia Canyon is a rock-cut, inaccessible and well-preserved monastery complex located in the south of Georgia. It is a monument of twelfth–thirteen centuries. Vardzia played a major role in the political, economic and cultural life of the country. In 1938, it was declared a museum-reserve.

The length of the monastery complex, which includes 13 floors, exceeds 500 m. There are several large temples and churches, numerous sources are mentioned, an internal communications system has been set up, up to 3 dozen cellars (wine storage, with a capacity of 100 thousand liters) and several hundred living rooms have been identified.

There are several dozen monuments of historical and economic significance in Vardzia Canyon, which is 3 km long and 600 m deep. Among them are the monastery complex of Vani (Vahan), Khertvisi fortress and traditional farming forms (vineyard and orchard terraces). Vani (Vahan) monastery complex was created in eighth–sixteenth centuries, in the northern part of Vardzia canyon. It was destroyed several times (by earthquakes, wars) and repaired. It has several temples and churches, a defense building, dozens of agricultural buildings and up to 200 living rooms. Vani Monastery Complex is an excellent example of Georgian history, architecture and culture.

Khertvisi Fortress is located in the extreme northern part of Vardzia Canyon, at the confluence of two important rivers (Mtkvari, Paravani). It has served as the most important defensive function for centuries. It is considered to be a fourteenth-century monument, although various historical sources point to buildings dating back several hundred years. The castle naturally blends in with the relief, which enhances its defensive and architectural significance. Currently, several historical monuments of Vardzia Canyon are included in the list of possible nominees for UNESCO World Heritage. The areas around the Vardzia Canyon are known throughout the Caucasus for their ancient specimens of agricultural. Here, in the conditions of difficult terrain, terrace farming was developed—it brought vines, fruits, grains and vegetables. Several dozen vine varieties have been spread here, which the locals are intensively trying to restore. Among the terraced agricultural centers (China, India, Syria, etc.), Georgia is also included (Fig. 6.6).

Fig. 6.6 Varzia Canion and Vardzia Monastery Complex

6.9 Conclusion

Tourism is considered to be one of the important directions of socio-economic development of Georgia. Most of the world-famous resort-recreational resources are also represented in Georgia. Such potential is a precondition for the development of various directions of tourism in our country. Nomination of separate territories of Georgia for the nomination of cultural landscapes will make its tourist attraction even more interesting. This will require a detailed knowledge of world experience and relevant scientific research. There is already an interest in such topics in the Georgian educational space, which can become a solid basis for the creation and popularization of cultural landscapes.

The allocation, study and establishment of cultural landscapes in Georgia should preferably be carried out in cooperation with scientific, academic and state organizations. For these activities, the proposed facilities could become pilot units. The natural and historical-cultural diversity of Georgia makes it possible to distinguish any form of cultural landscape that is even known in the world.

References

Beruchashvili N (1986) Landscape map of Georgia, Tbilisi, Georgia, p 1

Beruchashvili N (1995) Caucasus: Landscapes, models, experiments. Tbilisi, Georgia, p 355

Beruchashvili N, Zhuchkova V (1997) Methods for complex physical-geographical investigations: the manual. Moscow University Press, Russia, p 320

Biological and Landscape Diversity of Georgia (2000) WWF Georgia Country Office, Tbilisi, Georgia, p 312

Elizbarashvili N (2005) Geo-ecological basics of landscape planning. Tbilisi, Georgia, p 300

Elizbarashvili N (2010) Landscape planning of the Javakheti protected areas. Universal, Tbilisi, Georgia, p 123

Elizbarashvili N (2012) Fundamentals of geography. Universal, Tbilisi, Georgia, p 187

Elizbarashvili N (2013) Nature of Georgia. Palitra L, Tbilisi, Georgia, p 287

Elizbarashvili N (2016) Fundamentals of applied geography. Tbilisi, Georgia, p 376

Elizbarashvili N et al (2009) Landscape planning: methodology and experience, TJS-Caucasus, Tbilisi, Bonn, Irkutsk, p 188

Elizbarashvili N, Seperteladze Z, Elizbarashvili R (2015) Basic characteristics of stability of mountain landscapes of Georgia. Conference proceedings: landscapes: perception, knowledge, awareness and action, Addleton Academics Publishers, New York, pp 244–254

Elizbarashvili N, Meessen H, Kohler T, et al (2018). Sustainable development of mountain regions. Tbilisi, Georgia, p 291

European Landscape Convention. (2000). European Treaty Series—No. 176. Florence. http://conventions.coe.int.

Geography of Georgia. (2000). Tbilisi University, p. 300. https://www.icomos.org/centre_documentation/bib/2013Biblio_Cultural_landscapes.pdf

Kharatishvili, M. (1990). Physical geography of Georgia. Tbilisi, Georgia, p 215

Methodology of Landscape-geophysical research and mapping of conditions of natural-territorial complexes. (1983) Tbilisi, Georgia, p 255

Piloting landscape planning in the countries of the Southern Caucasus. (2009). Baku, Tbilisi, Erevan, p 121

Part II
Placemaking: Identity and Exposition

Chapter 7
Ayodhya, India: Placemaking and Transformation of Historic Urban Landscape

Rana P. B. Singh and Sarvesh Kumar

Abstract Places are the repository of the human imprints on the landscape and the true representative of the changes and transformations that took with time, through the channel of placemaking. The studied city, Ayodhya is one among the holy-heritage cities in Hindu cosmogony that is known for bestowing salvation. Of course, it had a history of the Babri Masjid—Ramajanmabhumi issue—which once recorded an example of the contested landscape; however, it is resolved smoothly through the jurisprudence process and public awakening. The transformation of the landscape and the issue of placemaking, with case studies of the Ramajanmabhumi temple and Queen Huh memorial park, are illustrated here as examples of making and re-creating the cultural landscapes in the holy-heritage city of Ayodhya in the frame of historical processes and events. The historical growth and landscape changes in Ayodhya are narrated, emphasizing the re-creation of the Ramajanmabhumi temple. Moreover, the recent development and expansion of Queen Huh Memorial Park, honouring Queen Huh who went to Gimhae (Korea) and married King Kim Suro in CE 48 as narrated in the thirteenth century memoir—the *Samguk Yusa*—represent an example of Indo-Korean interfaces of placemaking and will promote cross-cultural integration and cultural tourism. The narration and interpretation of the case studies will further be compared with a similar case of holy-heritage cities in India.

R. P. B. Singh (✉)
Cultural Landscapes and Heritage Studies, Banaras Hindu University, Varanasi, India
e-mail: ranapbs@gmail.com

Asian Cultural Landscape Association (ACLA), Varanasi, India

Present Address:
Usha Apartment (Flat 101-102), Lane 5—Janaki Nagar Colony, Flyover BLW end—Badi Patia Road, PO: Bazardiha, Varanasi 221106, UP, India

S. Kumar
Department of Geography, D.D.U. Gorakhpur University, Station Road, Gorakhpur, UP 273009, India
URL: https://bhu-in.academia.edu/RanaPBSINGH/Papers

R. P. B. Singh et al. (eds.), *Placemaking and Cultural Landscapes*,
Advances in Geographical and Environmental Sciences,
https://doi.org/10.1007/978-981-19-6274-5_7

Keywords Contested landscape · Placemaking · Historical events · Ramajanmabhumi · Ayodhya · Riverfront · Queen Huh Park

7.1 Ayodhya: The Geographical Personality

Ayodhya is counted among one of the seven most sacred and salvation-endowing cities of India (i.e., Ayodhya, Mathura, Maya-Haridvar, Kashi, Kanchi, Avantika-Ujjain, Puri, Dvarka) and is referred to in the *Puranic* literature of ca CE tenth century (e.g., the *Garuda Purana*, 49.114). Ayodhya is situated on the right bank of the Sarayu (Ghaghara) River at a distance of 7 km east of Faizabad city. Ayodhya (population: 55,890 in 2011 Census) was the part of Faizabad Metropolitan city and both were known together as Ayodhya-Faizabad twin city (Urban Agglomeration/Metropolitan City, population: 256,624 in 2011 census) extending between 26° 47′ North to 26° 80′ North Latitude and 82° 12′ East to 82° 20′ East Longitude (Fig. 7.1), but in May 2017 Faizabad was merged into Ayodhya Municipal Corporation. Of course, these two twin cities are divided by a pilgrimage route of Panchakroshi Yatra, and the entire sacred territory is roughly demarcated by the Chaudahkroshi Yatra.

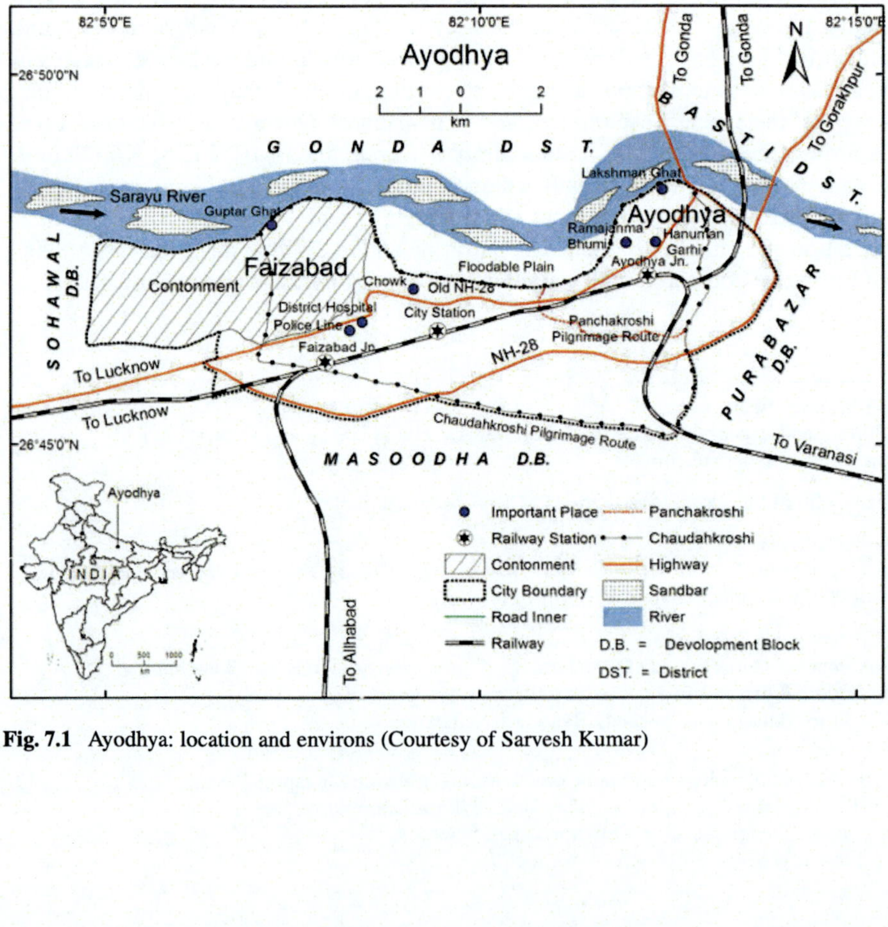

Fig. 7.1 Ayodhya: location and environs (Courtesy of Sarvesh Kumar)

The city of Ayodhya lies in the central part of the Ganga valley, north India, at a distance of 130 km west from the state's capital Lucknow, at a distance of 221 km in the south-east exits the holy-heritage city of Varanasi, and two other important nearby cities are Gorakhpur at 145 km in the east, and Prayagraj (Allahabad) lies at a distance of 167 km in the south. The Vedic mythology of the ancient era refers to Ayodhya as a city of King Manu (Hindu progenitor of mankind) and eulogized it as the birthplace of Lord Rama, the seventh incarnation of Vishnu. Among the ancient sixteen territorial kingdoms, *Mahajanapadas*, Ayodhya was prominently narrated as the first capital of the powerful Koshala Kingdom (Law 1944: p. 424; Chakrabarti 2000: pp. 378, 387). According to ancient epical sources, Ayodhya for a period of over two thousand years has borne witnessed the presence of Vaishnavism, Shaivism, Jainism, Buddhism, and Islam too; therefore, Ayodhya consists of the sacred and religious places for Hindus together with Jains, Buddhists, Sikhs, and Muslims too (Shaw 2000: p. 698).

Over time, Ayodhya was invaded and destroyed many times during the twelfth century under the sultanate rule at Delhi and Mughal rulers. The notable invasion and destruction were by the order of the Mughal invader Mir Baqi Tashkandi who demolished the famous Rama temple Ramajanmabhumi, belonging to Pratihara the Gahadavala period, i.e., the birthplace of Rama; moreover, in the following period of fifteen months, he built a Muslim monument (called Babri mosque) using the debris of the temple. Since its foundation, Ramajanmabhumi has been a controversial and sensitive place for centuries and even today. Muslims have never performed prayer (*namaz)* at this site, despite the construction of a mosque (*so-called*). Of course, it has been the place of Hindu-Muslim riots, the main site was not allowed for devout Hindus till 23 February 1857 when the East India Company (Britain) constructed a separating wall and ceased the entry of Hindus through the mosque-like built space since 5 January 1950 under the law, and only restricted entry was granted (Singh and Rana 2002: p. 301).

7.2 Historical Context

According to ancient history, Ayodhya was one of the holiest cities where the religious faiths of Hinduism, Buddhism, Jainism, and Islam are united together, converging into a place of enormous sacred importance. The history of Ayodhya is a chequered one (see Table 7.1).

In the Vedic text *Atharvaveda*, dated ca thirteenth century BCE, this place was described as a city that was made by gods and was as prosperous as heaven itself:

> *Aṣṭacakra navadvāra devānām puryodhyā.*
>
> *Tasyam hiraṇyah koṣah svargo jyotiṣavṛtah.* ...... *Atharvaveda* (10.2.31)

The powerful kingdom of ancient Koshala had Ayodhya as its capital. This city was also a significant trade centre in 700 BCE possessing layers of NBPW (Northern Black Polished Ware) and records continuity of prosperous settlements from then up

Table 7.1 Ayodhya: chronological history

Period	Archaeological and historical events
Seventh-sixth centuries BCE/BC	Northern Black Polished Ware levels Nalatila and Janmabhumi in the 'Ramkot' area. References in Buddhist texts such as *Digha Nikaya*
Fifth century BCE	Annexed into the kingdom of Magadha
Third century BCE	Part of Maurya Empire; Massive burnt brick wall on the west side of 'Ramkot'; coins of local rulers found in early historic levels at Hanumangarhi
Second century BCE/BC	'Sunga' inscription found in early historic levels at Hanumangarhi; coins of early dynastic kings (Vayudeva, Muladeva, Dhanadeva, Visakhadeva, Naradatta, and Sivadatta)
CE/AD 48	Mythic voyage of princess Suriratna (Huh Hwang-Ok in Korean) from Ayodhya to Gimhae/Kimhae to get married to King Kim Suro
CE 78	Kushana invasion; many Buddhist monuments were destroyed
CE 150–300	Mitra dynasty
CE 4th-5th cent	Gupta empire; Jain community developed in Ayodhya and Mani Parvat as ruins of the Buddhist site seen by Chinese pilgrim Faxian (Fa-Hsien) in ca CE 400
CE 320	Chandragupta I
CE 455–67	The capital was moved by Skandagupta/Vikramaditya from Pataliputra to Ayodhya to 'restore' Ayodhya's glory and revivification five important temples of Ayodhya, i.e., Ramajanmabhumi, Ratnasinhasanas temple, Kanakabhavan, Sheshavatara temple, Devkali temple, and bathing ghats developed at the riverfront of Saryu (Ghaghara)
CE sixth century	Huna invasion, followed by political fragmentation
CE eleventh century	Gahadavalan kings' period; fortification of Ramajanmabhumi, revivification of five Hari temples—Chandrahari, Vishnuhari, Dharmahari, Chakrahari, Guptahari, and Ayodhya—also became famous for three pilgrimage journeys, i.e., Panchakroshi, Chaudahkroshi, and Chaurasikroshi
CE 1192	Ban on Hindu temple construction under Muhammad Ghori
CE 1226	Ayodhya becomes the capital of Oudh Province under Delhi Sultanate, the debris of the Adinatha temple famous as a Shah Juran Tila
CE 1527–1528	Invasion by Mughal king Babar; destruction of Ramajanmabhumi temple
CE 1658–1707	Aurangzeb, the bigoted Mughal King, was known for destroying Hindu temples

(continued)

Table 7.1 (continued)

Period	Archaeological and historical events
CE late eighteenth century	The political centre moved from Ayodhya to Faizabad under Nawabs; Ramaite appropriation of Hanumangarhi temple and revivification of Nageshvarnatha temple
CE 1855	Communal violence breaks out at Hanumangarhi and spills over to Ramajanmabhumi
CE 1949	Beginning of police protection order at Ramajanmabhumi
CE 1992	6 December, destruction of dilapidated sixteenth century Babri building (mosque-alike) by the supporters of VHP
CE 2000	On 27 February, Ayodhya is nominated as the sister city of Gimhae by its mayor
CE 2001	06 March, the Queen Huh Memorial stone was laid in a park, near New Ghat at Ayodhya
CE 2018	06 November, Queen Huh Memorial Park, renovation and extension inaugurated by Mr. Do Jong-hwan, Minister of Culture, Korea, and the Chief Minister of Uttar Pradesh, Mr. Yogi Adityanath
CE 2019	9 October, the Supreme Court ordered the land to be handed over to a trust to build the temple of Ramajanmabhumi at its historical site
CE 2020	5 February, a trust known as 'Shri Ramajanmabhoomi Tirth Kshetra' was created by the Government of India
CE 2020	5 August, the site worship of constructing Ramajanmabhumi by Hon'ble Prime Minister Narendra Modi
CE 2021	4 November, the festival of light 'Dipavali'; the renovated and new Queen Huh Memorial Park is partially inaugurated but open to the public
CE 2023	26 December: Margashirsha Shukla 15th Samvata 2080 'Purnima', expected to complete the inner sanctum of the Ramajanmabhmi temple
CE 2024	17 April: Chaitra Shukla 9th Samvata 2081, planned to finally open the Ramajanmabhmi temple for devout people

Source Shaw (2000: p. 695), and Kumar (2018: pp. 70–71); updated

to CE twelfth century. That's now considered as one of the oldest living sacred sites in India (see Mani 2022). Historians have identified this place to be Saketa, a key Buddhist centre during the fifth century BCE (it is a widely held belief that Buddha visited Ayodhya on several occasions) which remained till the CE fifth century (cf. Sitaram 1932). Faxian (Fa-Hien, CE 377–422), the Chinese monk, kept a record of several Buddhist monasteries that he saw here (see Bakker 1986). Ayodhya has a historical significance for the Jain community too. This is the birthplace of two important Jain *Tirthankaras* who were born in the early centuries CE. Jain texts also stand testimony to the visit of Mahavira, Jainism's founder to this city. In the

CE seventh century, the Chinese monk Xuanzang (Hsüan-Tsang, 602–664) recorded spotting many Hindu temples in Ayodhya. In the epic *Ramayana*, the city of Ayodhya is cited as the birthplace of Lord Shri Rama, a Hindu deity who was worshipped as Lord Vishnu's seventh incarnation.

7.3 Rāmajanmabhumi: Placemaking and Rejuvenation

According to ancient Purāṇic tales, Manu, the first primordial divine progeny of Bhārata (India), founded this city in ca 7135 BCE (coinciding with the birthday of Lord Rāma, 5114 BCE). Ikshvāku was the eldest of Manu's nine sons, by whose name the race and clan were known. In the early period of development in and around Ayodhya (ca 1000 BCE–CE 500), Buddhism and Brahmanism alternatively dominated the landscape and culture. By the turn of the CE second century, the city of Ayodhya was well established and known as a pilgrimage centre, and by the turn of the Gupta period (CE fourth- sixth centuries), many temples and *ghats* along the Sarayu River were built up. Presently, the city possesses 1161 Hindu temples and shrines, 5 Jain temples, 2 Buddhist shrines, 7 Sikh *gurudvāras*, 2 churches, and 186 Muslim shrines (Kumar 2018: p. 136). The Sarayu River, the backbone of the city, is also perceived and eulogized as a goddess-like many other great rivers of India. Every *ghat* (stairway) possesses individual historical, mythological, religious folktales, and spiritual importance and is attached to distinct rituals and celebrations (cf. Singh 2020: p. 90).

With its general decline all over India from the sixth century onwards, Buddhism lost its position in Ayodhya too and appears to be virtually extinct thereafter CE 1000. It is evident through literary and archaeological evidences that in the twelfth century there were five important Vishnu temples located respectively at Guptar Ghat, Chakratirtha Ghat, and Rāmajanmabhumi, and the western and eastern sides of Svargadvara Ghat. Three of these temples were demolished and superimposed by mosques and one was swept away by the Sarayu River. The fifth one is perhaps occupied by the Chakrahari temple (cf. Fig. 7.2).

Like other places in north India, Ayodhya was also invaded by Muhammad Ghori in CE 1193. Under his patronage, his army officer Makhdum Shah Ghori reached Ayodhya in 1194 and destroyed the famous Jain temple of Adinātha. In the successive periods, under the Sultanate's rule at Delhi and the Mughal rulers, Ayodhya again seized and many temples were demolished. Over time, in 1528–1529, Mir Baqi Tashkandi, the army chief of Mughal invader Babur, had pulled down the famous Rāma temple of Pratihāra and the Gahaḍavāla period—existing at the birthplace of Rāma (Janmabhumi). Following the successive period of 15 months, he had built a Muslim monument (so-called mosque?) upon the rubrics of the temple (see Mani 2022: p. 101). Consequentially, this site of ancient temple turned to be a site of contestation, controversy, and sensitivity. Afterwards, Muslims have never performed prayer (*namāz*); however, the notion of being a 'mosque' (*Maszid*) has been widely propagated. This site and the environs have also been an issue of Hindu-Muslim

Fig. 7.2 Ayodhya (*Source* Singh and Rana 2002: p. 302)

conflicts and riots. The main site (Janmabhumi) was always open for Hindu devotees till 23 February 1857 when the East India Company (Britain) made a dividing wall and prohibited the entry of Hindus through the passage of mosque.

Taking into view the story of glories, contestation, and controversies, a recent book established successfully with inscriptional and archaeological evidences of the existence, importance, and continuity of the Rāmajanmabhumi Mandir and other auxiliary temples in its environs (Pandey 2019; cf. Fig. 7.3). The cultural landscape of Ayodhya possesses a rich heritage of icons of divinities, rubrics, monuments, sacred architecture, pilgrimage, and *parikramā* routes and the auxiliary elements and attributes; these altogether make the landscape a 'wholeness' and 'mosaic' of Indian culture, where the Rāmajanmabhumi serves as the *axis mundi*—as a node from where all the *saṅkalpa* (taking vow) initiated and after finishing get *purṇāhuti* (completion). This tradition of *darshan* (glimpsing) and *yātrā* (pilgrimage and circumambulation) goes long back in history. These get a formal built structure around ca CE fourth century by Vikramaditya Chandragupta II that finally enriched with superimposition and grandeur by Gahaḍavāla king Chandradeva around CE 1093 in the form of Vishṇuhari temple (Pandey 2019: p. 51). This temple was demolished and replaced by an Islamic-built structure (so-called mosque) in CE 1528 by Mir Baqi (Baqi Tashqandi), who was a commander of the Mughal Emperor Babur (r. 1526–1530).

In the holy territory of Ayodhya, Rāmajanmabhumi has been the pivot point that served as the initiation and completion of the pilgrimages and was surrounded by the fort structure, known as Ramkot (see Fig. 7.4). According to ancient texts, Ramkot was the ancient fort of Rāma. This fort covered a large extent of ground and according to ancient manuscripts, it was surrounded by twenty bastions, each of which was commanded by one of Rāma's famous generals, after whom they took the names from which they are still known. The guardian deities surrounding the Ramkot protect the holy places and their visitors; these guardian deities were the guards of the Rāma's Fort and it is also a great example of the coexistence of multi-cultural sites.

A recent book finally ends with the rational and ethically justified conclusion about the antiquity and spatial expansion of the Ramkot:

> Built in the period of Chandragupta II- Vikramāditya and further incorporated and spread over by Yashovarman and again by Meghasuta and Ayusyachandra in the Gahadavala period [CE 12th century], which was destroyed by the Mughal rulers; and re-construction (revival) of the same sacred *kshetra* (holy territory) as received by Savai Raja Jaisingh from the 16th century Mughal rulers will be the only solution for the controversy of the 'Rāmajanmabhumi Temple'. No way the footprints of history can be belied; this has to be accepted and this will be the best way to re-establishing the moral ethics of culture in terms of cultural rejuvenation. (Pandey 2019: p. 86)

Ramkot (Rāmadurga) is the main historical site that has a strong tradition of pilgrimage since the ancient past. It has been eulogized as a fort that had four gates in four directions, each one protected by guardians or demigods (Fig. 7.4). Of course, many of the spots and manifested icons are lost, and still, some of the prominent ones exist in the form of local deities, like Mattagajendra which lies in the basement of a mound in the north-eastern from the Rāmajanmabhumi, but nearby holy spots are

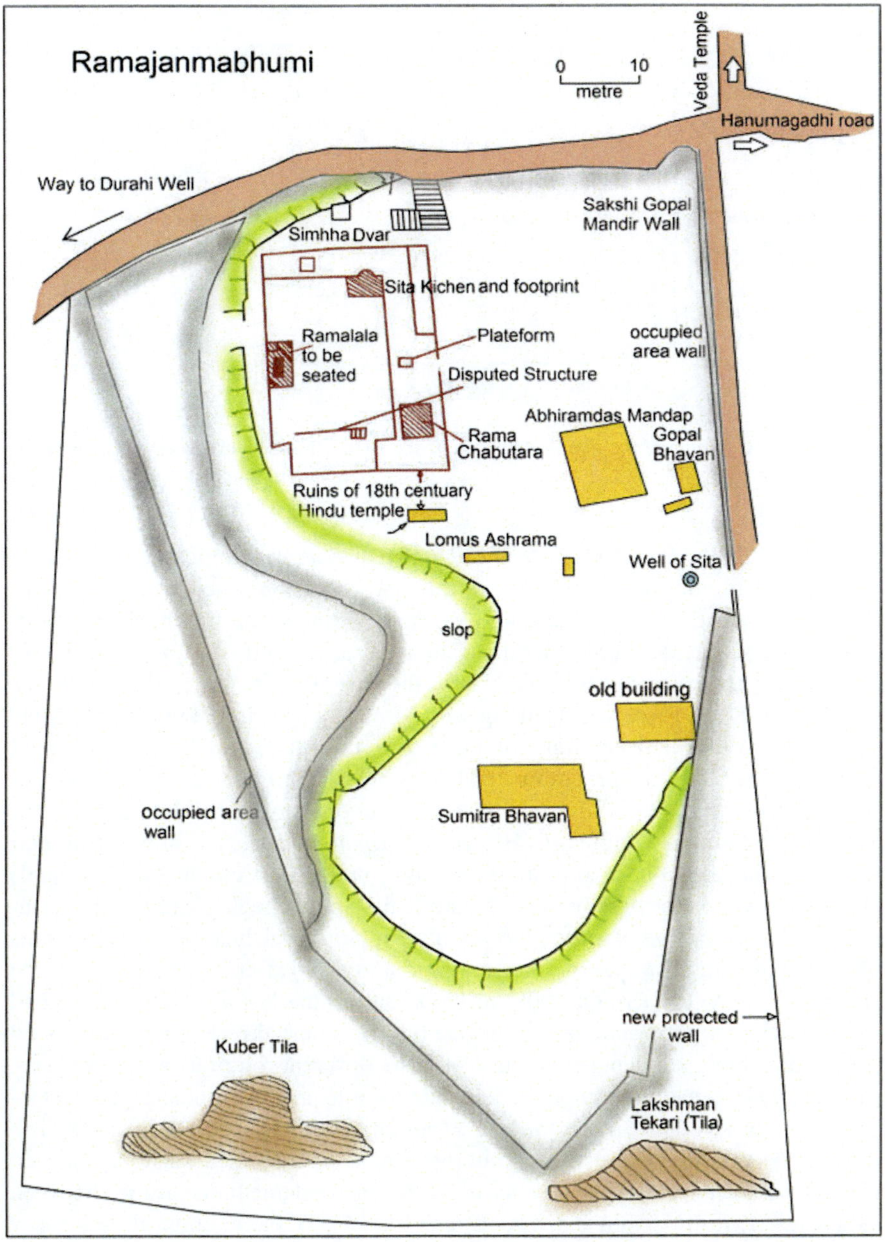

Fig. 7.3 Rāmajanmabhumi, Ayodhya: the notional map (*Source* Kumar 2018: p. 95)

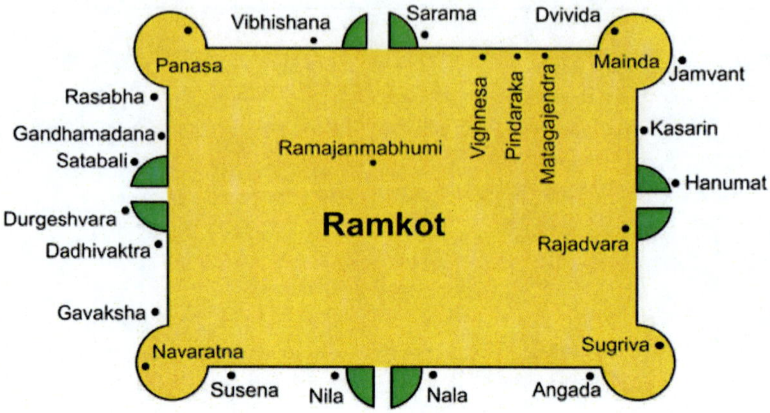

Fig. 7.4 Ramkot, Ayodhya: the manifested map of the Rāmajanmabhumi (*Source* Kumar 2018: p. 124)

disappeared (see Bakker 1986, II: pp. 117–118). The archaeological investigations approve the continuity of settlements in this part since the seventh century BCE, illustrated with massive brick walls, and Vaishnavite images of divinities; of course, there are pieces of evidence of a Buddhist stupa there, presently designated as Kuber Tila/mound in the west and Sugriva Tila in the east (Bakker 1986, II: p. 119; see Fig. 7.4). These two sacred spots along with Mattagajendra in the north demarcate the holy territory (*kshetra*) of Ramkot. Over time, the whole area became occupied by unplanned settlements, predominated by shrines and temples (see Fig. 7.5).

On 6 December 1992, a mob led by Hindu nationalists, the right-wing activist from the World Hindu Congress (VHP, *Vishva Hindu Parishad*), ultimately in their last attempt succeeded in razing the sixteenth century Babri monument (mosque!) in Ayodhya, which was already proved historically and archeologically as the birthplace of Lord Rāma, and was the glorious site with lofty Vaishnavite Hindu temples in the early twelfth century, but converted into a mosque after its demolition in the early sixteenth century (Singh 2008: p. 131). During the last 490 years, there had been several attempts to remove the mosque through court, direct action, or planned attacks. After India's independence in 1947, the different religions and their associated monuments had largely co-existed side by side and thus encouraged contestations, e.g., in Ayodhya, Varanasi, and Mathura (Singh et al. 2021: p. 245). The Ayodhya crisis must also be seen within the climate of increased tensions between India and Pakistan over the last few decades, and the fundamentalist groups between Hindus and Muslims within India itself (see Elst 2003). Here, cultural diplomacy could not succeed to resolve the issue, despite detailed archaeological evidence (Mani 2022).

In India, religion is more like a cultural symbol that fully fits the Hindu psyche, and politicians often use it for their vested interests and to maintain their authority. But as a pluralistic society, many times these strategies conflict (Singh 2011: p. 291). Since Prime Minister Narendra Modi and his party, BJP, took power in 2014, the

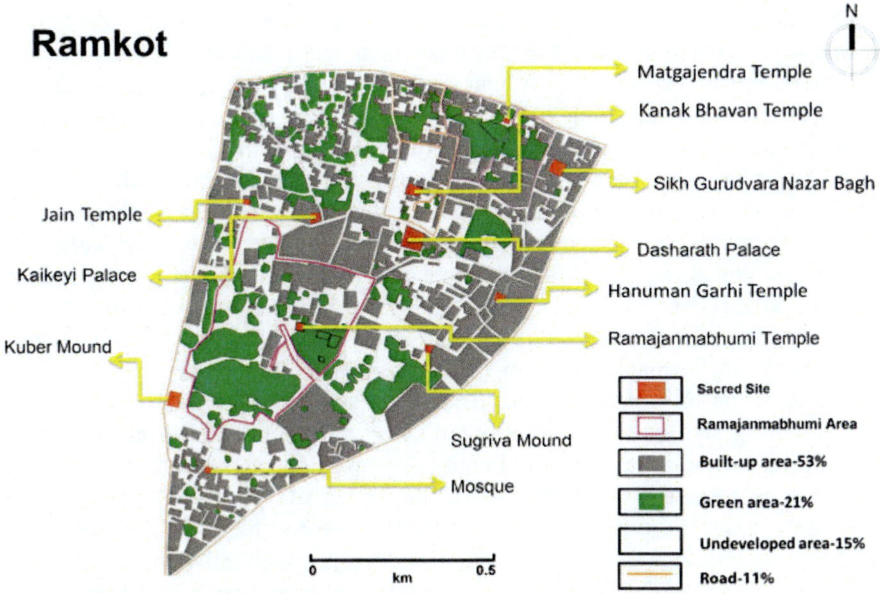

Fig. 7.5 Sacred site of Ramkot, Ayodhya: Land use and morphology (*Source* Developed on personal surveys by the authors)

rebuilding of a Rāma temple at Ayodhya has been at the forefront of their Hindutva agenda of political campaign and cultural diplomacy (see Singh and Rana 2021).

The Supreme Court Verdict in favour of a Rāma temple on 9 November 2019 that first began in 1990 was lodged on 30 September 2010 in the Allahabad High Court, but was ended almost after two decades; it has been one of India's longest land disputes that faced so much of the jurisdictive complexities that continued for decades. The Supreme Court vacated the previous decision and ruled that the land belonged to the government based on tax records, and further ordered the land to be handed over to a trust to build the Hindu temple. This way on 5 February 2020, the Government of India announced a trust named 'Shri Rāmajanmabhūmi Tīrtha Kshetra' to reconstruct a Rāma temple complex on the ancient site. As mentioned earlier, by early 2024 this temple will be ready to be opened to the public. This temple trust will also look after the revival and rejuvenation and maintenance of the pilgrimage journeys through their schemes of heritage inclusive development strategies, with emphasis on the promotion of pilgrimage tourism and spiritual tourism under the diplomatic policy of Hindu revivalism and renaissances of the Hindu nationalist party of the state of Uttar Pradesh.

The above judgement has put the town of Ayodhya on the cusp of a new identity, new makeover, and new development within the ideology of the present government's agenda for rejuvenating the ancient cultural identity. It is to be noted that the recent observation that 'religious travel to Ayodhya commemorates the obliteration of one site of India's Islamic heritage and the promise of the reclamation of India from Muslims, other minorities, secularism, and democracy—all threat to the Hindu nationalist vision' (Anand 2018: p. 112), is a completely superficial, biased, malicious, and fabricated conclusion without spot studies (see details, Kumar and Singh 2017a, b, also Singh and Rana 2021). To get momentum to plans, Ayodhya Tirtha Vikash Parishad (Development Authority of Ayodhya Tirtha, ATVP) is recently formed under the directives of the Supreme Court of India, which will take care of the construction of the 'grand and magnificent' Rāma Mandir at the ancient site, within a territory of 31.16 hectares (77 acres) of land in a period of four years, i.e., by 2024; and this territory will house various religious institutes and institutions for spiritual practices, like institute of Vedic knowledge, and also an ambitious 'theme park' showing model of historical events and Rāma Katha Museum. The ATVP will be operated by the Rāmajanmabhumi Trust, consisting of 15 members, out of which nine are the chief of monasteries (all Brahmins), a member from the royal family (also Brahmin), and the rest from political allies; no scholar who directly worked on Ayodhya was nominated in the board. The levelling of the land was already completed, followed by initiation rituals of the sacred site (*bhūmi-pūjana*), and the laying of the foundation stone on 5 August 2020 by the hon'ble PM Narendra Modi, thus declaring initiation of the construction of Rāmajanmabhumi Temple.

Of course, as a part of political strategy, having emotional support (or blackmailing in a way!) of the Hindu masses, a mega project of installing 251-m-tall bronze-made Rāma's image (including plinth and umbrella), surrounded over an area of 100 acres (40.47 hectares), is in process at the budget of Rs. 2,500 crore (US$ 330 million at the current rate), i.e., 20% of the total Ayodhya development project cost—assumed to be the tallest statue in the whole world. This results in the avoidance of developing infrastructural and health and civic services, environmental protection, and unwanted consequential repercussions that pervade the culture and societal development in Ayodhya. This is an example of using political power and cultural diplomacy of '*dharmacracy*' (use of '*dharma*' for protecting the political games) for placemaking and making false propaganda taking the god-fearing mindset of innocent Hindu masses. Of course, the basic foundation of Indian philosophy is '*dharmacracy*', which refers to '*Yato Dharmah Tato Jayah*', which means 'True victory comes from righteous conduct only'; however, politicians use this as tactful means for populism and emotional blackmailing.

There is an urgent need that the Government of India should review its archaeological laws and make suitable changes concerning its sacred sites, particularly those are pilgrimage places, like Ayodhya, Varanasi, and Mathura—the holy-heritage cities facing contestation, claims from different religious groups and superimposition of so-called modern and smart development that are not suited there. For this, they can use the guidelines under UNESCO World Heritage Site that refers to 'cultural heritage', 'cultural landscapes', and 'intangible resources'—all of these pertain to the continuity of traditions and performances that evolved from the historical past (Singh 2011: p. 297).

Among the five dimensions of making and maintaining the 'good city' form, viz. 'vitality', 'sense', 'fit', 'access', and 'control', Lynch (1981) emphasized the notion of 'access' that includes access to open space, friends, jobs, services, and information, while 'control' is an appraisal of how various forms of public space are rationally governed. On this scale, Ayodhya has not yet been successfully marched, despite several development projects concerning inclusive heritage development, promotion of pilgrimage and spiritual development, and the smart city plan; of course, they are mentioned in the Master Plan 2041 in progress. Lynch has further recommended rights of presence (who gets access), action (what we can do there), appropriation (the right to exclude others), modification (the right to change the design or move the loose parts), and disposition (sale, demolition) (Lynch 1981: Ch.11). He was right to say that the right of presence or access is the pre-condition for the following three: action, appropriation, and modification, which are relevant for the maintenance of public-space quality (see Lynch 1981: pp. 232–243). He notes that 'rights to action and appropriation are social and political freedoms of expression associated with the vitality and intensity of public space – while access does not ensure them, they can mean nothing without access' (Dovey and Pafka 2020: p. 236). The time shortly will answer these strategies in the case of Ayodhya.

7.4 Queen Huh Memorial Park: Cultural Strategy and Placemaking of Ayodhya

The legendary tale of Queen Huh was first mentioned in the *Samguk Yusa* (삼국유사, 'Memorabilia of the Three Kingdoms'), a Chinese-language memoir of an old historical tales book of Korea dated ca 1281 and compiled by a group of Buddhist monks as a collection of allegories and historical accounts connected to the Three Kingdoms of Korea that included Baejke, Silla, and Gogureyo, and was written by the monk Il-Yeon (일연– 然; 1206–1289) [cf. Ha and Mintz 1972]. This book narrates the story of Queen Huh Hwang-Ok (wife of King Kim Suro), who founded the ancient Karak Kingdom (Geumgwan Gaya) and was believed to be born in the city of Ayuta/Ayodhya (India), which was named Saket in the ancient period. This book further describes that in the year CE 48 (27 July?), an Indian princess Suriratna (Korean name Huh Hwang-Ock/'Heo' 허황옥), the daughter of King Padmasen and Indumati, who was 16-year-old at that time, voyaged by the sea to Korea from Ayodhya and later married with King Kim Suro of the ancient Korean Kingdom of Kaya (Gaya). This kingdom is now represented by the Gimhae city (Kimhae city). King Kim Suro had 10 children, those established lineages of the Kim or Karak clan. Prof Byung Mo Kim, a Professor Emeritus of Cultural Anthropology at Hanyang University and famous national archaeologist from Korea, shares a 'genetic connection' with Ayodhya (see Choong 2011: pp. 115–136).

Over time, a delegation led by Gimhae city Mayor Song Eun-Bok paid a visit on 27 February 2000, to their ancient ancestral site and proposed Ayodhya as a sister city of Gimhae; resultantly, a plan was enhanced to set up a memorial building in honour of Queen Huh in the future (Kumar and Singh 2017a, b), as an icon representation of the Tomb of Queen Huh in Gimhae. Proceeding further, in 2000, the Korean government accepted Ayodhya as the sister city of Gimhae. In the memory of Princess Huh, the

Fig. 7.6 Aerial view of the Huh Memorial Park, Ayodhya (Based on Google Image)

Korean government with the support of the Queen Memorial Society, constructed a monument in the city of Ayodhya at the riverfront on 6 March 2001 (see Fig. 7.6). The plaque at the monument inside the courtyard reads, 'Queen Huh was a princess of the kingdom. Her father, the king of Ayodhya, on receiving a divine revelation, sent her on a long sea voyage to the Karak kingdom in southern Korea to marry King Suro'.

Taking in view the narratives of a Korean tale recorded in 'Memorabilia of the Three Kingdoms', *Samguk Yusa* (dated ca 1281), the senior author has visited (Nov. 2017, Feb. 2019, Nov. 2019) those places related to the events, referring arrival of Queen Huh, who was from the royal family of Ayuta (identified as Ayodhya, the headquarter of the same-named district of Uttar Pradesh, India, firstly by Prof. Kim Byung-Mo, an anthropologist from the Hanyang University, based on his study of phonetic similarity; see Choong 2011: p. 34). Some historians, however, believe that the voyage of Queen Huh Hwang-Ok of Ayodhya is only a myth (cf. Tayal 2015: p. 23). This idea is supported by the fact that no records of this legend in Indian history have been traced yet (BBC 2001), as proclaimed on the occasion of unveiling the memorial stone of Queen Huh on 06 March 2001. Some scholars trained in the west are sceptical thinking that *Samguk Yusha*'s narration is more fiction and less historical evidence; contrarily, it is argued that this is related to old classical languages of Chinese and Korean, therefore to be interpreted in that context. Using myth and classical traditions of old languages and their reliability is justified by some other scholars (McBride II 2006: pp. 1–2). Based on his six times visits to Ayodhya and close interactions with royal families and historians and personal investigation, Prof. Kim Byung-Mo's proclaims have been widely propagated through the newspapers in India and Korea; however, no substantive research paper, or book, has yet been published (Singh and Kumar 2019: p. 17).

The revitalization and expansion of Queen Huh Memorial Park in Ayodhya are depicted as a symbol of Indo-Korean cultural links; the extended zone adjacent to the earlier one was recommended in March 2016, by a 38-member South Korean delegation who paid the visit for evaluation. Its present foundation was laid on 6 November 2018 by South Korean First lady Kim Jung-sook. In the same year, on 24 October 2018, Hon'ble Prime Minister of India Narendra Modi was nominated for the highest civilian award of Korea, Seoul Peace Prize, for his efforts in making 'regional and global peace' and future efforts in creating 'progressive Asia' under Indo-Korean venture (TOI 2018). In this continuation, on 16 June 2021, the Korean Ambassador expressed Huh Memorial Park as a place of pilgrimage for 'Kim' clan descendants of Korea, which consists of about ten per cent (i.e., 5.13 million) of the Korean population (est. 51.3 million in 2021). On the recently held sacred festival of light, 'Dipavali' (which fell on 4 November 2021), this park is partially inaugurated in its present form. This park, already in high speed for completion, will be inaugurated in its present form (see Fig. 7.7), and on the line of previous celebrations by the Korean delegation, a delegation from the Korean embassy in New Delhi has attended as a witness to the revival of Indo-Korean cultural interconnectedness. The main monument was built in Korean traditional style, using a 3-m-high and 7,500-kg heavy stone shipped from Korea. The Korean Pavilion reflects the architectural style of Aelyeonjeong at Changdeok Palace—listed as the UNESCO World Heritage site in Seoul. This Pavilion embodies the simple beauty of Joseon Palace Korean architecture of the seventeenth century. Under the bilateral agreement, half of the construction cost is borne out by the state government of Uttar Pradesh, i.e., worth Rs. 250 million (equal to US$ 3.36 million at the present rate), and the rest by the Korean government under the patronage of Gimhae municipality and the Karak Clan Society.

These cultural links and programmes are driven by the contemporary economic and geopolitical agenda, i.e., the Indo-Korean vision of new Asia that has a deeper history. Both countries were formerly colonies of empires and saw parallel journeys to have freedom in the twentieth century. Philosophers of idealism, Buddhists, socialists, revolutionaries, and peaceful protesters from both countries took each other as a point of reference as they carried their nation to independence on 15 August, of course, South Korea in 1948, i.e., after one year of India's independence. Nonviolent mass protest presents one key example of similar struggles and articulations of freedom.

In a recent development, a strong strategy between India and Korea has been formulated through an MoU (Memorandum of Understanding), and 11 broad and basic issues have been formulated (see Table 7.2). This MoU has paved the path for the Indo-Korean joint programme for sustainable development, including special consideration for developing and enhancing Queen Huh Memorial Park and related cultural heritage in the environment (MoU no. 11); this new monument will be a tribute to the shared cultural heritage and long-lasting friendship between India and Korea.

Fig. 7.7 Coming up Queen Huh Memorial Park, Ayodhya (Photo courtesy Sarvesh Kumar, 26 November 2021)

7.5 Indo-Korean Interfaces in Placemaking Through Queen Hue Memorial Park

For developing sustainable development of cultural and responsive tourism and also to support in preparing the dossier for an inscription of the two cities in the UNESCO WHL, as discussed between the two sides of scholars and authorize (Ayodhya and Gimhae), the following issues are considered for follow-up action programme on the line of Ayodhya-Gimhae Interfacing Model of Cultural Landscape and Heritage Planning (Singh and Kumar 2019: p. 20):

1. Interfacing cross-cultural study of cultural landscapes between Ayodhya and Gimhae.
2. Development of Tangible and Intangible cultural heritages and related tourism in both the city through exchanges.
3. Construction and building replica of the tomb of Queen Huh Hwang-Ok in Gimhae (see Fig. 11.6 and 11.7), and representation (mural, frescoes, sculptures related to the Queen's story) in the Korean Park, Ayodhya.
4. Development of museum and theatre to show the life stories and incidences of the Queen's voyage and marriage.
5. Advancing and advocating integrated messages from Sanatana Hinduism and Buddhism in making global peace and human welfare, taking into account of ancient past, esp. ca CE first century.
6. Preparation of common platform, guideline, and concerned criterion of heritage properties for making valid and justified Dossiers to get both the cities, Ayodhya (including Queen Huh Korean Memorial Site) and Gimhae (the two tombs of

Table 7.2 Indo-Korean MoU for reciprocal cultural and economic developments

Se	MoU Focus	Objectives
1	Comprehensive Economic Partnership Agreement (CEPA)	Upgrading the India-RoK CEPA by identifying key areas for trade liberalization
2	Trade Remedies	Cooperation in the area of trade remedies through strong Cooperation
3	Future Strategy	Development of technologies and commerce to reap the benefits of the Industrial revolution
4	Cultural Exchange Programme, 2018–2022	To deepen cultural and people-to-people relations
5	Cooperation in scientific and technological research, between CSIR (India) and NCST (Korea)	Cooperation in scientific and technological research, transport system
6	Research Design and Standards Organization (RDSO)	Cooperation in railway research, exchange of railway-related experience
7	Cooperation in the field of biotechnology and bio-economics	Adoption of biotechnology and Bio Big-Data in health, medicine, Agro-fishery products, etc.
8	Cooperation in the field of ICT and Telecommunications	Expansion of state-of-the-art Telecommunication/ICT services
9	Cooperation in the field of Micro, Small, and Medium Enterprises	Cooperation in the development of micro, small, and medium enterprises
10	Govt. of Gujarat and Korea Trade Promotion Agency (KOTRA)	Enhance industrial and investment relations between South Korean companies
11	Promoting Queen Suriratna Memorial Project	Upgrading and expansion of Queen Huh Memorial Park, and promoting cultural heritage

Source Embassy of India, Seoul, RoK, https://www.indembassyseoul.gov.in/page/list-of-mous-agreements-signed-during-the-visit/; dated: January 2020

King Suro and Queen Huh), inscribed in the UNESCO World Heritage List (of course, firstly 'Tentative List').

7. Reconstructing the 'imagined' cultural landscapes, routescapes, and mythical landscapes, taking into account the tale of *Samguk Yusa*, field studies, and marriage rituals of north India.
8. Media productions and publications of research papers/books on these aspects, as knowledge-addition to Indo-Korean studies, which will further promote cultural tourism.

Special festival of 'oil lamp' (Dipavali) celebration, a mark of the revival of an ancient tradition, recently started at a grand scale along the riverfront of Sayaru called 'Dipotsava', when 1.2 million oil lamps are decorated along the *ghat* as a symbol of welcoming god Rama who after killing the demon returned to Ayodhya as the ancient myths referred to. This was celebrated before the eve of the festival of light

'Dipavali' and held recently on 19 October 2017, 6 November 2018, 26 October 2019, 13 November 2020, and 3 November 2021 (following the Lunar calendar of Hindus, i.e., Karttika dark fortnight 15th 'no moon day'). This is now taken as a good example of 'placemaking' and promoting Indo-Korean cultural interaction. Already in this continuity, a grand celebration was held in 2018 and 2019 in which a special delegation of artists and members of the Karak clan (Gimhae and other places) have taken part and made a grand show of the dramatic part of the voyage of Queen Huh to Korea. This programme is led and supported by the Karak Clan Society, Gimhae. Initiations are in process under the MoU to further take part and represent India (through the myth and practices in Ayodhya) in the forthcoming Gaya Cultural Festival, which was started in 1976 and celebrated every year in high esteem. This way Indo-Korean cultural ties will be more active. Unfortunately, due to the pandemic situation in 2020 such celebrations were held in abbreviated forms. The recent one held on 3 November 2021, revived the old tradition and made a world record (as noted in the Guinness Book of World Records) with lightning 1.2 million oil lamps, and was attended by a delegation from the Korean Embassy at New Delhi. However, the afterward result deteriorated the environment and turned to different degrees of pollution and chaos in a lack of post-festival sustainable planning for the consequences.

In October 2005, the Korean government organized an international festival at Gimhae city to mark the 2000 years of the visit of Queen Huh, where the officials and artists of the U.P. Cultural and Tourism Department participated in the festival. This has been celebrated on a grand scale in 2017. Now, a similar celebration is planned to hold at Ayodhya and also to make a landscape model of Huh's tomb as in Gimhae. On India's republic day celebration, 26 January 2010, Korean President Hon'ble Lee Myung-Bok had been the chief guest; this indicates further promotion of strong ties between the two countries.

The recent visit of Korean President Hon'ble Moon Jae-in with his wife Kim Jung-sook and a delegation of officials in India during 8–11 July 2018 has opened a new step in the Indo-Korean cultural interaction. On 10 July 2018, Korean President agreed to grant a fund of US$ 860,000 for infrastructural development; it includes the establishment of the memorial of Queen Huh (Suriratna) at Ayodhya, a symbol of Indo-Korean cultural relationship having a history of more than 2000 years. This is on the line of Hon'ble PM of India Narendra Modi's visit to Korea in May 2015, now resulted in MOU (10 July 2018) for the bilateral agreement and MoUs for Cultural Exchange Programme for the period 2018–2022, including 'Queen Suriratna Memorial Project' (Table 11.2); this is 'to deepen cultural and people-to-people relations by providing for institutionalized cooperation in fields of music and dance, theatre, art exhibitions, archives, anthropology, mass media programmes, and museum exhibits' (see Singh 2018; Singh and Kumar 2019: p. 21). This agreement and development plan will promote cultural tourism (Sharda 2018a), where Ayodhya will serve as the nexus for this programme. The background has already been prepared through a novel telling the magical story of Princess Suriratna who became Queen Huh (Parthasarathi 2015, who was an Indian Ambassador in Korea during 2005–2008). Also, a film by

Indian producer Jayanth Sinha is in progress to make the story a medium to awaken the Indian mass knowing their historical-cultural links with Korea (Sharda 2018b).

7.6 Prospecting Placemaking: Reviving the Essence of Holy-Heritage Cities

It is rightly remarked: 'India has always led the world spiritually. Divinity here has attracted pilgrims and travellers since time immemorial. Most of our pilgrimage places are displaced, with their original sacred geography distorted. Ayodhya gives us a unique opportunity to recreate the sacred geography that leverages modern technologies but stays rooted in the Indic philosophy of *yātrā* (pilgrimage)' (Goyal 2021). Restoring, reviving, re-establishing, and re-creating the lost sacred geography in the past have transformed the image of Ayodhya and opened several avenues for varying attributes of placemaking, which will serve as a model at a successive period: 'succession-sustenance-sustainability'—'evolution-existence-expansion'. This study will help for similar studies of holy-heritage cities, e.g., Varanasi and Mathura—propagated as cities of contestations and conflicts in recent history, especially between Hindus and Muslims.

India and Korea had significant space in the mental cognizance of each other in the past. However, more than the material exchange, the mutual relations were based on ideational and cultural commonalities and mythological beliefs. The direct or indirect way of Buddhism, from India to Korea, along with the legend of the princess Huh of the Kaya kingdom was the initial route of interactions between the two cultures. Thanks to the Korean people (especially the Karak clan group) to get revived the old cultural linkages, thus using the old legend of the 13th-century book of *Samguk Yusa*, to create an additional arena of placemaking in Ayodhya. Through such structural and memorial developments, the Queen Huh legend (?) has been transformed into a built-up space, resulting in the opening of cross-cultural interaction and the promotion of cultural tourism. Queen Huh Memorial Park is now already an impartial part of heritage walks, cultural tourism, and sacred scenic spot in Ayodhya.

Under the present government rule, the use and role of heritage and associated placemaking are used as a political strategy in restoring and reviving the history of the ancient past under the vision of Hindu ideology—*dharmacracy*. These are illustrated in the recent transformations and regeneration activities and residuals that taking place in Ayodhya. This political-ideological march reinforces the point that while there is a strong correlation between conservation, regeneration, placemaking, and economic development, other motivations might be in play and help in image-making and fulfilling the sentiments of the dwellers and pilgrims. These interventions follow the top-down approach in a peculiar way to propagate the heritage values and place-identity—of course, several times following the ways to manipulate and manifest the place-identity.

For a religious heritage to become a resource for development, first it needs to be documented, then protected, and finally utilized according to specific heritage guidelines and legislation, within the purview of historical-cultural images and indigenous traditions (see, Singh et al. 2021: p. 243). Only then, with increased stakeholders' awareness and participation, will policy efforts and interventions become sustainable—environmentally, socially, and culturally. We may separate ourselves from the web of our heritage in the pursuit of modernity and secularism, but it will always be at the cost of our hearts and our souls (Singh 2015: p. 120).

Under the increasing pace and pressure of the modern way of pilgrimage/religious tourism, predominated by huge masses of visitors, the pilgrimage landscapes are on the verge of threatening and deterioration, ecologically and culturally both. This tendency helps to increase the pilgrimage economy in a more materialistic frame and violates the sanctity and solace of the environment. A recent study of the holy-heritage site concludes that 'the faith of devotees in maintaining these places as sacred is central to this process: they travel to these places to fulfil their religious and spiritual needs, perform pilgrimage rituals, invest in symbolic structures, and as patrons along with other social actors contribute in making the environment. However, these needs are manifested differently in different periods and therefore it is necessary to rethink the issues surrounding the physical environment that arise over time, especially when pilgrimage sites are used for multiple purposes including pilgrimage and tourism' (Shinde 2012: p. 126). These issues should be given priority in the developmental strategy for holy-heritage cities like Ayodhya. The way outside touristic forces and the overload that are grabbing the city of Ayodhya, the eternity and serenity are in danger. This will ultimately loss the indigenous culture, if not responsive and sustainably development strategy is taken care of at all the corners.

The emerged heritagescapes through placemaking that survive will be those we preserve and make that constantly live, functional and transformative in the span of time. 'Let us rejoice, for we are alive, and life is good; let us participate in the celebration of renewal and rejuvenation, caring for one another and our cosmos through lived placemaking' (Singh 1995: p. 210). The place is considered as the crucible element of 'space and function' of cultural and economic interactions, and at the other end, cultural identity and economic order condense out to emerge as the monopoly power of place. Hence, it is important to recognize the symbiotic inter-relationships between place, culture, and economy to protect the spatial root of cultural-economic aspects and maintain it with the ever-changing forces of placemaking. It is rightly observed that 'the linkages among spatiality of time, the temporality of space, sacrality of space and further their counter networking with the spirit of placefulness would open a new site of pilgrimage-tourism and spirituality' (Singh and Jaiswal 2018: p. 67).

Acknowledgements The fieldwork in Gimhae (1917, 2018, 2019) was supported and guided by (*late*) Prof. Sung-Kyun Kim (SNU Seoul, Korea, b. 1956–d. 2020) under the aegis of NRF International Project—'Comparative Study of Cultural Links and Ritual Landscapes between Gaya (Korea) and India', ['General Research Collaboration Support Project', 20 Dec. 2018–19 Dec. 2019, Project No.: 500-20180233]. Many of the general and supplementary documents and information brochures in Korean were translated by Prof. Sung-Kyun Kim and incorporated here. We express our gratitude and thanks for his assistance and support; in fact, this paper is a memorial tribute to him.

References

Anand D (2018) Approaching Ayodhya on "Valor Day": Hindu Nationalism and pilgrimage as politics. In: Pinkney AM, Whalen-Bridge J (eds) Religious journeys in India: pilgrims, tourists, and travelers. State University of New York Press, Albany, NY, pp 91–116

Bakker HT (1986) Ayodhyā: the history of Ayodhyā from the 7th century BC to the middle of the 18th century [Three parts]. Egbert Forsten, Groningen

Chakrabarti DK (2000) Mahajanapadas states on ancient India. In: Hansen MH (ed) A comparative study of thirty city-state cultures. Danske Videnskabernes Selskab, Copenhagen, pp 375–391

Choong SK (2011) Voices of foreign brides: the roots and development of multiculturalism in Korea. AltaMira Press, New York

Dovey K, Pafka E (2020) Mapping the publicness of public space: an access/control typology. In: Mehta V, Palazzo D (eds) Companion to public space. Routledge, London, pp 234–248

Elst K (2003) Ayodhya, the finale: science versus secularism the excavations debate. Voice of India, New Delhi

Goyal A (2021) Ayodhya, an opportunity for Indic tourism model. NewIndiaExpress E-paper, November 8. https://www.newindianexpress.com/opinions/columns/2021/nov/08/ayodhya-an-opportunity-for-indic-tourism-model-2380703.html. Accessed 13 Nov 2021

Ha T-H, Mintz GK (ed & trans) (1972) Il-Yeon's Samguk Yusa. Yonsei University Press, Seoul

Kumar S (2018) Cultural landscape and heritage of Ayodhya-Faizabad: A geographical analysis. Unpublished Doctoral Dissertation in Geography, Banaras Hindu University, Varanasi [Supervisor: Rana P.B. Singh]

Kumar S, Singh RPB (2017a) Ayodhyā, the holy city of India: riverfront and a Place of Indo-Korean interfacing symbolic Landscapes. Sthāpatyam, Indian J Architect Art 3(6):61–74. ISSN: 2349-2368

Kumar S, Singh RPB (2017b) Muslim cultural landscapes of Ayodhya-Faizabad (India): a geographical scenario. The Geographer (Geog. Society, Dept. of Geography, Aligarh, India) 64(2):95–109

Law BC (1944) Ayodhyā in ancient India. J Ganganatha Jha Res Inst (Allahabad) 1(4):423-444

Lynch K (1981) A theory of good city form. MIT Press, Cambridge MA

Mani BR (2022) Ayodhya revisited: experiencing an unprecedented excavation in 2003. In Tiwari MNP, Sinha SS (eds) Kalpavriksha. Dr. R.D. Choudhury felicitation volume, vol I. B.R. Publishing Corp, New Delhi, pp 97–103 and p 223 plates

McBride II RD (2006) Is the Samgukyusa Reliable? Case studies from Chinese and Korean sources. J Korean Stud 11(1, Fall):163–189

Pandey DB (2019) Rāma Janmabhūmi Mandir Ayodhyā kī Prāchīnatā [Based on pieces of evidence from archival and archaeological excavations]. Standard Publishers, New Delhi (India) [in Hindi]

Parthasarathi N (2015) Silk and empress/legend of Indian princess in Korea. National Book Trust, New Delhi

Sharda S (2018a) Contest to decide a design for Queen Hu's Ayodhya memorial. The Times of India (a daily from Lucknow), June 19. http://timesofindia.indiatimes.com/articleshow/64642152.cms?utm_source=contentofinterest&utm_medium=text&utm_campaign=cppst. Accessed 9 July 2018

Sharda S (2018b) Tale of Ayodhya princess to come upon the silver screen. The Times of India (a daily from Lucknow), May 1. http://timesofindia.indiatimes.com/articleshow/63981410.cms?utm_source=plus.google.com&utm_medium=social&utm_campaign=TOIDesktop&utm_source=contentofinterest&utm_medium=text&utm_campaign=cppst. Retrieved 9 July 2018

Shaw J (2000) Ayodhyā's sacred landscape: ritual memory, politics, and archaeological 'fact'. Antiquity 74(285):693–700

Shinde K (2012) Place-making and environmental change in Hindu pilgrimage site in India. Geoforum 43(1):116–127

Singh RPB (1995) Heritage ecology and caring for the earth: a search for preserving harmony and ethical values. Natl Geogr J India 41(2):191–218

Singh RPB (2008) The contestation of heritage: the enduring importance of religion. In: Graham B, Howard P (eds) Ashgate research companion to heritage & identity. Ashgate Pub, London, pp 125–141

Singh RPB (2011) Politics and pilgrimage in North India: Varanasi between Communitas and Contestation. TOURISM, an Int Interdisc J 59(3):287–304 (Institute of Tourism, Zagreb)

Singh RPB (2015) Heritage value of religious sites and built archetypes: the scenario of hinduism, and illustrating the riverfront Varanasi. Aatmbodh, J Knowl Self 12(1, Spring):1–23 (U.P. PG College, Varanasi)

Singh RPB (2020) Sacrality and waterfront spaces in India: myths and making of place. In: Ray C (ed) Sacred waters: A cross-cultural compendium of hallowed springs and holy wells. Routledge, London, pp 80–94

Singh RPB, Jaiswal S (2018) Kardameśvara (Kāśī) temple, a religious heritage from India: sacred landscape, architectural designs, and perspectives. Sthāpatyam, J Indian Sci Archit Allied Sci 5(3):47–70. ISSN: 2349-2368

Singh RPB, Kumar S (2019) Interfacing cultural landscapes between India and Korea: Illustrating the memorial of Korean Queen Heo in Ayodhya. The Geographer (Aligarh Muslim University) 66(2):12–23

Singh RPB, Rana PS (2002) Ayodhyā. In: Singh RPB, Rana PS (eds) Banaras region: a spiritual and cultural guide. Pilgrimage & Cosmology Series: 1. Indica Books, Varanasi, pp 277–285

Singh RPB, Rana PS (2021) Cultural diplomacy in India: dispersal, heritage representation, contestation, and development. In: Niglio O, Lee EYJ (eds) Transcultural diplomacy and international law in heritage conservation. Cultural Diplomacy & Heritage series. Springer Nature Pte Ltd., Singapore, pp 267–294

Singh RPB, Rana PS, Kumar S (2021) Hinduism, heritagization, and holy cities in India. In: Kim C, Kim H (eds) Great transition in Indian Humanities. Institute of Indian Studies, Hankuk University of Foreign Studies, Seoul. Seoul- Korea, Dahae Design, pp 227–270

Singh S (2018) Here is the list of MoUs signed between India and South Korea. The CNBC TV News (a daily from New Delhi), July 10. https://www.cnbctv18.com/politics/here-is-the-list-of-mous-signed-between-india-and-south-korea-285091.htm. Accessed 11 July 2018

Sitaram (1932) Ayodhyā Ka Itihas. Arya Book Depot, New Delhi (in Hindi)

Tayal SR (2015) India and the Republic of Korea: engaged democracies. Taylor & Francis, New York

TOI (2018) PM Narendra Modi awarded Seoul Peace Prize 2018. Times of India, October 24. https://timesofindia.indiatimes.com/india/pm-modi-conferred-seoul-peace-prize-2018/articleshow/66341708.cms. Accessed 15 Dec 2018

Chapter 8
Placemaking Approach in Revitalizing Cultural Tourism in Temple Towns: Case of Melukote, Karnataka

Surekha Ramineni, Monalisa Bhardwaj, and H. Tejaswini

Abstract Heritage landscapes are a direct reflection of human actions, natural and ecological processes, and prevalent culture (Hayden in J Archit Educ 41:45–51, 1988). Delconte et al. have observed that the creation of a cultural substrate of rituals, culture, festivals, and processions plays an important role in enriching people's experiences (Delconte et al. in J Herit Tour 11:324–335, 2016). Tourism can revitalize the intangible cultural heritage of the historic town, and preserve the continuity and relevance of heritage. Through a coordinated approach, new opportunities can be created to enhance the image, and identity of the place and help the users interpret, understand, and experience the tangible and intangible qualities the place offers (Maiti and Bidinger in J Chem Inf Model 53:1689–1699, 1981). This study looks at placemaking approaches for revitalizing cultural tourism in Melukote, a temple town in Karnataka. Melukote once flourished under royal patronage as an important pilgrim center, but now lacks the resources to maintain its character and significance. The study discusses the need for revitalizing cultural tourism in Melukote, and the role of the placemaking approach in terms of—spatial analysis of ritual landscape and pilgrim paths, activity mapping of locals & tourists, interviews with stakeholders and subject experts on placemaking and cultural landscapes. The study concludes with strategies for effective placemaking and possible design solutions to enhance tourism.

Keywords Placemaking · Temple towns · Rituals landscape · Cultural tourism · Melukote · Holy-heritage town · Revitalizing

---

S. Ramineni · M. Bhardwaj (✉) · H. Tejaswini
Ramaiah Institute of Technology, Bangalore, Karnataka, India
e-mail: monalisa@msrit.edu

S. Ramineni
e-mail: surekha.ramineni@msrit.edu

H. Tejaswini
e-mail: tejaswinibaskar@msrit.edu

R. P. B. Singh et al. (eds.), *Placemaking and Cultural Landscapes*,
Advances in Geographical and Environmental Sciences,
https://doi.org/10.1007/978-981-19-6274-5_8

8.1 Introduction: Temple Towns as Cultural Landscapes

In India, religion and culture are regarded as inseparable. Myriad cultural heritages and landscapes exist in India, of which, one unique type is temple towns. South India has many of these temple towns, principally—Kumbakonam, Tiruchirappalli, Thanjavur, Melukote, Hampi. They are the living connections between the past and the present. For Indians, visits to such temple towns are part of their religious duties, a kind of cultural tourism that is bound to a pilgrimage; is self-motivated; and has an intense influence on one's senses and relationship with these places. Places can be regarded as cultural manifestations of intangible meaningfulness and associations that are created by both physical and metaphysical aspects that develop organically as a response to human needs from time to time (Mohamed et al. 2020). They not only constitute monuments and landscapes but also rituals, lifestyles, experiences, and memories (Alexandrakis et al. 2019). In a mostly democratic world with a highly dynamic demographic composition, the cultural memories are somewhat kaleidoscopic, fluid, and ever-evolving, while at the same time imparting identity to cultural and historic places.

Melukote is a well-known temple town in Karnataka and a place of unique cultural heritage and religious prominence. Melukote once flourished under royal patronage as an important pilgrim center, but now lacks the resources to maintain its character and significance. The study is focused on understanding the need for revitalizing cultural tourism in Melukote, and the role of the placemaking approach toward this end. The study attempts to understand the role of placemaking as a tool to create hybrid options that acknowledge coexisting relevance of multiple identities of a place and to revitalize cultural tourism in temple towns of South India in the context of Melukote, a temple town in Karnataka.

8.2 Methodology

The study defines the context in terms of—spatial analysis of ritual landscape and pilgrim paths, activity mapping of locals and tourists, feedback from stakeholders and subject experts on placemaking, and cultural landscape potential in Melukote. It concludes with identifying effective placemaking strategies and recommendations to enhance tourism. The study is structured to firstly identify the need to revitalize the temple towns. Through literature review, it aims to understand the concepts and theories of 'Placemaking' in the context of heritage sites, specifically temple towns. This is further developed by collecting contextual data of the study area that allows for the identification of cultural and heritage assets of Melukote.

Based on observational studies in the field—spatial mapping of activities by tourists, pilgrims, and locals has been developed. Further, on-site informal interviews with the local communities to understand the values they attach to the heritage of Melukote, their acceptance of Melukote as a tourist destination, their interest in the development of cultural tourism of Melukote, and their role in and expectations from tourism. Online interviews with the tourists to understand their motivations, expectations, and their willingness to revisit and engage in the public realm of the study area post-placemaking. The findings are analyzed to identify placemaking opportunities based on existing theories. The study concludes with placemaking strategies in the case of study context.

8.3 Need to Revitalize the Temple Towns as Heritage Sites

Temple towns are the living heritages with their roots embedded deeply in the local architecture and the intangible heritage from different centuries that is simultaneously alive in memories, day-to-day acts of rituals, performances, and skills passed from one generation to the next. As opposed to many other heritage sites, the uniqueness of these temple towns (as stated above) is that they continue to be inhabited and valued in continuity. The continuous nature of the use and the notion of cultural traditions in a highly diverse society add a critical dimension to the plurality of perceptions of the place, plurality of the memories, and multiple representations of the past.

Revitalization of the heritage sites with a vision from a local perspective can be achieved by bridging these binaries, identifying the new typologies, imagining new futures evolving from informed pasts, and developing sustainable strategies to suit contemporary needs. The complex layers of myriad contemporary cultures and times impart distinctiveness to the Indian context as opposed to the Western context. Therefore, the norms and practices adapted to revitalize cultural heritage in temple towns cannot be completely based on Western models. It is imperative to develop a more sensitized toolkit to foster the cultural image and identity of the place in the Indian context.

To preserve and celebrate cultural heritage, many national and international institutions like UNESCO and the Government of India have rolled out focused programs like—UNESCO World Heritage Cities Program and HRIDAY (Heritage City Development and Augmentation Yojana). The main objective of these programs is the inclusive development of heritage cities to improve quality of life, promote sustainable tourism, and revive the selected city's distinctive cultural identity.

8.4 Placemaking as a Strategy for Revitalizing Temple Towns and Foster Cultural Tourism

Place and Heritage: Place is the core concept in environmental psychology (Lewicka 2008) and can be defined as a space endowed with meaning and identity due to either distinctiveness or sameness. Heritage comprises the physical and metaphysical that change across time, space, and people in a given geographical setting and is ever-transforming due to anthropological factors (Anjali and Vinod 2017). The physical includes the built whereas the intangible includes the symbolic manifestation of culture like practices, traditions, rituals, customs, art, festivals, beliefs, etc. People develop an attachment with the place which is affective, cognitive, and behavioral (Jorgensen and Stedman 2001). Place attachments are particularly strong when the place in question has heritage value (Delconte et al. 2016). People could be the visitors who engage with a place for a shorter duration or the local communities who share the identity of the place. These communities are constituted by members of various socioeconomic strata who are diverse and multivocal and carry a plurality of memories of a place yet view heritage more inclusively and expansively with a holistic perspective. The connections of people with places and known surroundings impart a sense of trust (Rodman 2003), safety, and stability to the local communities.

In temple towns, the lives of people are centered around religious practices and rituals which impart a unique identity to the place and are rubbed off on residents and tourists (Uzzell 1996). The religious heritage emerges where the spirit of nature and culture meet and is highlighted by the rituals and festivities and lingering memories (Singh 2015). Temple towns are considered epicenters of spiritual energy with innumerable rituals focusing on worship and pilgrimage where the interaction between human actions and sacredscapes has generated an environment of faithscapes (Singh 1997). Although the temples are most revered, there exists parallel sacred urbanism that can be witnessed in various elements like anointed shrines on the sides of the road, venerated trees, paths of religious processions, etc. (Bharne 2013). In the backdrop are the living settlements whose lifestyle imparts an identity to the place.

Association between people and place gives rise to multiple concepts like place identity, place attachment, space identity, community identity and dependence, and placemaking (Twigger-Ross and Uzzell 1996; Franzén 2004; Seamon and Sowers 2008; Proshansky et al. 2014). Place attachments are particularly strong when the place in question has heritage value (Delconte et al. 2016). One of the principal functions of heritage interpretation is to enhance the visitor's sense of place and place identity. If this is to occur, the interpretation should be planned and designed with that outcome in mind (Twigger-Ross and Uzzell 1996; Kasimova et al. 2019). Though placemaking initiatives focus primarily on urban settings, they are equally relevant in rural and heritage sites (Markusen and Nicodemus 2010).

The first step of placemaking is to identify the major actors involved, i.e., the tourists and the local community and possible challenges and opportunities during their engagement; impact of modern developments; and the dangers of routine placemaking.

- Tourists are varying in age groups, backgrounds, and interests, maybe local or international. They may have little in common, little to no shared perspective on the importance of heritage. Their motivation may lie in religion, history, art, science, lifestyle, heritage, and culture of the community or region (Silberberg 1995). Such travel is focused on experiencing cultural environments, including landscapes, the visual and performing arts, and special lifestyles, values, traditions, and events both passively and actively. When tourists encounter the local communities in non-staged environments, they absorb the everyday life of the locals (Sinha 2017). Developing an understanding of tourists' motivation and catering to it is important due to its two-pronged benefit to both the tourist and the local economy.

- The locals play a crucial role in the success or failure of any form of tourism. The communities may either exhibit a positive attitude toward tourist potentials and attractions or may not be motivated to be involved in providing complementary services to promote tourism. It is necessary to involve communities in decision-making processes and the promotion and maintaining tourism destinations. By doing so, they can end up having an improved quality of life, improve their sense of belonging, develop social networks, and inculcate a greater appreciation and understanding of the value of the local area. It is necessary to tap the locals as unique resources for tourism vowing to their immense knowledge, skills, and products. This helps to reduce their investments and increase profits. Locals could be the trained guides and cultural performers as part of the conducted tours.

Modern Development: Heritage and modernity are continuous spatiotemporal processes of the society that benefit from archaic wisdom and must co-exist to understand humanity's link with nature (Singh 1997). Dr. Singh suggests that the sacredscapes contain the roots and prospects of our existence. Sacredscapes are an integral part of the heritage ecology. INTACH has accepted the concept of a Cultural Heritage Zone around built heritage including artifacts. The design of these zones is based on the guidelines laid out by ICOMOS. The basic idea behind this approach is 'Placemaking.' The main objective of placemaking is to strike a balance between conservation, authenticity, tourism, and economic development.

Dangers of route placemaking: Urbanization results in development that can either be in harmony or in conflict with the local heritage. Any disregard for historical context, will result in disruption of the cultural fabric and change the character of the place. All insertions and interventions must respect the pre-existing built heritage and follow suitable placemaking guidelines. Heritage sites and open spaces in historic areas provide opportunities for regeneration schemes. The cultural integration potential of these sites can be increased with placemaking by balancing the scales between heritage conservation and modern development. The idea of placemaking may lead to the manifestation of the design principles with the introduction of physical elements which lead to default urbanism. This may again lead to damage to the local identity. It is important to identify the layers of history and traditions which reflect the context of topography, climate, and culture of the specific neighborhood. The original placemaking philosophy which was based on the appreciation of diversity and the recognition of the place is intrinsically tied to culture. Facing the danger of routine application, placemaking needs to be looked into with a different concern with the physical and economic aspects of vitality. The narrow focus on walkability and mix of functions without reflection on values of history, heritage, and culture would present a missed opportunity to produce uniquely different and beautiful places (Fig. 8.1).

Fig. 8.1 Keys of inclusive placemaking (*Source* Author)

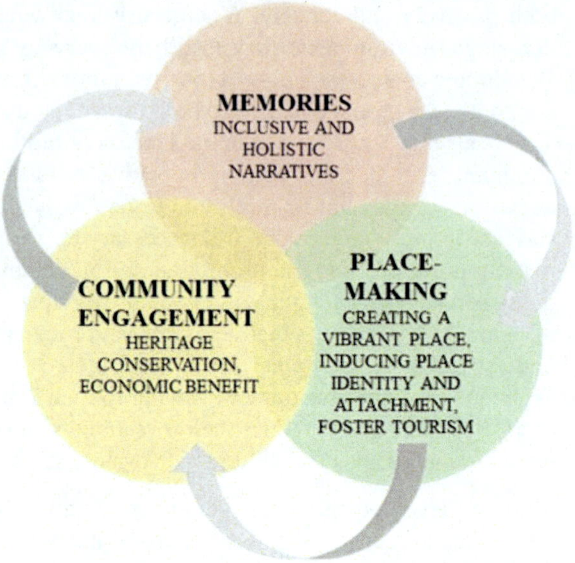

For placemaking to be successful, it is imperative to adopt a participatory approach by opening up a dialogue with the local communities, building consensus, and developing a model of co-design. By this, placemaking fosters not only individual identity but also community identity and creates a strong civic fabric. If the meaning of a site is largely defined by experts that differ in meaning and value ascribed by the locals, the communities disengage from the development module and considerable tension may arise when hosts and guests meet. This discord threatens the heritage sites, particularly the sacredscapes. A participatory, community-centered approach encourages tourists to stay longer and/or to spend more money and improve the socioeconomic conditions of the local communities. The prerequisite to the success of placemaking lies largely in the local community engagement. The coexistence of developmental programs and cultural milieu forms the foundation of integrated sustainable development of any historic site especially temple towns. The sentiments and values that locals attach to the heritage though different are no less than the values assigned to by the tourist, historians, and archaeologists. Through research-based placemaking, a context-sensitive, integrated, and inclusive approach can be developed for the effective conservation of historic landscapes where temporally distant events define the present meanings and experiences.

8.5 Context Study of a Cultural and Heritage Site—Melukote, Temple Town South Karnataka

This study is focused on understanding the need for revitalizing cultural tourism in Melukote, and the role of the placemaking approach toward this end. Melukote is a fortified temple town located on an elevated hillock 3589 feet high in Pandavapura Taluk of Mandya District, Karnataka. An important historical and religious center, this is also referred to as Theerthakshetrapura. It is believed that during the twelfth-century Sri Ramanujacharya, a Sri Vaishnava pontiff, spent 14 years of his life in this sacred place during which time he recovered the idol of Cheluvanarayana Swamy and built the temple with the help of the then Hoysala king Vishnuvardhana. A local legend says that Melukote, the residence of God Narayana, was Narayanadri in Kritayuga, Vedadri in Tretayuga, Yadavadri in Dwaparayuga, and Yatushaila in Kaliyuga. Various records indicate that this place was referred to with different titles and received continuous patronage from kings starting from Hoysala Vishnuvardhana (1008–1148 A.D.) up to the kings of Mysore who built various temples and water bodies, generously contributing to various temple activities. Melukote is a house to many substantial architectural and archaeological sites such as Cheluvanarayana Swamy Temple, Yoganarasimha Temple, Raya Gopura, Panchakalyani, Akka Tangi Kola. The unique architecture of Melukote, rituals and practices, centered around the temples, ingenious water system; the town planning strategies all constitute the metaphysical charm of this temple town.

Melukote is a magnificent ancient town considered to be the epitome of culture throughout history. It is this heritage along with the seamless manifestation of its spirit over the physical structure of Melukote which makes it a unique entity worth exploring. Every structure and fragment in Melukote is a true document of cultural and technical knowledge systems. The evidence of their history is preserved as 'layers' of built fabric, making them highly readable entities. Coherence is added by the specific nature of heritage components, historic building typologies, and their co-relations.

The living communities around the temple, the practices, rituals, processions, literature, and a lip-smacking dish puliyogare form a cultural ensemble. The Temple Town of Melukote pulsates with rituals of Cheluvanarayanaswamy and Yoganarasimha Swamy endlessly throughout the year. The temples and related structures around the Panchakalyani retain their ancient pride with the huge traditional pond, rich architectural edifices, and spiritual ambiance of Vedic chanting still resonating in its walls. Due to the geographical proximity of Melukote to the state capital Bangalore and cultural capital Mysore, it is a popular destination among others for thousands of tourists, both as a pilgrimage site as well as a weekend getaway.

8.5.1 Tangible and Intangible Layers of Melukote

Melukote was reigned by various dynasties which have contributed to its unique architectural achievement. The lion pillars and the two-tier cave temples of Melukote are believed to date back to the Ganga period of the tenth century. Many distinct characteristics of the Vijayanagara architecture and Hoysala architecture can be seen in Melukote in the form of temples, mathas and 76 mantapas that though have their unique identity have blended into the cultural heritage of Melukote. The settlement geography of Melukote is embellished with temples of various sizes and scores of anointed idols, venerated trees, and water bodies associated with the central spirit of the place that carve out beautiful spaces within the landscape. The singular dominating culture and function of the town have been the focal identity of this historic town. The functional character of the town has allowed for a continuing living heritage character of the town. Socio-urban character is defined by an intimate human scale given the horizontal skyline. The distinctive architectural ambiance and social pattern of the various communities reflect their unique cultural and social identity. The plan of the houses of each community complements their occupation and lifestyle. A definite fragmentation of social spaces can be seen in the settlement layout.

Archaeological studies and recently discovered line drawings on a boulder at Melukote as well as the burials called Pandavara gavi on the hill range show that this was a habitation site of the Megalithic people (Karnataka 2019). The caves to the left of the road to the Yoganarasimha temple are called Pandavara gavi on the hill range show that this was a habitation site of the Megalithic people (Karnataka 2019). The caves to the left of the road to the Yoganarasimha temple are called 'Pandava Gavi' and the worn-out boulders, as the plates used by the Pandavas. There are also relics of the Megalithic culture. Raya Gopura, the south gate of Melukote, has been recognized as an archaeological site by the Archaeological Survey of India (ASI).

The ingenious water system of 108 water bodies dotting the landscape of Melukote, developed in response to existing climate and geography, can undoubtedly be deemed as a perfect marriage between nature, social, spiritual, and sacred (Srijayachar and Natampalli 1998). These remarkable characteristics of the water systems have always attracted many researchers, academicians, and professionals. These were constructed in different periods by donees with religious inclinations for multipurpose uses and they show the significance of Melukote. Panchakalyani/Pushkarini is the most important due to its religious association with Cheluvanarayanaswamy Temple, this is where Dattatreya taught Vedas to his disciples. Teppa kola is where the float festival used to be conducted in the earlier days but is now in a dilapidated condition owing to ignorance. Veda, Darbha, Padma, Yadava, Palasha, Narayana, and Vaikunta Theerthas are considered Astha Theerthodakas (eight sacred ponds) (Filliozat 1998). The water bodies form the core of social interaction at various levels. The myth and stories behind each of them add to one's curiosity.

Melukote is undoubtedly a planning masterpiece with chronological layering and growth pattern with the gradual addition of diverse communities onto its canvas

(Keskar 1998). This is substantiated by various building typologies and architectural styles. The hierarchy of roads, strategic location of water bodies, and judicious positioning of functional areas. Melukote covers an area of 1.5 sq.m. and is a linear development in the north–south orientation. Raja Beedhi forms the central spine connecting temples and related areas and various landmarks like Rayagopura in the South and the Pete Bagilu (entrance gateway) in the North. The secondary streets branch out from Raja Beedhi connecting to various settlements. Yatrinivas and commercial activities are mostly centered around the Raja Beedhi. On reviewing the cultural context of town planning, the settlement pattern indicates caste-based society and Brahmin supremacy with a well-defined hierarchy of settlements. Brahmins occupied the top rung while communities supporting the ancillary temple activities like weaving, pottery, and farmers occupied the lower rung. The privileged Brahmins settled around the temples whereas the non-Brahmin communities settled toward the periphery.

Melukote is an abode for heritage studies institutions. The Academy for Sanskrit Research established in 1853 A.D. by the Mysore king Mummadi Krishnaraja Wodeyar is the second earliest Sanskrit college in the country, the other one being the Sanskrit college at Kashi (Karnataka 2019). It has started the studies in Sanskrit literature, collection of manuscripts and its studies as well as publishing them. The center houses a rich Library with 25,000 books and 10,500 ancient manuscripts that are preserved through scientific methods. This institute is a bridge between the past and present as well as a beginning for the future (Fig. 8.2).

Melukote is vibrant with festivals and fairs being held throughout the year. Festivals of the deity Cheluvanarayana are carried out according to Pancharatra agama, and the reciting of Divya Prabhandas (Dravida Veda) during the worship is something special (Karnataka 2019) (Fig. 8.3).

Melukote has sheltered many artisan communities since ancient times. Many ancillary communities like weavers, potters, peasants, leaf plate makers, and Thiruman makers moved to the Melukote to aid in the temple activities (Srijayachar and Natampalli 1998). With time and modernization, these professions that once reflected the culture of Melukote have lost their identity and are on the verge of disappearance due to a lack of appreciation and funds. Melukote's distinctiveness as a temple town is defined by the traditional communities and cultural manifestations. With a change in socioeconomic circumstances over a course of time, the strong spiritual connections and the cultural bonding that once existed are gradually getting homogenized and resulting in loss of associations, identity, memory, and attachment. If we recognize that the processes that people use to engage with the pasts in a heritage context are as revealing as those pasts themselves, substantial emphasis can be laid on culture as tradition and culture as communication. Together they bring about a sense of place, continuity, security, distinct identity, and new opportunities. A collaborative approach to placemaking can aid in recreating an authentic sense of the place, bonding the end-user with the genius loci, and developing Melukote as a cultural tourism destination attracting tourists from all over the world.

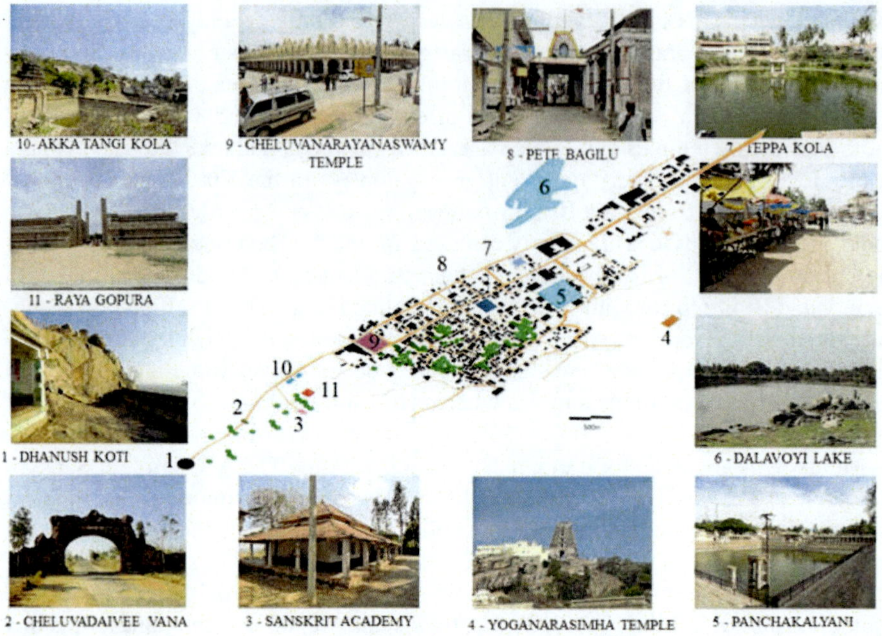

Fig. 8.2 Major landmarks of Melukote (*Source* Author)

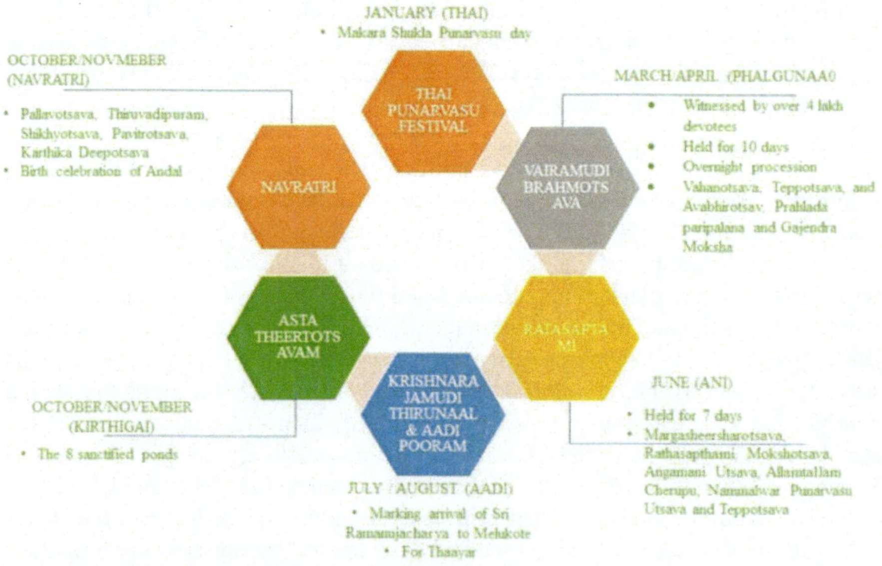

Fig. 8.3 Rituals conducted in Melukote (*Source* Author)

8.6 Context Study and Analysis

To understand and evaluate the scope for cultural integration through placemaking interventions; informal interviews with the locals, online interviews with the tourists, and site study were carried out.

The purpose of the interviews with local communities was to understand their level of awareness of local history, their sense of place attachment, place identity, and their views about Melukote as a tourism destination. Semi-structured face-to-face interviews were held with local communities as the focus of this research was on obtaining insight and understanding. Interviews were conducted in Kannada to create a sense of trust with the locals, ease communication, and obtain clear inputs. About people were approached out of which only 35 were forthcoming to engage in conversation with the authors. Documentation of conversation excerpts was done in notebooks which were comprehensively analyzed to put forth the common observations. The views varied between different communities. Brahmins were aware of the history of Melukote and exhibited a sense of pride and a strong sense of association with Melukote. Most Brahmins continuously engage in temple activities and view religious tourism as important. The non-Brahmin community, i.e., the weaver community, the peasant community, and the potter community are busy making ends meet. They are aware of the history of Melukote and associate the place differently. Nevertheless, in terms of tourism, the interviewees were neither embracing nor rejecting tourism in its current state. Neither do they participate nor interfere with activities of tourism. They do not see any direct benefit from tourism. Upon enquiring about their willingness to participate in activities to promote facilitated cultural tourism and their interest in interaction with tourists, they showed interest in seeing this opportunity as a platform to promote local culture and to seek economic benefit.

The interviews with tourists were based on the hypothesis that there exists a notable correlation between people's awareness of heritage and their willingness to return to visit it. The questions in interviews thus focused on people who had a contextual understanding of the pilgrim character of Melukote and their experience of the town. The questions further explored the willingness of people to use the public realm of the town for cultural interaction from a tourism perspective. Another factor explored through interviews was the people's perception of the use of vehicular traffic on historic streets. The interview of both potential tourists and the tourists who have visited Melukote was conducted. A total of 177 people were interviewed. The sample included participants from various categories and age groups. The findings of the study have been summarized below (Table 8.1).

The results indicate that though Melukote is a culturally valued town, it has reasonable cultural tourism potential. The unique ritual setting, food culture, and arts and crafts are related to the experience of the temple town of Melukote. The public realm and settings are integral to the cognitive image of the town and integral to how tourists recall the image of the town. Thus, they play a significant role in designing tourism

Table 8.1 Summary of survey findings

	Factors	Outcomes		
Historical context awareness				
		Yes	No	
1	Are you aware of the history of Melukote?	62.7%	37.3%	
2	Are you aware of the food and traditions in and around Melukote?	92.0%	8.0%	
3	Are you aware of the arts and crafts of Melukote?	71%	29%	
Perception of cultural interaction through tourism in the public realm				
		Yes	No	Not aware
1	Does the place engage your interest due to its cultural uniqueness?	88.7%	3.4%	7.9%
2	Do public places in Melukote allow for cultural interaction?	72.9%	16.4%	10.7%
3	Is it easy to reach one landmark to the other by walking?	54.8%	33.3%	11.9%
4	Have you visited major landmarks of Melukote?	94.3%	5.7%	
5	Are there any places to sit and relax in the public areas?	63.8%	24.9%	11.3%
Perception of use of vehicular traffic on historical streets				
		Yes	No	Not aware
1	How do you personally feel about the use of vehicles in the historic streets?	39.5%	60.5%	
2	Are there sufficient parking areas?	74%	15.8%	10.2%

(continued)

Table 8.1 (continued)

Tourism potential

1	What is the motivation behind your visit to Melukote	Pilgrimage	Getaway	Cultural Richness	Other	Not visited
		55.9%	16.4%	47.5%	18.1%	6.8%
2	Is it important to preserve cultural heritage?	0 to 1	1 to 2	2 to 3	3 to 4	4 to 5
		2.3%	1.7%	1.7%	9%	85.3%
3	How many times, have you visited Melukote?		0	1 to 3	4 to 5	More than 5
			8.5%	62.7%	10.2%	18.6%
4	How much time did you spend in Melukote?		Half a day	Full day	More than a day	Not visited
			52.5%	26%	14.1%	7.3%
5	Have you stayed overnight in Melukote?				Yes	No
					23.2%	76.8%
6	Would you be interested in staying longer than a day in Melukote?				Yes	No
					74%	26%

The willingness of tourists to revisit

1	Are you interested to participate in any of the following activities?	Pilgrimage tour	To learn about ingenious water systems	Interact with the local community	Attend a food festival	Archaeological tour	Nature trek	Architectural tour
		48.6%	44.1%	29.4%	42.9%	53.7%	54.8%	49.2%

interests. The public realm, experiences, and cultural integration are most successfully stitched together through placemaking. One may conclude that placemaking holds tremendous value in revitalizing the cultural tourism of temple towns given their historical settings.

As part of site study of Melukote, a normative approach was adopted to map the pilgrim paths, rituals and procession routes, linkages, street life, evening use, and neighborhood. The study and mapping revealed that Raja Beedhi (central street) is the common element in each of the aspects studied. The picturesque streetscape is created by the facades of the vernacular houses which are enlivened by the jagali (the raised platform between the house and street) used as a sit-out and facilitated social interaction. Jagalis blur the boundaries that define individual territories and comfort the human height. The highly decorated entry doors; terracotta roof tiles define the vernacular character of buildings. The shops on Raja Beedhi and the frontage provide a stage for social interaction. The current houses lack the concept of Jagali and the typical vernacular character and disturb the urban aesthetic and the sense of place. Raja Beedhi has ample opportunity to promote social and cultural integration on daily basis for both locals and tourists when designed for lingering and walkability (Figs. 8.4 and 8.5).

Based on the analysis and site study, it could be inferred that placemaking can be a successful tool to improve the quality of historic places and communities by improving the urban fabric; improving the aesthetic appeal and use of the public realm; and activating underutilized public spaces; facilitate a wide range of entertainment, recreation options for both tourists and locals; promote local art and culture and create economic competitiveness; empower local communities and promote art, retain talent.

8.7 Proposed Placemaking Strategies for Temple Town—Context Melukote

Cultural tourism can be successful if it is sustainable. Both tourists and resident communities should build a sense of attachment to the cultural values that strengthen the identity of the place. The objective of the placemaking strategy in temple towns is to enrich the cultural knowledge of the tourists without changing the heritage character of the town or disrupting the day-to-day life of locals while also building sensitivity to the local context. Placemaking interventions can, therefore, contribute to building creative expression and a coherent understanding of the historic place.

The survey undertaken for the study indicates visitors' interest to learn more about the less-known facts of Melukote, cultural and built heritage. Constructed themed itineraries like religious walkthroughs linking temples and other religious buildings, rituals and ceremonies, events, and pilgrim places; cultural trails featuring literature, music, local art, Sanskrit, and food; archaeological trails connecting all well-known and not-so-well-known archaeological sites; historical trail focusing on architectural

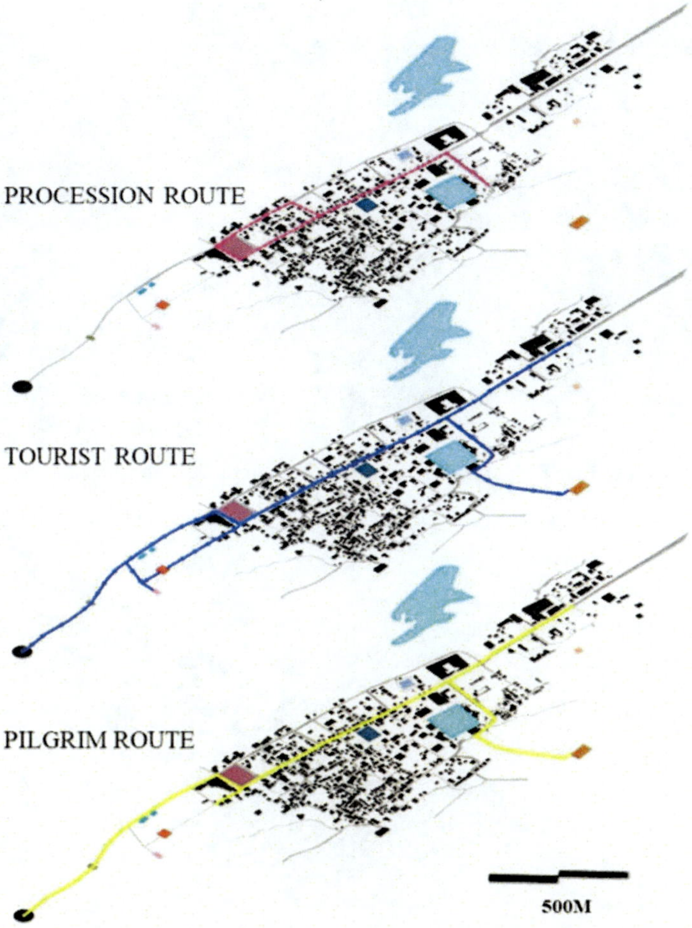

Fig. 8.4 Routes used by pilgrims, tourists, procession (*Source* Author)

attractions; nature walks through the natural features; water walks to include a tour of the 108 water bodies can be organized to enhance visitors' experience. Holy dips at the water tanks at an affordable price can be arranged and the income generated can be used to revive the lost or dilapidated water bodies (Fig. 8.6).

The experiences can be reinforced with interactive activities by organizing events like weekly workshops to learn local art like pottery, weaving, etc.; food melas; story (katha) telling events from mythology, reading sessions from scriptures, spiritual discourses by scholars in the field, dance, music recitals, and play.

Although the culture of Melukote is the strongest magnet that attracts visitors; physical inserts include architectural and public realm infrastructure like signages; street furniture; lighting; signboards or narrative boards at each site of importance;

Fig. 8.5 Character of Raja Beedhi during various events and times of the day (*Source* Author)

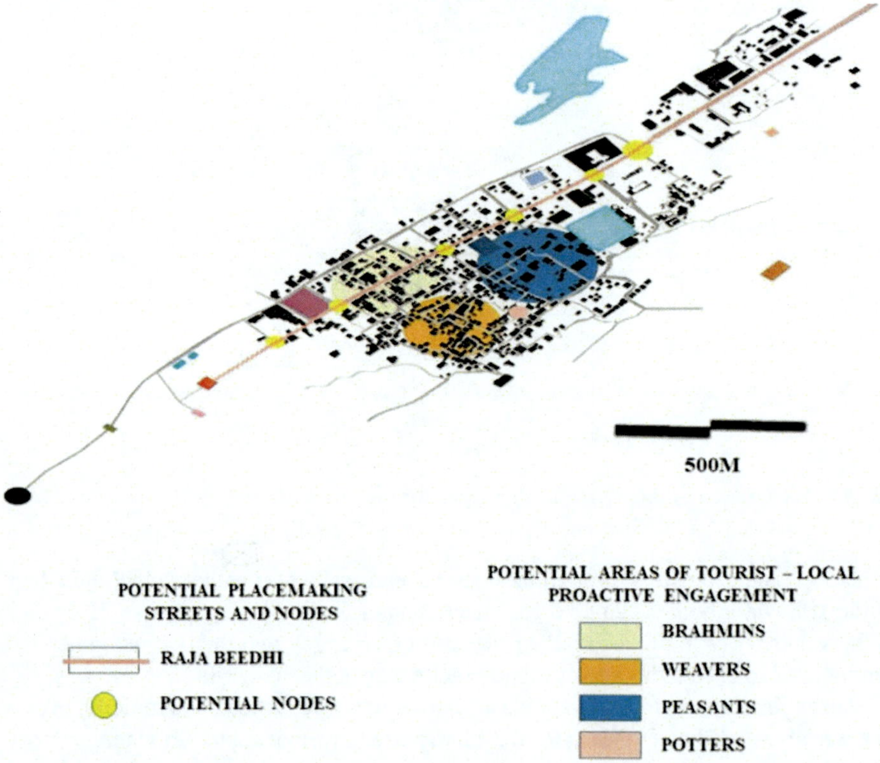

Fig. 8.6 Potential sites for placemaking (*Source* Author)

ample parking space; redirecting vehicular traffic on days of festivities and celebrations; and preserving facades of the street to reflect vernacular style can act as catalysts to improve visitors' experience.

Ample tourism design opportunities can be created by integrating existing streets into interpretive trails that can knit together the scattered heritage precincts into an interwoven system. These trails especially along the Raja Beedhi, which is the major path of tourist and pilgrim movement between landmarks, can be intercepted by relevant public spaces like performance areas, souvenir and craft shops, melas, resting and viewing points which can act as platforms of social interaction between tourist and local community. These trails can be embellished with native trees to provide shade and promote walking. Considering the results of the survey indicating visitors' interest in spending more than a day in Melukote, homestay programs may be promoted which would allow the visitors to absorb the culture of Melukote and the lifestyle of the locals in its entirety. This may prove beneficial to both the visitors and the locals, where the former gains through the intangible experience and the latter benefits economically.

8.8 Conclusion

The temple towns in South India are a unique and distinctive type of cultural landscape characterized by monoculture urbanism with a living tradition. These landscapes have a significant role to play in cultural memory. The pilgrim value makes them a tourist destination. These towns provide an opportunity for cultural integration engaging tourists and local populations—the custodians of their unique cultural heritage. Placemaking is an effective strategy to achieve this goal. The role of the public realm and its urban design, architectural conservation, and engagement opportunities through interaction and experience are likely to build a sense of association and memory of the place. These can contribute to revitalizing the purpose of historic towns in a more inclusive and democratic society, which was once sustained by royal patronages and religious donations.

References

Alexandrakis G, Manasakis C, Kampanis NA (2019) Economic and societal impacts on cultural heritage sites, resulting from natural effects and climate change. Heritage 2(1):279–305. https://doi.org/10.3390/heritage2010019

Anjali C, Vinod K (2017) The Indian historic cities: Exploring the cultural capital, pp 1–6

Bharne V (2013) The emerging Asian city: concomitant urbanities and urbanisms. In: Bharne V (ed), Oxon, Routledge

Delconte J, Kline CS, Scavo C (2016) The impacts of local arts agencies on community placemaking and heritage tourism. J Herit Tour 11(4):324–335. https://doi.org/10.1080/1743873X.2015.1088019

Filliozat DV (1998) 'Melukote through the ages. In: Astatirthotsavam. Melkote, Academy of Sanskrit Research, pp 241–246

Franzén M (2004) Anthony M. Orum and Xiangming Chen (2003) The World of Cities. Places in Comparative and Historical Perspective. Malden Ma: Blackwell. 180 pp. ISBN: 0631210253. Hous Theory Soc 21(3):143–144. https://doi.org/10.1080/14036090310017032

Hayden D (1988) Placemaking, preservation and urban history. J Archit Educ 41(3):45–51. https://doi.org/10.1080/10464883.1988.10758490

Jorgensen BS, Stedman RC (2001) Sense of place as an attitude: Lakeshore owners attitudes toward their properties. J Environ Psychol 21(3):233–248. https://doi.org/10.1006/jevp.2001.0226

Karnataka G (2019) Places of interest: Europe. Mandya District Gazetteer. https://doi.org/10.4324/9781315172828-12

Kasimova RR, Ziyadullayev AR, Ziyadullayeva AA (2019) Comparison of ceremonies and legends associated with the belief in reverence for water. https://doi.org/10.15863/TAS

Keskar J (1998) Melukote through the ages. In: A town planner's view. Melkote, Academy of Sanskrit Research, pp 141–144

Lewicka M (2008) Place attachment, place identity, and place memory: Restoring the forgotten city past. J Environ Psychol 28(3):209–231. https://doi.org/10.1016/j.jenvp.2008.02.001

Maiti and Bidinger (1981) 済無No Title No Title. J Chem Inf Model 53(9):1689–1699

Markusen A, Nicodemus AG (2010) community development investment review 35 Federal Reserve Bank of San Francisco creative placemaking: How to do it well. (October), pp 35–42. http://engagedscholarship.csuohio.edu/urban_facpub/1162

Mohamed AMRM et al (2020) The role of placemaking approach in revitalizing AL-ULA heritage site: Linkage and access as key factors. Int J Sustain Dev Plan 15(6):921–926. https://doi.org/10.18280/ijsdp.150616

Proshansky HM, Fabian AK, Kaminoff R (2014). Place-identity: Physical world socialization of the self (1983). The People, Place, and Space Reader, pp 77–81. https://doi.org/10.4324/9781315816852

Rodman M (2003) The anthropology of space and place: Locating culture. In: Empowering place: Multilocality and multivocality, pp 204–223.

Seamon D, Sowers J (2008) Place and placelessness (1976): Edward Relph. Key Texts in Human Geography, (May), pp 43–52. https://doi.org/10.4135/9781446213742.n5

Silberberg T (1995) Cultural tourism and business opportunities for museums and heritage sites. Tour Manage 16(5):361–365. https://doi.org/10.1016/0261-5177(95)00039-Q

Singh RPB (1997) Sacredscape and urban heritage in India: Contestation and perspective. In: Shaw Brian J, Roy Jones (eds) Contested Urban Heritage: Voices from the Periphery, vol 1. Ashgate Publis., Aldershot, pp 101–131. https://s3.amazonaws.com/academia.edu.documents/37783955/118-97_Ur_Heri_Con_Ra_Asgat.pdf?AWSAccessKeyId=AKIAIWOWYYGZ2Y53UL3A&Expires=1502003762&Signature=qKlR4KhS6W%2FWbPlTbSDHnCJpOBc%3D&response-content-disposition=inline%3Bfilename%3D118-97_._Singh_Ran

Sing RPB (2015) Heritage value of religious sites and built archetypes: The scenario of Hinduism, and illustrating the riverfront Varanasi. Aatmbodh, the Journal of Knowledge of Self, 12(1):21–40. http://www.academia.edu/download/37777472/430.15_a._Hindu_Rel_Sites-Rana__Aatmbodh_12-1__Spr_2015.pdf

Sinha A (2017) The public realm of heritage sites in India: Sustainable approaches towards planning and management. Cultural landscapes of South Asia studies in heritage conservation and management. Routledge, Oxon, pp 227–240

Srijayachar, Natampalli M (1998) Melukote through the ages. In: Home industries. Melkote, Academy of Sanskrit Research, pp 375–382.

Twigger-Ross CL, Uzzell DL (1996) Place and identity processes. J Environ Psychol 16(3):205–220. https://doi.org/10.1006/jevp.1996.0017

Uzzell DL (1996) International journal of heritage studies creating place identity through heritage interpretation creating place identity through heritage interpretation. Int J Heritage Stud Publ 1(4):219–228

Chapter 9
The Challenges of Integrated Conservation and Development in Historic Rural Landscapes; Case Study: The Historic Villages of East Azerbaijan, Iran

Farnaz Faraji and Elham Masoumi

Abstract The notion of Historical Rural Landscape was introduced in 1990 by National Park Service. This approach studies the relationship between man as a cultural agent, nature as a physical intervener, and the built environment due to the coordination of these two in historic places. In recent years, the conservation of historic rural landscapes has evolved and strains to reestablish the balance between protecting the local values in historic villages and development in these settlements. Contemporary demands of rural communities in Iran for development and the increasing speed of change have encountered rural heritage and protection of equilibrium among them with many challenges. Therefore, with the aim of quantitative approach and field surveys, this study tries to identify and classify the existing challenges in the historical rural landscapes of East Azerbaijan villages. Based on the findings, these challenges are categorised into global and local challenges and sub-sections related to nature and human beings. Additionally, the results indicate that many of the challenges are rooted in humanistic issues, for example, inadequate technical and professional knowledge, inadequate legal structures, and residents' tendency to adopt living urban cultures. All of which result in a failure to establish a balance between integrated conversation and development.

Keywords Integrated Conservation and Development · Challenges · Historic Rural Landscape · Azerbaijan villages · Management of change

---

F. Faraji (✉)
School of History, Classics and Archaeology, Newcastle University, Newcastle Upon Tyne, UK
e-mail: farnaz.faraji@newcastle.ac.uk; farnaz.faraji@ncl.ac.uk

E. Masoumi
Architecture Department, College of Fine Arts, Tehran University, Tehran, Iran
e-mail: elham.masoumi@ut.ac.ir

R. P. B. Singh et al. (eds.), *Placemaking and Cultural Landscapes*,
Advances in Geographical and Environmental Sciences,
https://doi.org/10.1007/978-981-19-6274-5_9

9.1 Introduction

The Challenges of conservation vs development have risen mainly after World War II with emphasis on the importance of preservation of historic and cultural proprieties as well as the necessity of providing the essential needs for the human population in living environments (Hanachi and Koosheshgaran 2011). In response to the raising conflict, conservationists seek to limit the effects of climate change on an urban settlement and enhance its cultural significance. While simultaneously, development seeks to intrigue the context for changes growth (Fadaei Nezhad and Ehsrati 2015).

From the 1980s onward, cultural heritage conservation debates have placed emphasis on the importance of preserving heritages assets. Thus, both tangible and intangible heritages have contributed to the development of heritage conservation principles over the last few decades. Moreover, the post-war international debates on heritage conservation show that the subject of historical monuments conservation has changed to the preservation of the sites through different approaches, e.g., landscape conservation, international charters, discussions, etc., led to the importance of values protection within communities (Habibi and Maghsoudi 2013).

In the twenty-first century, the most recognised approach to cultural heritage preservation discourse has been continued with the introduction of new concepts such as management of change. The new approach of "change management" mainly prepared the setting to introduce the approach of *Historic Urban Landscape* (HUL) introduced by UNESCO (2011) as a holistic, integrated, and flexible approach respecting sustainable development in the urban environments. Based on this approach, the HUL is the urban area understood as the result of a historic layering of cultural and natural values and attributes, extending beyond the notion of "historic centre" or "ensemble" to include the broader urban context and its geographical setting (UNESCO 2011, art I. 8).

A recent study has focused on the case of HUL in rural contexts by identifying the challenges that stand in the way of conserving the historical value of the rural fabric. According to this new framework, building a mechanism that allows the integration of conservation and development goals within these areas in a coordinated manner becomes essential.

Therefore, the primary objective of this study is to incorporate historical research conducted by the National Park Service in the 1990s about rural historic landscapes. This will help to identify and integrate patterns and trends across literature. Cigdem Asrav (2015) and Di Fazio and Modica (2018) have introduced the topic briefly on the challenges and the importance of conserving rural historic landscapes. Similarly, this study aims to analyse the challenges of conservation and development of rural landscapes in five villages in Iran. For the methodology, the authors used an overview of the evolution of the HRL concept. To identify the pre-used framework and classify the challenges observed in the World Heritage Sites (WHS), the scholars have referenced a study conducted by ICOMOS. Eventually, based on the case studies, a comparative study among these villages shows the most significant challenges to the integrated conservation and development of the Historic Rural Landscape.

9.2 Historic Rural Landscape

Having reached the new century, the concept of sustainable development now entails the conservation of cultural properties (Bandarin and Oers 2015). Francis Nocca (2017) argues that economic, social, cultural, and environmental systems are interconnected. He highlights heritage as a "glue" that can link different aspects of sustainable development (Nocca 2017). The notion of conservation is thus seen to be a holistic approach that involves the notion of "change management" concerning the different dimensions of legacy development, including maintenance, care, and sustainability (ICOMOS 2005; Smith 2015).

A decade ago, a similar designation of heritage conservation was applied to rural landscapes. First, before the UNESCO World Heritage Sites discussions on the inclusion of a new concept of "Cultural Landscapes" in the list, there was a widespread debate on the importance of the preservation of the rural landscape. The outcome of this discussion resulted in the integration of rural landscapes into a wider concept of cultural landscapes. Resulting of the international meetings on the importance of leaving the division between nature and culture in UNESCO criteria on the inscription of universal heritages in the list, the new convention of 1992 on "Protection of Cultural Landscapes" was released (Harrison 2013). This late action of UNESCO put an end to the issue of rural landscapes. However, it was not enough and led to forgetting the importance of the issue for a decade later.

Simultaneous with the discussion on rural landscapes preservation in Europe, and the UK, in the USA, McClelland and Melnick (1999) in the *National Park Service* (NPS) of the U.S. published a leaflet concerning the importance of documenting the rural landscape's properties.

It is necessary for you to understand before diving into the history of the rural historic landscapes that the planning process of a rural landscape conservation plan involves two main steps. First, it is necessary to recognise and understand the primary components of rural properties and understand their cultural significance to the owners. Then, understanding and formulating this plan based on the historical layers they constitute through the years as the layers of rural context. Thus, historic rural landscapes are subtle cultural landscapes whose fundamental concepts are driven by cultural landscape definitions and characteristics applicable to all mutual historical sites.

The *Rural Historic Landscape* (RHL) concept was first introduced in the 1990s by the National Park Service (NPS) of the United States. NPS defines the concept of Rural Historic Landscape as "a geographical area that historically has been used by people or shaped or modified by human activity, occupancy, or intervention, and that possesses a significant concentration, linkage, or continuity of area of land use, vegetation, buildings and structures, roads and waterways and natural features" (McClelland and Melnick 1999). The main question here is how a HRL can be recognised and differentiated. To answer this question, NPS U.S. clarifies the HRL's most crucial feature by introducing this concept as "structured landscapes of daily activities of the people and concerning traditional activities, such as work in mines, fishing, and all kinds of activities related to the field of agriculture" (McClelland and Melnick 1999, p. 5). Ownership, land uses, policies, strategies, and social-cultural, economic, and natural resources constitute other visible features in a historic rural landscape.

Also, NPS defines and categorises the types of historic rural landscapes based on four main processes that have a reciprocal affection between rural areas' context and the formation of the rural landscapes. These four processes are land use and activities, patterns of spatial organisation, response to the natural environment, and cultural traditions. However, these four processes are not considered the only drives of historic rural landscape creation. The circulation networks, boundary demarcations, vegetation related to land use, buildings, structures, objects, clusters, archaeological sites, and small-scale elements are other components that affect the essence of various rural historic landscapes formation (McClelland and Melnick 1999, p. 4).

Despite the definition introduced by NPS on the concept of HRL and its preservation techniques, this perspective was criticised by Rottle (2008). Regarding the recent changes in international and national scales and to resolve the conflicts of these challenges, she claims the necessity of a "balanced continuity of the exploitation of rural lands" and "codification of process-based management systems in historical protection" (Rottle 2008). Other scholars in recent years have proposed to change the lenses. They admire the importance of a change in the conservation practice of rural historic landscapes from a "monumental", "rigid", and "non-inclusive" perspective to an approach consisting of "landscape" in the conservation of historic environments. Therefore, studies by Di Fazio & Modica (2018), Cigdem Asrav (2015), and Zarif and Jokilehto (2016) have introduced the concept of "Historic Rural Landscapes", based on the recently introduced approach of "Historic Urban Landscape" to apply in the rural environments with multilayered histories and significance of the place.

In the new definition of this concept, the historic rural landscape is a subjective–objective concept that compiles from triple interaction between humans, nature, and the environment. In this tripartite interaction, the identity characteristics of local society shape the products of this triple interaction as an absolute unity that bears the historical-cultural significance of the place to other landscape aspects in rural settlements. According to the new definition, recognising the existing challenges is the first step that can be taken to provide a comprehensive management plan and preserve the historic rural landscape (Cigdem Asrav 2015) (Fig. 9.1).

9.3 Challenges in the Conservation of Historic Rural Environments

From the second half of the eighteenth century up to now, at an increasing speed, the development of new energy sources, technology innovation, transport means and infrastructures, communication/information technologies, and media has progressively allowed communities to achieve greater independence from place constraints.

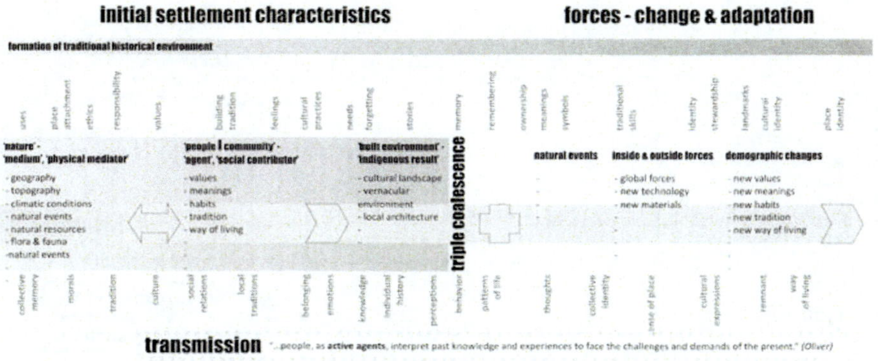

Fig. 9.1 The formation of historic rural landscapes (*Source* Cigdem Asrav 2015, p. 43)

This evolution has accompanied a new social and cultural climate characterised by widespread unconditional faith in technological development and a typical ethical horizon (Di Fazio and Modica 2018).

Villages as a parallel settlement alongside cities face some mutual challenges in the world globally, such as Climate Change, Globalisation of cultures, Demographic rapid changes, Modernisation, the Tourism Industry and its mismanagement, consummation, and consumption, and other issues like gentrification by economical-social points, pressures risen from political relations, economic development that has been recently raised affected villages as much as it did on cities (ICOMOS 2005; UNESCO 2011).

Agricultural and livestock activities, mainly based on agriculture, have been dominant economic activities in historical rural landscapes. Thus, it encourages migration to urban centres for better alternative economic revenues, thus raising unemployment or lower income levels. Moreover, it decreased the rural population, primarily the young generation, with its strong human capital.

The consequence of this is the disruption of population balance in rural areas. The contemporary pressures of urbanisation, modernisation, and globalisation manifest themselves in heterogeneous and non-controlled developments throughout the world, threatening the physical, social and cultural attributes of rural environments. These threats also emerge from widening urban borders into rural areas and introducing new materials and methods unsuited to existing rustic patterns. As such, the tourism industry is also a sinister force; when providing tourism facilities to meet the current needs of rural residents, it has a higher priority. Essentially, this encourages residents to adapt traditional buildings to tourist functions such as hotels, shops, and so on, shifting rural affairs into commodities for display in a way that they are not normally viewed (O'Donnell 2015) (Fig. 9.2).

From 1995 to 2004, ICOMOS guided a study focused on the threats and challenges against world heritage sites and cultural landscapes that, based on that, categorised following threats into eight groups consisting of (A) Deterioration, (B) Development,

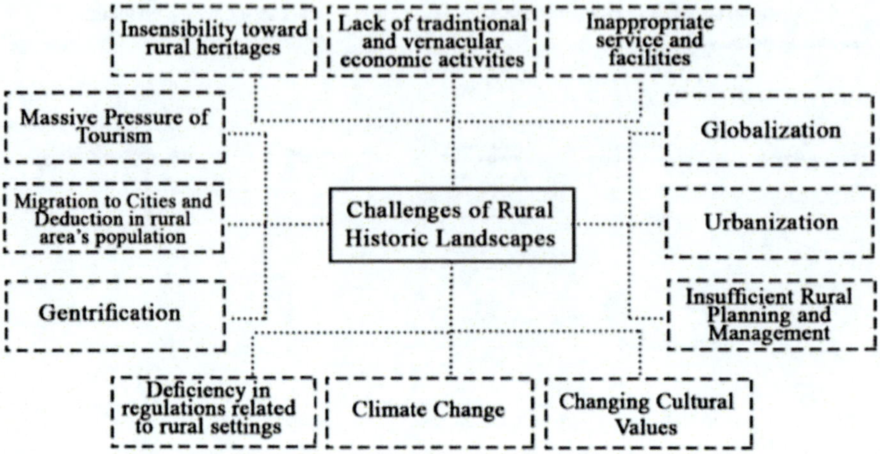

Fig. 9.2 Global challenges meeting rural settlements

(C) Extraction of resources, (D) Large-scale development projects (e.g., energy-transport), (E) Tourism, (F) Local on-site management deficiencies, (G) Cultural changes of deficiencies, and (H) Socio-economic, national infrastructure context (ICOMOS 2004).

In Table 9.1, international challenges for cultural heritage and cultural landscapes are defined to help to define the challenges of study through determining the challenges for village studies, merging them with national challenges, and then conducting a case study; the first classification of challenges for the sample studied, and then the degree of comprehensiveness of all challenges as they affect the sample studied. It aims to identify which challenges are at what level and how their solutions may be applied in the future.

9.4 Challenges of Rural Development and Conservation in Iran

In the early nineteenth century, late Qajar era, and early Pahlavi kingdom, the Modernism movement and theory helped to bring about significant changes in Iranian law and regulation. In rural areas of Iran, these laws have significantly reformed traditional Iran into a modernised country. Changing demographic patterns in rural areas, changes in land ownership policies, changes to village planning, and the move of production cities from villages influenced the centre features of rural landscapes, understanding of villages' importance for people, and their interpretation of villages. However, the change process has not been fully implemented in Iran, particularly in rural areas, where the economy does not completely rely on agricultural products. Thus, cities became a new attraction for the youth population of villages to live and

Table 9.1 An analysis of threats to world heritage sites 1994–2004 (*Source* ICOMOS 2004)

Description	Category
Deterioration	*Man-made aggression and disaster* 1. Pollution (Air Pollution, Ecological disaster) 2. Vandalism 3. Looting 4. Armed conflicts *Natural aggression and disaster* Climate change Natural deterioration Natural disasters (earthquakes, floods, hurricanes, drought, etc.)
Development	1. Urban Pressure (destruction of traditional building/construction of large buildings, high rise, modern houses, incinerators, demographic growth) 2. Commercial activates 3. Encroachment 4. Agriculture pressures 5. Illegal construction/ destruction 6. Garbage 7. Old or new water supply system/pumping station/drainage
Extraction of resources	Legal or illegal extraction of resources (natural resources inside and outside the site). Oil gas; Water; deforestation
Large-scale development projects (e.g., energy-transport)	Construction of bridges/roads/underground transport/airports/pipelines Industrial complex Dam Power lines Lack of impact assessment
Tourism	1. Vandalism 2. Leakage Tourism/unsustainable tourism 3. Illegal construction of infrastructures for tourists within or outside the site 4. Over-visiting/tourism pressure 5. Garbage 6. Uncontrolled tourism flow or development/lack of tourism management plan/regulation about the management plan of the site 7. Lack of impact assessment on tourism 8. Inadequate tourist facilities/infrastructure/projects 9. Overuse of the resources for the tourists/commercial use 10. Increase of tourist vendors within or outside the site

(continued)

Table 9.1 (continued)

Description	Category
Local on-site management deficiencies	*Management issues* 1. Lack of financial resources/human resources/inadequate management of funds 2. Lack/Inadequate management planning process 3. Lack/Inadequate security system 4. Weak or no implementation of regulations/laws/international conventions 5. Illegal/inappropriate archaeological excavations/lack of research 6. Inadequate/lack of management strategy/priorities/Plan (conservation and management plans) monitoring/mechanisms 7. Inadequate/lack of interpretation/presentation 8. Inadequate/lack of documentation 9. Inadequate/lack training/unsustainable training 10. Inadequate/Lack of institutional/project coordination 11. Unclear boundaries/needs to be updated 12. Inadequate/lack of rehabilitation 13. Inadequate/lack of infrastructure 14. Inadequate/lack of preventive mechanisms 15. Inadequate/lack of techno-administrative structures 16. Inadequate international collaboration 17. Incapability/unsustainability between national or international development projects and conservation *Social dimension in management issues* 18. Lack of participation/consultation/engagement with the local population. Lack of democratisation of culture 19. Incompatibility between development needs (ex: poverty) and conservation issues. Need to link management issues, find sustainability in the people, human resources/institutions 20. Challenges of combining conservation and cultural changes 21. Inadequate promotion, educational awareness to protect Cultural Heritage (the tangible and intangible heritage) 22. Difficulty in dealing with conflicts that emerge from the different values and uses assigned to the site/ or coming to a consensus *Conservation issues* 23. Inadequate/Lack of conservation/projects/planning/priorities 24. Use of inadequate material for restoration inadequate techniques 25. Lack of impact/risk preparedness 26. Inadequate/lack of maintenance/restoration

(continued)

Table 9.1 (continued)

Description	Category
Cultural changes or deficiencies	1. Lack of awareness of respect for the outstanding universal values of the site 2. Loss of authenticity 3. Loss of integrity 4. Loss of knowledge in traditional construction techniques or processes 5. Lack of common language about concepts/loss of significance 6. Loss of social/community connection with property 7. Loss of traditional or religious associations 8. Conflicts between different values and uses associated with the site 9. Changes in values and uses/ loss of significance 10. Changes in values and services/ loss of significance 11. Abandonment of the site 12. Loss of qualities that contribute to outstanding universal values
Socio-economic national infrastructure context	1. Weak socio-economic situation of the country 2. No legal framework to control development/land use 3. No urban planning/operational instruments 4. Failure to apply laws 5. Lack of impact assessment 6. Geopolitical conflicts between countries/lack of collaboration 7. Lack of institutional coordination/collaboration/dialogue 8. Weak legislation/corruption 9. Socio-economic changes 10. Weak governance on culture/economy/social/politics/unstable political situation 11. Weak institutions/Lack of training 12. Globalisations 13. The inadequate legal framework to protect Cultural Heritage

work, which led to a massive transformation in both the physical and social fabric of villages in Iran (Hanachi and Koosheshgaran 2011).

Two main factors characterised this transformation period from the perspective of contemporary Iranian villages: firstly, the villages adopted themselves to the gradual changes. The immediate result of the first factor relies on the gradual entry of materials, technology, and new and non-native construction facilities in the village. Secondly, by entering the changes and even doing the planning and rural plans, it is impossible to witness the change in the original face of the village and its conformity with contemporary conditions. In this sense, new housing areas stand beside and on the outskirts of the "old fabric" of the village. The formation of this new residential area has no relation to population demographic changes. Still, it is due to the arrival of public services and their establishment in the margin of the main village. Relatively, the old fabric of the village remains intact, but it gradually faded away from its inhabitants and is abandoned. Instead, the new context is more likely to be compounded by the lack of coordination with the characteristics of the local context and the inability of new materials and techniques.

In the context of Iran, major challenges significantly affect the deterioration of rural land ranging from economic instability, environmental issues, and cultural and social changes, e.g., lifestyles, to religious and lack of proper regulations and policies. Among these challenges are factors such as the loss of indigenous economic activities, a lack of young people ready to work, modifications in rural population balance, erosion of homes and other historic rural structures, changes in the physical, cultural, and social environments, pressures from globalisation, industrialisation, urbanisation, and modernisation, unrestrained development of tourism, lack of rural infrastructure, etc.

The loss of indigenous economic activity, a lack of young people willing to work, adjustments in rural balance, and erosion of buildings have contributed to these challenges. Furthermore, phenomena such as globalisation, urbanisation and modernisation have exerted pressures, as have unrestricted tourism development and lack of proper rural infrastructures. In fact, most of these challenges, with their direct and indirect effects on other micro-domains, operate in a reciprocal loop. Strengthening one can strengthen the others. When indigenous economic activities disappear, the unemployment rate will go up, and villagers will migrate to the cities. In rural areas, this process results in undesirable changes in population balance and the loss of young labourers (O'Donnell 2015).

In fact, most of these challenges, with their direct and indirect effects on other micro-domains, operate in a reciprocal loop. Strengthening one can strengthen the others. When indigenous economic activities disappear, the unemployment rate will go up, and villagers will migrate to the cities. In rural areas, this process results in undesirable changes in population balance and the loss of young labourers (O'Donnell 2015).

In Iran, the lack of appropriate regulations and policies is a major challenge. In the 1960s, for example, the "Law on land reform", also known as the "White Revolution", was the first government-approved programme for rural development. In the years following the implementation of this law, the characteristics of rural areas have changed spectacularly (Hojat 2006).

After the Islamic revolution in Iran, organisations like "Jihad of Construction" and the "Housing Foundation of Islamic Revolution" have managed plans and activities to provide services for rural areas based on the rehabilitation plans and rural guide plans (Azkia and Ghafari 2013). Also, in 2005, with the enactment of the "Law on Territory and Buffer Zone of Towns, Cities, and Villages", the responsibility of making decisions about the rural area's land use was given to the village administrator. Due to a lack of specialised knowledge, this practice often led to uneven spread of rural constructions, major changes in noble characteristics of rural landscapes, and, consequently, accretion of problems and challenges. In general, less attention has been paid to villages than cities in terms of adopting legal measures and tools. The current regulations are also not adequate and have no executive guarantee.

Therefore, rural areas are susceptible to any intervention that can lead to the loss of identity and tangible and intangible values. Accordingly, challenges in Iran villages are mostly related to the rapid and uncontrolled changes and the imbalance and balance between the activity balance and the previous relations between the villages and the country's cities. Nowadays, these challenges do exist in most African and Asian villages. On the other hand, lack of managerial knowledge and adequate legal measures, besides ignoring the rural issues, are some of the factors that have led to the reduction of the quality of life in the villages of Iran.

9.5 Case Study: Four Villages in the East Azerbaijan Province, Iran

The four villages that constituted this study, *Ushtabin, Kandovan, Zonuzaq,* and *Jeraghil,* are in the northwest part of Iran, politically considered under the management of East Azerbaijan province. The geopolitical importance of this region has resulted from its vicinity to the neighbourhood countries, e.g., Azerbaijan, Turkey, and Armenia (Fig. 9.3). This geographical closeness and their unity within Iran during history have had a great impact on the formation of local characteristics of the rural areas. These similarities of villages can be traced nowadays to rural neighbourhood countries. An outstanding example of this simulation is in the form of ceilings used in the stables built in Ushtabin village called "*Hazarashen*" and the same form of the ceiling in many Armenian local churches built in the vicinity through the first centuries of Christianity in Armenia and Iran. However, these are not the only reasons for this research to conduct on the four selected villages.

East Azerbaijan province has about 3,000 villages, out of which about 22 villages have been considered historical villages by the Cultural Heritage, Handicrafts, and Tourism Organization of Iran (Ministry of Cultural Heritage, Tourism and Handicrafts 2015). Among all these villages, four selected villages, *Kandovan,*

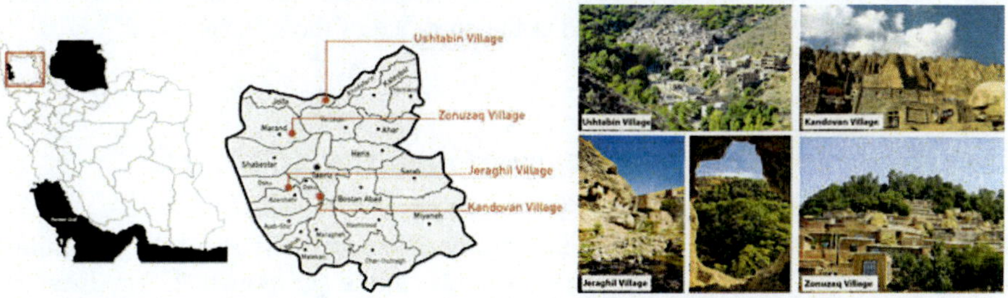

Fig. 9.3 Location of case studies of this research in East Azerbaijan province, Iran

Jeraghil, Ushtabin, and *Zonuzaq,* have historical, cultural, and environmental-physical features that have distinguished them from the remaining 18 villages. The rocky settlements with stepped patterns, staircase formation of these villages in which each row of houses has been built on top of the other, as well as the construction of rural settlements of various historical periods (esp. In the case of Jeraghil with the dug mountain as the primary rural settlement, and Kandovan village with three different types of human settlements in forms of dug, built like a sugarloaf, and recently constructions), same socio-cultural background, the existence of ancient languages, etc., are some major characteristics of these villages.

Nevertheless, each of them has unique features that are going to be introduced briefly:

9.5.1 Historic Rural Landscape of Ushtabin, Jolfa, Iran

The historic village of Ushtabin is one of the stepped villages of *Azerbaijan* located near *Jolfa* city, beside the *Aras* River and on the common border of Iran and Armenia. Based on available references, the village dates to the *Sassanid* era, but a significant portion of its current structures (historical houses) are attributed to the *Safavid* era. The values of this village include its cultural significance as a birthplace of a regional poet, *Abol-ghasem Nabati* (a well-known Turkish Poet who lived within the seventeenth and eighteenth centuries), special culture and traditions, the history of its traditional *Tati* language (an ancient language mainly used in the Northwest of Iran and *Qafqaz* region), historical values, its historic structures, its environmental values as a habitat of rare flora and fauna. The constructions of the village can be summarised in four main periods. First, constructions of the *Safavid* era, constructions of the *Qajar* dynasty, constructions built after the 1920s, and constructions that have been built in the last two decades.

The historic fabric of the village that is mainly located in the centre has been demolished completely, and one or two houses have good condition in it. During the *Qajar* dynasty, the fabric of the village is created around the central core of the village. The constructions which have been built in the third period are mostly composed of rural structures located in the vicinity of the *Ushtabin* River, in the east–west direction, in the north of the old village. Finally, the fourth-period fabric is composed of new structures located at the entrance of the village.

9.5.2 Historic Rural Landscape of Kandovan, Osku, Iran

The historic village of Kandovan is an internationally renowned rocky village, similar to *Cappadocia* in Turkey. This village is located near the *Osku* town, 22 km distance far from Tabriz, the capital of East Azerbaijan province. The outstanding features of this village go back to the architecture of its rocky houses, its special culture and customs, and its relationship with nature. It is believed that the mountaineers founded the village in the early Islamic centuries when Islam hadn't been brought to Iran by Arabs. However, a considerable part of this village's current structures has been attributed to the *Ilkhanid* era.

In the past, the inhabitants of this village lived in the surrounding mountains in habitats dug in the mountain called "*Hileh-var*". Then they migrated to the rocky regions. The village fabric can be divided into four areas: the *Hilevar* village that its constructions are located on the "*Sahand*" Mountain slope; the rocky residential structures of Kandovan, the constructions that are the result of the resident's migration to the new houses which have been built mainly in the 1980s and 1990s, and the constructions located at the other side of the river built during the last two decades, as a result of truism development in the region.

9.5.3 Historic Rural Landscape of Jeraghil, Azarshahr, Iran

The historic village of Jeraghil, like Kandovan village, is a rocky village in the region. In fact, this village is located along the ridge that forms the context of the Kandovan and is situated on the other side of these mountains. Jeraghil village fabric is divided into three areas: The oldest part of the village is composed of a series of houses and a mosque which have been excavated in the mountain, facing the surrounding rivers and gardens; another part of the village is located on the *Sahand* Mountains slope, and its wooden and stone structures have formed part of its characteristic. Besides, there is another part that includes the constructions which have been built during the last two decades. This area has been distinguished from the other two areas by using modern materials and patterns.

9.5.4 Historic Rural Landscape of Zonuzaq, Marand, Iran

As an old and large stepped pattern village, the historic village of Zonuzaq is in the *Marand* city. Some of the existing historic structures in this village, such as the residential spaces excavated in the mountains and its historical mosque, date back to the *Ilkhanid* era. Based on the region's climate and its topographical features, the village has a dense fabric and a stepped pattern. Since the 1940s, this village has also undergone new construction because of regional rural development planning. Therefore, the ancient part of the village has a stepped structure and unlike the later built section in the fabric of the village, which reside considerable population in it. The new fabric of the village is divided into two areas: an area that extends along the river and its mainly brick structures, mostly located on the mountain slope in a linear form, and an area that includes residential structures along the access road to the village. These structures are constructed with materials commonly used in urban constructions and have disrupted the integration of the village landscape.

9.6 Existing Challenges in Historic Rural Landscape of East Azerbaijan Villages

According to this newly developed conceptualisation of Historic Rural Landscape, conservation is not passive or conservative. Acceptance of change is a calculative process in the context of the conservation of historical monuments, sites, and heritage areas. Monitoring changes and managing them helps to prevent the further negative impact of new socio-economic relations on originality and cultural significance. In addition, it helps to distinguish the character of urban heritage as another essential requirement for preservation. Due to the limited effects of changing contexts on historical contexts, the implementation of regulatory tools and planning is required to meet this challenge (ICOMOS 2005).

In Table 9.1 of this chapter on the four historical villages of East Azerbaijan province in Iran as a case study, the eight significant challenges for world heritage sites are caused by national and international dilemmas affecting the development of the four case studies villages their own unique characteristics. These factors affect a rural landscape's overall image, along with the pressures placed on the urban areas.

Here, the challenges in the four villages surveyed are classified into three general categories: "Challenges rooted in natural events", "Challenges rooted in human interventions", and "Challenges affecting the environment". In the classification, the challenges arising from natural events are those that occur without human intervention and are based on the physical environment and geographical location of each village.

Challenges resulting from human interventions often occur over time and in the form of complex human interventions in the natural environment of the rural context and often are related to the different aspects of the social, economic, and cultural life of the residents and to the management approaches in the national and local levels. In comparison, challenges affecting the environment are, in fact, part of human interventions in the natural environment that has impacted the process of livelihoods in historical rural landscapes but have been considered as a separate category due to the growing importance of the environmental activities and interventions of human in the nature (Table 9.2).

9.7 Concluding Remarks

Conflicts between conservation and preservation of historic-cultural assets, and the endless need for societies to develop economically, challenged the historic rural environment. Although, in recent years, the conservation of historical monuments and sites has evolved to include an integrated approach to this duality. Concepts such as management of change, historic urban landscape, and cultural landscapes conservation have contributed to the resolution of conflicts between the necessity of saving cultural heritage for the future and the necessities of communities to improve living standards.

Table 9.2 Challenges of integrated conservation and development in the historic rural landscapes of four villages, Jeraghil, Ushtabin, Kandovan, and Zonuzaq

Primary Category of Challenges	Secondary Threats	Description	*Ushtabin*	*Zonuzaq*	*Jeraghil*	*Kandovan*
Nature-based challenges and threats	Unexpected Natural Events	e.g., earthquakes, floods, hurricanes, drought, etc	–	*	-	–
	Gradual Natural Events	e.g., Natural deterioration, seasonal rains, snowing, etc	*	*	*	*
Human Activity-based challenges and threats	Weakness in rural management	Illegal/inappropriate restoration of rural structures by local institutions	*	*	*	*
		Lack of participation/consultation/engagement with the local population	*	*	*	*
		Fragmented rural management systems affected by local governmental authorities	*	*	*	*
		Illegal constructions/destructions/changes in land use	*	*	*	*
		Imposing plumping lines in villages and destroying the integration of rural fabric	*	*	–	*
		Imposing urban/regional development planning plots for rural areas	*	–	*	–
		Rural sprawl	*	*	*	*
	Insufficient Local/National regulations		*	*	*	*
	Weakness in Tourism Management	Insufficient/lack of tourism infrastructures	*	*	*	–
		The commodification of the rural landscape as an effect of mass tourism	–	*	–	*
	Extensive pressure on rural development	Second-home settlers in the village	*	*	*	*

(continued)

Table 9.2 (continued)

Primary Category of Challenges	Secondary Threats	Description	*Ushtabin*	*Zonuzaq*	*Jeraghil*	*Kandovan*
		New construction with modern materials & techniques, e.g., concrete	*	*	*	*
		Lack of rural infrastructures & services for saving harvests	*	–	*	–
	Threats arising from Socio-Cultural issues	Weakness/lack of proper education and informative centres for locals and other stakeholders	*	*	*	*
		Change in lifestyle of locals	*	–	–	*
		A large portion of elderlies in rural demography	*	*	*	–
		Change in values/uses; loss of significance/qualities that contribute to the OUV	*	*	*	–
		Globalisation and Urbanisation trends	*	*	*	*
		Loss of authenticity and integrity of place and significance of tangible and intangible values	*	*	*	*
		Loss of social/community connection with cultural properties	*	*	*	–
		Loss of traditional/religious associations, festivals, and events	–	–	–	*
	Threats arising from economic issues	Loss/Lack of traditional economic activities in the villages	*	*	*	*

(continued)

Table 9.2 (continued)

Primary Category of Challenges	Secondary Threats	Description	*Ushtabin*	*Zonuzaq*	*Jeraghil*	*Kandovan*
		Lack/Loss of young workers by migration to cities	*	*	*	*
		National economic situation/ raise in inflation rate	*	*	*	*
		Lack of financial resources/investments. Inadequate management of funds	*	*	*	*
	Human Demolitions (e.g., destruction of heritage values, use of valuable materials in new buildings, destruction of historic flora and fauna)		*	*	*	–
Threats and challenges affecting the environment	Weakness in environmental management	Unsuitable garbage/sewage disposal	*	*	–	*
		Lack of proper management of valuable environmental habitats of under threat floras and faunas	*	–	–	*
		Pollution of water, soil, and air resources (rivers, spas, etc.)	*	*	*	*
	Climate Change		*	*	*	*

Historic rural landscapes are cultural landscapes that embody triple interactions between human beings, nature, and the environment in rural settlements. A rural landscape consists of the traits and characteristics of rural life in a natural environment. Prior to the recent emergence of changes and transformations, highly valued villages interacted with their surroundings. The challenges were met in this way. In rural landscapes, however, increasing changes in social, cultural, and economic circumstances, as well as insufficient laws and regulations at local and national levels, have doubled the challenges. It aims to address the problems faced by the historically valuable villages of East Azerbaijan province.

This study shows that challenges arising from Historic Rural Landscapes can be categorised into three categories: common global challenges of historic rural landscapes, national challenges, and challenges arising from local human-nature characteristics of rural settlements. Global challenges include issues such as urbanisation and global culture. On the other hand, the lack of efficient national policies and regulations or a weak rural management programme is classified as a national challenge. In addition, some of the categories, such as environmental issues associated with village natural resources, low levels of tourist hostility among villagers, and low awareness of heritage preservation, are inherent challenges of this research.

The existing challenges in Ushtabin, Kandovan, Jeraghil, and Zonuzaq show that each of these challenges needs a strategy appropriate to its scale. Solutions and strategies for the global challenges of rural property conservation should be comprehensive and inclusive and emphasise the local potential. Local solutions are better for tackling global challenges.

Some of these solutions may refer to the national challenges of rural settlements that turn out to be inappropriate and unsuitable for rural management. Others, on the other hand, can be managed by governmental institutions to resolve weaknesses or lack of economic activities besides the village's dependence on urban areas. By establishing management plans, careful planning, and continuous monitoring, these challenges can be addressed by creating creative business opportunities or a revival of traditional economic activities.

References

Azkia M, Ghafari G (2013) Rural development; with emphasis on Iranian rural community, 5th edn. Nashre Ney Publication, Tehran

Bandarin B, Oers R (2015) Reconnecting the city: The historic urban landscape approach and the future of urban heritage. Wiley, NewYork

Cigdem Asrav E (2015) Place and community driven conservation and empowerment in historic rural landscapes: Principals and strategies for Taskale Village, Turkey. Ankara. The Degree of Master of Science in Restoration in Architecture, The Graduate School of Natural and Applied Sciences of Middle East Technical University

Ministry of Cultural Heritage, Tourism and Handicrafts (2015) Most valuable historic village in East Azerbaijan of Iran. Tabriz: Cultural Heritage, Handicrafts and Tourism Organization of Iran

Di Fazio S, Modica G (2018) Historic rural landscapes: Sustainable planning strategies and action criteria. The Italian experience in the global and European context. Sustainability 10:1–27

Fadaei Nezhad S, Ehsrati P (2015) A definition of authenticity concept in conservation of cultural landscapes. Int J Archit Res 9(1):93–107

Habibi S, Maghsoudi M (2013) Urban conservation. University of Tehran, Tehran

Hanachi P, Koosheshgaran A (2011) Conservation and development in valuable rural fabrics. Bonyad Maskan Emam Khomeini, Tehran

Harrison R (2013) Heritage: Critical approaches. Routledge, London

Hojat I (2006) Impact of land reform on the form of Iranian villages. Fine Arts Journal of University of Tehran-Architecture and Urbanism, pp 75–84

ICOMOS (2005) Xi'an Declaration on the conservation of the setting of heritage structures, sites and areas. Xi'an, ICOMOS

ICOMOS (2004) Threats to world heritage sites 1994–2004. ICOMOS

McClelland F, Melnick R (1999) Guidelines for evaluating and documenting rural historic landscapes. United States, NPS: National Park Service

Nocca F (2017) the role of cultural heritage in sustainable development: Multidimensional indicators as decision-making tool. Sustainability 9:1–28

O'Donnell P (2015) Historic urban landscape a new UNESCO tool for a sustainable future. In: Ken Taylor AS (ed) Cultural landscapes: Twenty-first century conservation opportunities and challenges, Routledge, pp 163–182

Rottle N (2008) A continuum and process framework for rural historic landscape preservation revisiting Ebey's landing on Whidby Island, Washington. In: longstreth R (ed) Cultural landscapes; Balancing nature and heritage in preservation practice, University of Minnesota Press, pp 129–150

Smith J (2015) Applying a cultural landscape approach to the urban context. In: Ken Taylor AS (ed) Conserving cultural landscapes: Challenges and new directions, Routledge, pp 182–198

UNESCO (2011) Recommendation on the historic urban landscape. Paris: adapted by the General Conference at its 36th session: UNESCO World Heritage Center

Zarif KB, Jokilehto J (2016) General framework for the conservation of historical rural landscape: A case study of Qadisha Valley in Lebanon. Dissertation. Nova Gorica, B. Zarif Keyrouz

Chapter 10
Chinese Cities as Sacred Landscapes: The Case of the Capitals of the Ming Dynasty

Norma Camilla Baratta, Mariacristina Giambruno, Giulio Magli, and Jing Zhou

Abstract The advent of the Ming dynasty marks a break through in the history of Chinese architecture. The first Ming rulers Hongwu and Yongle were indeed engaged in a huge building effort, aimed at showing symbolically their divine rights to power—the "Mandate of Heaven". This resulted in as much as three projected capitals in the course of a few decades: first, a re-planning of Nanjing; second, a completely new capital in Fengyang, which remained unfinished; and finally, of course, the planning and building of Ming Beijing. In the present paper, we explore and compare cognitive and symbolic aspects of these three "sacred landscapes of power" using the methods of modern Archaeoastronomy and Archaeotopography. In this way, we are able to highlight astronomical alignments but also topographical and magnetic orientations, which appear to be related to the feng shui tradition in both its "form" and "compass" schools. These results help in placing the architecture of the Ming capitals within the original perception of the landscape at the time of their planning, also with the aim of contributing to the preservation and fruition of this important Chinese cultural heritage.

Keywords Archaeoastronomy of Ming Capitals · Form feng shui · Compass feng shui · Ancient Chinese urban planning

N. C. Baratta · M. Giambruno
Department of Architecture and Urban Studies, Politecnico di Milano, Milan, Italy
e-mail: normacamilla.baratta@polimi.it

M. Giambruno
e-mail: mariacristina.giambruno@polimi.it

G. Magli (✉)
Department of Mathematics, Politecnico di Milano, Milan, Italy
e-mail: giulio.magli@polimi.it

J. Zhou
Architecture Department, Human Settlement and Civil Engineering School, Xi'an Jiaotong University, Xi'an, China
e-mail: zhoujing@xjtu.edu.cn

R. P. B. Singh et al. (eds.), *Placemaking and Cultural Landscapes*,
Advances in Geographical and Environmental Sciences,
https://doi.org/10.1007/978-981-19-6274-5_10

10.1 Introduction

The second half of the 14 century AD marks in China the beginning of the Ming dynasty, proclaimed in 1368 after the defeat of the Mongol (Yuan) dynasty. The first Ming Emperor Hongwu made his capital Nanjing, immediately after he started the project of a brand new capital, Mingzhong, in his homeland; construction works there lasted for some 7 years but were interrupted before completion, so that Nanjing returned to be the elected capital up to the advent of the third Ming Emperor, Yongle, formerly Prince of Beiping (1402). Later, Emperor Yongle, for uncertain reasons (discussed in Liu 1998), decided to elect the city of his principate, Beiping, as capital. The city, which had been previously the southern capital of the Mongol Yuan dynasty under the name Dadu, was then renamed Beijing (Northern Capital). With his return, Yongle decided to make significant urbanistic interventions to the city, which were carried out over a period of 13 years and concluded in 1420.

All in all, in the space of just 40 years, the Ming emperors planned three imperial capitals, two of them through a radical re-planning of important existing cities and one of them from scratch in an area that had never hosted an imperial city. The focus of our study here is on the symbolic aspects of the planning of these towns from a unified point of view. Indeed, several architectural solutions are repeated and re-enhanced in all the 3 projects. The unified approach we propose here follows the lines of modern Archaeoastronomy (Magli 2018) as recently applied specifically to the Beijing case (Baratta and Magli 2021).

An archaeoastronomical study aims at analyzing the relationships between the built environment and the celestial sphere: here, the celestial landscape is shown to play an important role in the formation of the sacred landscape. Through a similar approach, an investigation is carried out into the relationships that the city establishes with the terrestrial landscape, generating alignments linked to the traditional Chinese doctrine known as feng shui. In doing so, both contexts (terrestrial and celestial), with their alignments and relationships, are studied in a unified scientific approach This approach leads to a better understanding of several architectural issues which are inter-connected by the same political message: according to Chinese imperial ideology indeed, since the first emperor Shihuang the ruling dynasty was entrusted with the Mandate of Heaven. The ruler thus justified his power because he was chosen by Tian, the deified heaven: this was recognized as the celestial order and the regularity of the celestial cycles. The emperor therefore took upon himself the task of bringing this same order on earth. Urban planning, and especially the imperial capitals (the highest expression of ruling power) are the instrument, the "pivot of the four quarters" (Wheatley 1975), to bring celestial order on earth: they are the true cosmic center of the world (Krupp 1982).

In the first millennium AD, to these traditional ideas, others were added by the development of the feng shui tradition, based on the concept of *qi*—the "vital energy"—and on the search for favorable places for its flow. Of course, the emperors became interested in showing their explicit connection (and the connection of their tombs) with auspicious places, and so the doctrine started to influence imperial architectural projects (Magli 2019).

Two schools of feng shui can be recognized, traditionally called form and compass feng shui (Bruun 2011). While the first school is based on the investigation of the topography of the place, the second is based on the use of an instrument called Luopan (a magnetic compass).

The first school therefore involves an analysis of the morphological characteristics of the place: the presence of mountains and watercourses, their shape and position, and their relationship with the winds. The aim was to identify a "favorable place" characterized by the accumulation of positive energy: qi. The general picture of an auspicious place has as its first aspects the highest visible mountain located to the north, behind the site, and a river flowing to the south, in front of the site. To these are then added further features that increase the suitability of the place: the undulating shape of the mountain to the north, the presence of two protective reliefs, one to the east (Green Dragon Hills) and one to the west (White Tiger Hills) of the site, and a slight relief to the south. The second school of feng shui probably originated with the discovery of the magnetic properties of lodestone but did not develop until the end of the 1st millennium, with the introduction of a compass based on a free-floating needle (Needham and Ling 1956). The second school involves instead the use of a magnetic orientation instrument: the objective is certainly not magnetic north but rather to follow a favorable direction that the earth itself indicates and that can be read through the instrument. Added to this is a complicated investigation of the topographical elements that are assigned an auspicious value based on their magnetic azimuth. Of course, there is no such thing as "Qi" flowing on earth, so that Feng Shui theories are superstitions; however, their importance for the history of Chinese architecture and landscape is fundamental. Today, it is possible to study the use of magnetic orientation in ancient times using Paleomagnetic models, that is models of the behavior of the earth's magnetic field (which varies continuously in time and space).

10.2 The Project of Ming Nanjing

Hongwu ascended to the Dragon Throne in 1368, establishing his residence in Nanjing and immediately starting a series of building projects there. Shortly after however, these works were slowed down by the decision of building a brand new capital in Hongwu's homeland, modern Fengyang. According to the existing sources, he took this decision in 1370. Building activity in Fengyang lasted about six years, up to 1376 when the decision was taken to abandon the project, and to come back to Nanjing as the definitive capital, who took the name Yingtian (actually the name Nanjing—meaning South Capital—will enter in use only after the transfer of the Ming court in Beijing). The above-described succession of events makes it likely that at least the general inspiring ideas of the Ming Nanjing plan—and, in particular, that of the imperial residence—were conceived first, and then adapted to Fengyang, where the geographical situation was far easier.

For the Hongwu architects, making Nanjing the center of the universe, the place of residence of the Mandate of Heaven was not an easy task, both from the geographical and from the historical point of view. Geographically, Nanjing is located in a peculiar situation characterized by prominent natural features: to the west, the Yangzi River; to the north, the Xuanwu lake; and to the north/north-east, the Zhongshan hills (usually called Purple Mountain). Furthermore, historically, the city had already a millenary past. In particular, it had been the state capital of the reign of Wu and the imperial capital during the long and turbulent period of the so-called Six Dynasties, with the name Jiankang. With the advent of the Sui dynasty, the city was almost destroyed but raised again at imperial status with the name Jinling during the Southern Tang. In spite of such a long and undoubtedly glorious story, however, the city had never been the capital of a dynasty reigning over the whole country (contrary to Xi'an, the place of the capitals of the Qin, Han, and Tang rulers). In a sense thus, Nanjing history was inauspicious for a dynasty that claimed for a return of the epoch of the grand unifications.

To tackle this problem, imperial propaganda sat about to present Nanjing in its Ming re-foundation as precisely the right place for the residence of the Mandate of heaven. The choice was to distinguish Ming Nanjing as a "brand new" city, and to this aim nothing was left to chance: from literary writings of propaganda to a complete re-planning of the existing city in accordance with the new rulers' ideas and beliefs. Interestingly enough, there are intertwining connections between all these aspects which help us in understanding the symbolic framework in which the ideas on Nanjing (and actually all the three Ming capitals) were generated. In particular, in literature devoted to Nanjing and in a text written by Hongwu himself, it is explicitly stated that the Ming city was profoundly different from its predecessors at the same place, since the previous urban layouts were unable to harmonize with the surrounding mountains and rivers—as instead the Ming plan is claimed to do and for these reasons were inauspicious. As a matter of fact, as we shall now see, the Ming re-planning of Nanjing really was a huge and complex project conceived to address a series of symbolic issues.

10.3 The City Walls

The first and most striking feature of Ming Nanjing are the city walls. The two previous imperial towns had indeed been constructed occupying only a relatively small area of the above-described naturally bounded territory. The inspiring idea of the Ming project was to enclose the new town by remarking all the natural boundaries. The inner-city walls were thus planned with an unusual, extremely long perimeter, which follows the natural contours given by the mountain to the east, the lake to the north, and the river to the west/southwest. Of the pre-existing walls, only part of the southern fortifications was adopted and reinforced. The brand new core of the capital, the imperial residence city, was built in a virgin zone located to the east, inside a sort of natural bay formed by the west flank of Purple Mountain, with the Qinhuai River to the south (on the position and orientation of the imperial town we shall come back later).

Contrary thus to the traditional, rectangular layout of most Chinese cities (including previous settlements in Nanjing) the perimeter of the wall is irregular and encompasses a wide area which was actually reached by the city urbanization only in the twentieth century. Supporting strategic aims for such a huge project is at least dubious, and it has been proposed that the perimeter was symbolical, and intentionally planned to resemble the matching of the ideal shape of two constellations, the North Dipper (the seven stars of the Big Dipper of western astronomy) and the South Dipper (six stars of Sagittarius) so that the 13 stars forming the two constellations should correspond to the 13 gates of Ming Nanjing. The two shapes are supposed to connect at Tongji Gate in the southeast corner of the city wall and at Zhongfu Gate in the north-west corner, creating an ideal, transverse division of the city into two "halves" (Zhang 1985; Yang 1999) (Fig. 10.1).

The resemblance of the shape of the constellations is somewhat doubtful, and in absence of written sources, it is difficult to ascertain if this idea was really in the mind of the builders (similar ideas have been advanced for previous Chinese capitals of polygonal shape, notably Han Chang'an, again without definitive proofs).

What is certain, however, is that China is perhaps the unique example of a civilization in which a written historical text makes an explicit reference of a "image of heaven" actualized trough imperial architecture: it is the famous text by Sima Qian which parallels the position of the Qin imperial palace with respect to the Wei River with the position of the northern stars with respect to the Milky Way. Another argument in favor of the constellation's theory is the Taoist ideas which are known to have inspired Hongwu, and in particular his will of unification of Heaven and Earth (corresponding to the north and the south constellations) in cult practices.

10.4 The Imperial Palace

The area of the imperial palace is naturally individuated by the eastern sector of the town, between Tongyi gate and Taiping gate. To the left of this sector, the main axis of the town is skewed about 15.5° east of north, reflecting the orientation of the previous capital. The axis of the Ming imperial town is *not* parallel to this direction but bears an azimuth of about 5° east of north. Clearly, this was an important decision, introducing a radial skew between the main south-north streets of the town. Its reason is purely symbolic, as we shall see.

The imperial palace is today almost destroyed, but it is possible to get a general idea of the original project, which was similar—and certainly inspiring—for those of Fengyang first and of the Forbidden City in Beijing later. It was a huge, rectangular area, accessed by a winged (Meridian) gate to the south, flanked, as it will be in Beijing, by the Ancestral Temple and the Altar of Land and Grain. The center of

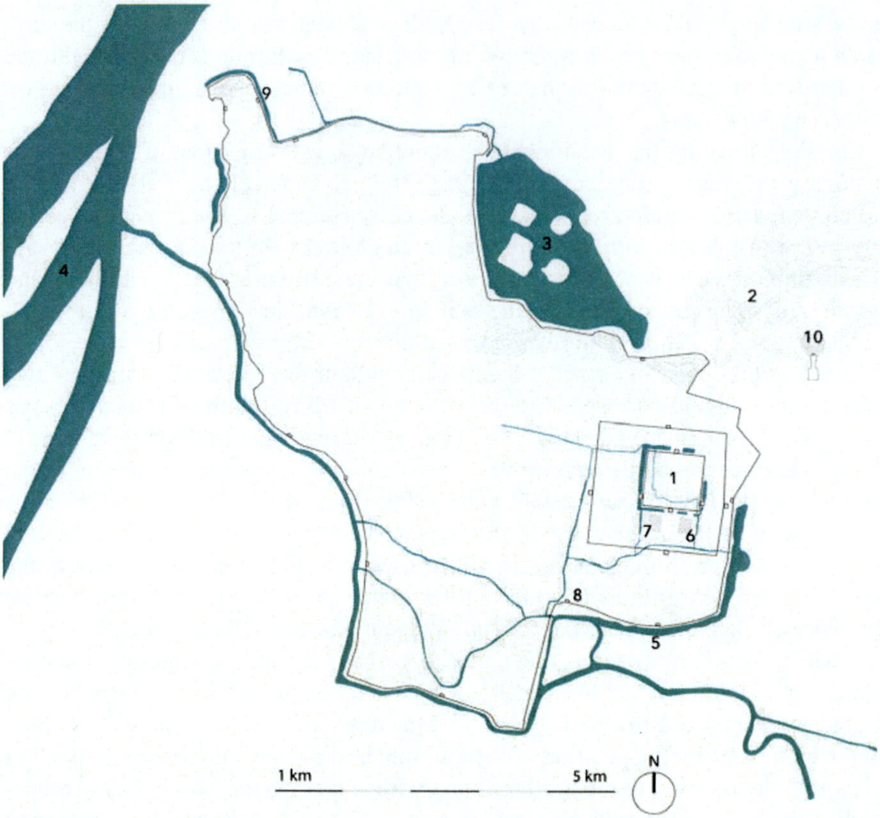

Fig. 10.1 Plan of Ming Nanjing. **1** Forbidden City, **2** Purple Mountain, **3** Xuanwu lake, **4** Yangzi River, **5** Qinhuai River, **6** Imperial Ancestral Temple, **7** Altar of Land and Grain, **8** Tongji Gate, **9** Zhongfu Gate, **10** Xiaoling mausoleum (Drawing by the authors)

attraction of the complex was a succession of three huge ceremonial halls, serving as a clear model for the future project of the Forbidden City. The halls do not exist anymore, but their location is marked in the park which has been established at the site, which also contains many scattered pieces of the original stone decorations.

Planning the palace in a virgin zone obtained different goals. First, the project avoided to overlap the old palace site of previous dynasties, which, as mentioned, were not considered auspicious for the Ming. Second, it allowed a direct, close connection with the Emperor's tomb, the magnificent Xiaoling mausoleum, located on Purple Mountain and easily accessed from Chaoyang Gate. Last but not least, it satisfied a series of symbolic canons which will be later applied also in Fengyang, only to become striking in Ming Beijing.

The first canon is form feng shui. According to this, as mentioned, a place is auspicious if there is a meandering river to the south, a mountain to the north, and gentle hills (Green Dragon and White Tiger) to the east and west, respectively. These elements were identified as follows, with a certain amount of ingenuity. The dragon hill clearly is Purple Mountain to the east, although it is a rather prominent, overwhelming presence on the whole city rather than a gentle hill at some distance. The mountain was however endowed with special, auspicious significance: the nickname "Purple Mountain" referred to flashes of purple light purportedly visible on the summit, associated with royalty—the zone of the heavens associated with the emperor was indeed called the Purple Enclosure. The function of white tiger, the hill to the west, is played by is the so-called Stone city. It is a low natural stone relief which was fortified during the Wu Kingdom, 1000 years before the Ming (today, only a huge wall remains). Again, it is a place traditionally endowed with special significance, as still today the face of a benign spirit is supposedly visible on it. The river to the south is Qinhuai River, which is meandering to the south of the Nanjing Forbidden city precisely as the "Golden Water" (artificial) River will be projected to do in the Forbidden City in Beijing. Finally, to obtain an alignment to a hill peak screen to the north, the project was located in such a way that the axis points to a low spur of the northern flank of Purple Mountain.

Interestingly enough, further to form feng shui, we have evidence that also compass feng shui was used. Indeed, reconstructing the magnetic field declination in Nanjing at the time of construction, the skew to the east qualitatively corresponds to the magnetic declination at the date of foundation. In the present paper, we use the Palaeomagnetic model CALS10k.2, a global geomagnetic field reconstruction, developed at the German Research Center for Geosciences in Potsdam: it allows us, knowing the specific coordinates of a place, to calculate the value of the magnetic declination in that place over time (Korte et al. 2011; Constable et al. 2016).

10.5 The New Capital in Fengyang

Immediately after the foundation of the Ming dynasty, the Emperor Hongwu conceived the idea of building a new capital in his homeland, Linhao (modern Fengyang in Northern Anhwei). In the same place, he ordered the construction of a magnificent tomb for his parents, deciding to honor his own ancestors giving them a posthumous imperial status. The imperial mausoleum—called Huangling—was brought to end. Its sacred road, endowed with magnificent stone statues, is still visible and reminds the grandeur of that of the Hongwu's own mausoleum, Xiaoling on Purple Mountain. Contrary to the Mausoleum, the new capital, named Mingzhong, was not doomed to last.

According to the chronicles, construction works went on rapidly, and in 1374 as much as 140,000 persons were conscripted to work on the walls; when the emperor visited the site in 1375, workers assembled and petitioned for better conditions; and the prime minister Li Shanchang overreacted and had thousands of people killed. Inhabitants rebelled and the emperor considered this a very bad omen. Apparently for this reason, he ordered to stop the project of the new capital and came back to Nanjing (Chen 2009). The city remained at the status of provincial town and is today known mostly for the imperial Mausoleum. However, although the construction of the imperial city was abandoned, its planning and layout had a profound impact on future Ming project (Chen 2019).

10.5.1 The Urban Plan

Contrary to the plans of Ming Nanjing and later Ming Beijing, which had to consider the pre-existing towns, Mingzhong was planned in an empty field and anew. As we shall see, the location was chosen carefully, in order to harmonize the town with the topography along the accepted canons of geomancy. Mingzhong was actually the first city planned on virgin soil after the Ming Dynasty established the Ming ritual system, and the commander-in-chief who presided over the formulation of both the ritual system and the Mingzhong Capital Construction Project was the first official Li Shanchang (Chen 2009).

The construction of Mingzhong was based on the so-called magic square plan: a square divided in nine equal sub-unities; together with the location of markets to the north and the construction of the Ancestral Temple to the east of the administrative city and of the Altar of the God of Land and Grain to the west of it, these features make the city conform to the recipes of the "Kaogong Ji" section of the Confucian classic Zhou Li. This text—written during the Han period—reports the Chinese traditional canons for town construction which were allegedly already existing during the Zhou (Schinz 1996; Xu 2019). Ironically, the Mongols—and thus foreign rulers—were the first to adhere rather strictly to the Chinese canons, more than 1200 years after their writing, when they planned their capital Dadu. The Ming project of Mingzhong follows similar guidelines: the square, whose size was determined by a basic unit of one and half li (like Mongols' Dadu), was composed by nine such units (the length of the sides was 13.5 km); there would have been in total 12 gates, three on each side. As in Dadu, the imperial palace was not in the perfect center but was shifted southward, leaving a space two and half li wide between the southern walls for the Thousand Pace Causeway in front of the palace (Schinz 1996).

Nevertheless, probably to make the defense of the city more effective, the east wall of the city was moved eastward, to include Dushan hill, and the southwest wall was protruded by a corner to the south, to include Fenghuangzui hill (Chen 2019).

As the east wall of the outer city moved eastward, the imperial city, built on the gentle slope south of the Phoenix Mountain, and originally in the center of the city, was slightly westward. The walls around, built with masonry, with turrets on the four corners, featured four gates, one on each side: the Meridian Gate to the south, the Xuanwu Gate to the north, the Donghua Gate to the east, and the Xihua Gate to the west. The imperial city was again surrounded by walls, built with masonry, enclosing the main peak of Phoenix Mountain and its associated Longevity peak, and featuring a gate on each side.

Fig. 10.2 Mingzhong, aerial view (Image used under license from Shutterstock.com)

The third ring of walls of the outer city had nine gates (initially thought to be 12, three on each side, the number was perhaps reduced following the strike that will finally lead to the abandonment of the works). The layout of the central city follows the principle of symmetry, with a central axis running through the city from north to south, extending from the Hongwu Gate of the Outer City in the south to the Zhengbei Gate of the Outer City in the north (not completed at that time), with a total length of nearly seven kilometers. The buildings are arranged symmetrically on both sides of the central axis. The Drum Tower and the Bell Tower are three kilometers apart, located at the two ends of Yunji Street, symmetrical to the east and west sides of the central axis (Zhao et al. 2018; Chen 2009).

The Imperial Ancestral Temple and the Altar of Grain and Soil were placed on the left and right sides of the central axis before the Meridian Gate, as will be later done in Beijing, emphasizing the position of the central imperial road, but also highlighting the hierarchy of the Mandate of heaven. The two temples were built in 1370, while the following year the emperor built the Circular Mound south of the city outside the outer walls, the Square Mound to the north, and the Altars to the Sun in the east and the Moon in the west (Steinhardt 1999) (Figs. 10.2 and 10.3).

10.5.2 The Sacred Landscape of the New Capital

The entire city was planned to be located within a natural context characterized by an accumulation of vital energy (qi): behind the palace city, enclosed within the second ring of walls, there is the Phoenix hill, whose peak is intercepted by the same axis that governs the entire urban layout. Next to it, slightly east of the north, is the Longevity hill and to complete the panorama in the north-east and north-west direction are the Moon hill and the Sun hill, respectively, all enclosed within the third ring of walls. A stream flows from east to west passing south of the city beyond the outer city walls, while to the north the city axis intercepts the West Lake and the Huai River. Finally, also in Fengyang we find a clear skew from the cardinal orientation, in this case 4° east of north, fully compatible with magnetic declination at the time of planning.

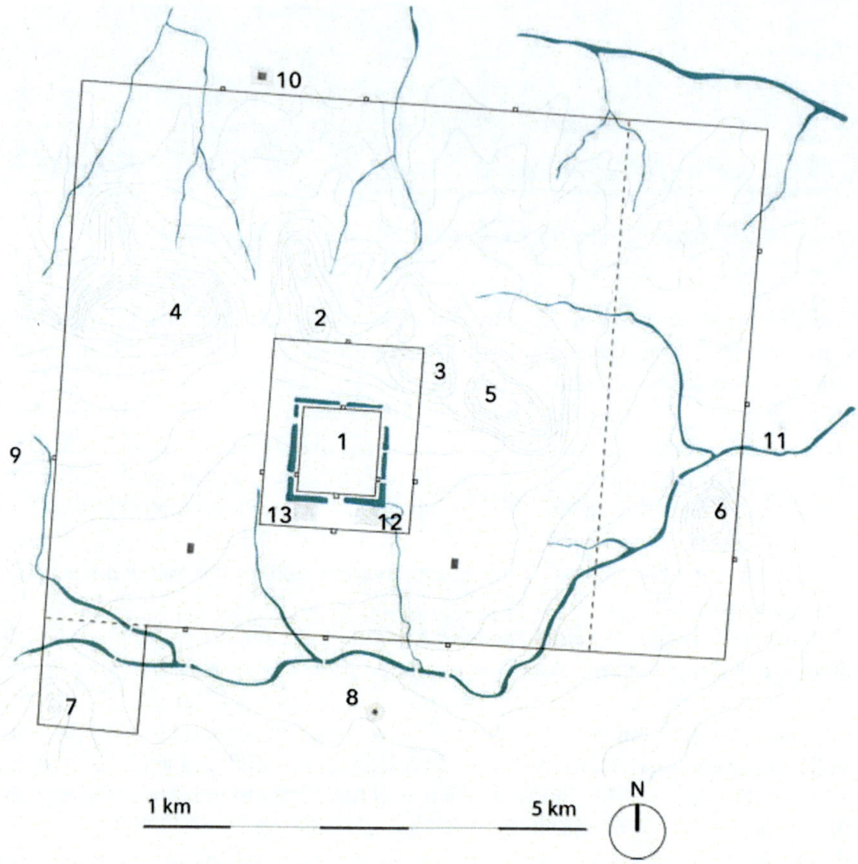

Fig. 10.3 Plan of Mingzhong. The original square is shaded. **1** Forbidden City, **2** Phoenix hill, **3** Longevity hill, **4** Sun hill, **5** Moon hill, **6** Dushan hill, **7** Fenghuangzui hill, **8** Round Altar of Heaven, **9** Altar of the Moon Temple, **10** Square Altar of Earth, **11** Altar of the Sun, **12** Imperial Ancestral Temple, **13** Altar of Land and Grain (Drawing by the authors)

10.6 Ming Beijing

The Ming urban program in Beijing brought to the extreme completion the ideas of "cosmic" city landscape, symbolically joining a series of cognitive aspects related to the Heavens and the celestial cycles as well as to feng shui. We have studied in depth all such aspects in a recent paper (Baratta and Magli 2021) to which the reader is referred to for an exhaustive treatment.

10.6.1 The Urban Plan

The first Ming intervention involves the reduction of the size of the urban plant by moving the northern limit toward the south of a few kilometers; at the same time the southern limit was somewhat moved, with the intention of making the Forbidden City the ideal center of the capital. It was only with the expansion works of 1550 that, with the addition of a further area to the south, the urban layout assumed the well-known double rectangle shape.

The Imperial City is anyhow located slightly south of the center. The layout, 2500 m from east to west and 2750 from north to south, features four doors, one on each side. In the area to the north, an artificial relief had been made with the excavated material of the moat surrounding the Forbidden City, located in the center along the meridian axis.

The name of the Forbidden City was actually *Zijincheng*, or "Purple Forbidden City": the reference is to the region of the circumpolar sky (the "celestial Purple Enclosure"), whose stars were associated, according to Chinese tradition, to the emperor and his entourage (Needham and Ling 1959).

The Purple Forbidden City was conceived as the earthly counterpart of the celestial one. Its construction took into account the experiences made in the planning of the two previous capitals. Its layout, approximately 1,000 m north–south by 750 m east–west, features an extremely rigorous geometry. The wall enclosure, surrounded by a moat, is characterized by the presence of towers at the corners and a door on each side. A east-wet division identifies a public area to the south and one dedicated to the emperor to the north; while the meridian direction was divided into 3 sections, of which the central one contained the path for the exclusive use of the ruler. In the southern area flowed the "Golden Water" River (an artificial canal) with its characteristic sinuous shape, which can be crossed, coming from the Meridian Gate, through five bridges. Continuing toward the north stands the Gate of Supreme Harmony which hides behind it a second square that prepares for the fulcrum of the complex, three consequential pavilions, the Hall of Supreme Harmony, the Hall of Central Harmony, and the Hall of Preserving Harmony, positioned on a raised terrace.

The design of the Forbidden City influenced the entire layout of Beijing: in fact, the main axis of the imperial city, after crossing the Tiananmen Gate, becomes the central axis of the entire city (Krupp 1983). Going south along this axis, there are the Altar of Land and Grain and the Ancestral Temple and, proceeding further south after Tianamen, the two huge complexes of the Temple of Agriculture and the Temple of Heaven. It is with the emperor Ming Jiajing (1521–1567) that the city system reaches a completion in its symbolic structure. The project's motivation may have stemmed from its adherence to Taoism, but its aim was likely to improve the Emperor's reputation. By constructing a number of buildings for the seasonal rites that would have been performed by the emperor personally, his position as the sole intermediary in charge of the connection with Heaven was undeniably strengthened. In particular, to the south of the Temple of Heaven and in axis with it, the Circular Mound Altar was added, where, on the occasion of the winter solstice, a very important ceremony of renewal took place (Krupp 1989). Three further temples were constructed to complete the symbolic system: the Temple of the Sun, the Temple of the Earth (also known as Fangze Temple), and the Temple of the Moon, which are situated east, north, and west of the Forbidden City, respectively (Fig. 10.4). The main entrance (and, by extension, the major direction of the temple) was distinguished by the presence of three stone-pillared doors. They were distinguished by the presence of a central open-air altar, which was surrounded by a fence with stone gates on either side.

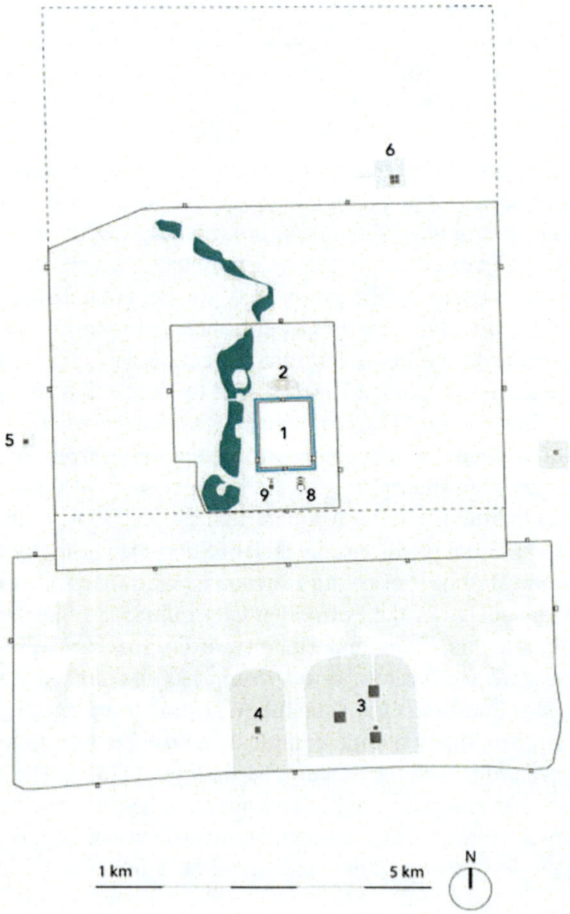

Fig. 10.4 Plan of Ming Beijing. The original borders the Yuan capital Dadu are shaded. **1** Forbidden City, **2** Jangshan hill, **3** Temple of Heaven, **4** Temple of Agriculture, **5** Temple of the Moon, **6** Temple of the Earth, **7** Temple of the Sun, **8** Imperial Ancestral Temple, **9** Altar of Land and Grain (Drawing by the authors)

10.6.2 Astronomy and Feng Shui of Ming Beijing

The ideas which inspired the projects of Hungwu and later of Jiajing appear to have been a complex combination of both feng shui schools, united with more ancient connections with the Heavens and the celestial cycles (Baratta and Magli 2021).

In brief, these ideas can be summarized as follows:

10.6.2.1 Compass Feng Shui

The orientation of the town's grid and of the Forbidden City was obtained through the use of a compass. We have indeed verified that while the original plan of Yuan Dadu followed in a rather strict manner the dictates of the Kaogong Ji, and was therefore oriented to the cardinal points, at the moment of planning Beijing the Ming architects used a magnetic orientation. This is visible through a small, but non negligible, deviation of 2.5° toward the west, which is in agreement with the magnetic declination error at the time of planning.

10.6.2.2 Form Feng Shui

There are various sources that study the location of the northern capital city analyzing it with reference to the theories of sacred geography and geomancy, an example is that of the "three Dragons" (Meyer 1991), according to which three mountain ranges could be identified that would ideally divide the entire country: Beijing, due to its location, would belong to the northernmost range. It is not easy to argue one or the other thesis, and thus whether it is a site that responds to the dictates of feng shui or refers to other cultural traditions; however, even the oldest sources referring to the subject were written after the site was chosen for the foundation of the capital Jin (Liu 1998). It should be noted that the capital area lacks the alleged typical characteristics that would make the site a "good" place according to the dictates of feng shui: in fact, they are all too far apart to be considered effective. Consequently, it is necessary to reflect on the "geomantic maps" that described the site of Beijing over time: these, either by exaggerating its morphological features or somehow falsifying the scale of representation, would have shown its auspiciousness (Yoon 2008).

Analyzing the specific context of the Forbidden City, the situation is quite different. Its design is certainly based on respect for the two main dictates of formal feng shui (i.e., hill to the north and meandering river to the south). As mentioned, both topographical elements were actually artificial, therefore, absent at the beginning of the project: they were designed and constructed to give the Forbidden City "a feng shui landscape where there is no feng shui landscape" (Magli 2020, p. 111). The hill to the north, known as Jangshan Hill, made from the excavated material of the moat, is more than 50 m high and looks like a five-peaked backdrop of the

Forbidden City, directly overlooking the area where the emperor's private gardens were located, while the River of Golden Water was obtained by using an existing canal connected to the lake area of the Western Park.

10.6.2.3 Astronomical and Topographical Alignments

Astronomy also played a role in symbolically linking the architectural designs of Ming Beijing. An example is the astronomical alignment that results when standing in the heart of the Forbidden City, in the center of the Hall of Supreme Harmony, and looking in the direction of the eastern gate: this, together with its western counterpart, is in fact not positioned along the median line of the inner wall but is slightly shifted southwards. The alignment is to the sunrise on the day of the winter solstice: as already mentioned, this is a day of great importance in Chinese imperial ritual (Krupp 1989), and the repetition of such a focus on this particular day of the year indeed reinforces its significance. Other astronomical alignments are those that were made during the phase of completion of the symbolic structure of the capital under the guidance of Emperor Jiajing. In particular, this is the location chosen for each of the new temples built. The center of the Temple of the Earth, north of the Forbidden City, is almost located in the same meridian that crosses the heart of the circular mound to the south (about 8 and a half kilometers away), while the parallel that passes through the center of the altar of the temple of the Sun, to the east, crosses the very heart of the Forbidden City and probably had to connect with the center of the Hall of Supreme Harmony, which is less than 150 m away; plus, the parallel on which the center of the Altar of the Temple of the Moon is located, to the west, crosses the Hall of Supreme Harmony with impressive precision.

10.7 Conclusions

The plans of all the three Ming capitals appears to have been inspired by a symbolism which, since the times of the first emperor, aimed at establishing an explicit connection between the design of the capital and the sky.

The key of this connection was in a person: the emperor. The capital was indeed the place of residence of the Mandate of Heaven, personified by the ruler, and consequently it was the pivot of the four quarters, the cosmic center. This tradition can be traced from the Han up to the Yuan dynasty, but in Ming capitals we observe a clear change and enlargement of focal point, from the sky in itself (and therefore, from cardinal and astronomical orientations) to the global geomantic properties of the architectural projects. In particular, we have shown that the planning of the imperial palaces was always based on magnetic orientation—and therefore, on ideas related to compass feng shui—while their siting in the landscape was always governed by form feng shui, either observing natural features (as in Nanjing and Mingzhong) or literally building the required ones ex novo (as in Beijing).

During the Ming dynasty, many other towns, some of which quite important (think, for instance, to Ming Xi'an), were founded (or re-founded) according to the imperial canons. In a sense therefore, the present research is only at the beginning, as we plan to apply our approach to a much wider sample of Ming towns in the near future.

References

Baratta NC, Magli G (2021) The role of astronomy and Feng Shui in the planning of Ming Beijing. Nexus Netw J 23:767–787. https://doi.org/10.1007/s00004-021-00555-y

Bruun O (2011) Fengshui in China geomantic divination between state orthodoxy and popular religion, NIAS—Nordic Institute of Asian Studies

Constable C, Korte M, Panovska S (2016) Persistent high paleosecular variation activity in southern hemisphere for at least 10,000 years. Earth Planet Sci Lett 453:78–86

Chen H (2009) Comparison of three capitals planning system in the early Ming Dynasty. Proc Forbidden City Soc China 5:233–244

Chen W (2019) On the construction and abolition of the central capital of the Ming Dynasty and its impact. J Jiangnan Univ (Humanities & Social Sciences) 18(4):66–74

Korte M, Constable C, Donadini F (2011) CALS10k. 1: A Holocene geomagnetic field model based on Archeo- and paleomagnetic data. 25th IUGG General Assembly, Melbourne, Australia

Krupp EC (1982) The Mandate of Heaven. Griffith Observer 46(6):8–17

Krupp EC (1983) Echoes of the ancient skies: The astronomy of lost civilizations. Harper & Row, New York

Krupp EC (1989) The cosmic temples of Old Beijing. In: Aveni AF (ed) World archaeoastronomy. Cambridge University Press, Cambridge, pp 65–75

Liu L (1998) The influence of Fengshui on the building of the city of Beijing in the Ming Dynasty. https://era.ed.ac.uk/handle/1842/28440

Magli G (2018) Archaeoastronomy: introduction to the science of stars and stones. Springer Verlag, New York

Magli G (2019) Astronomy and Feng Shui in the projects of the Tang, Ming and Qing royal mausoleums. A satellite imagery approach. Archaeological Research in Asia 17:98–108. https://doi.org/10.1016/j.ara.2018.10.004

Magli G (2020) Sacred landscapes of Imperial China. Springer Verlag, New York

Meyer JF (1991) The dragons of Tiananmen: Beijing as a sacred city. University of South Carolina

Needham J, Ling W (1956) History of scientific though. In: Science and civilisation in China, vol 2, p 361

Needham J, Ling W (1959) Mathematics and the sciences of the heavens and the earth. In: Science and civilisation in China, vol. 3. Cambridge University Press

Schinz A (1996) The magic square: history of Chinese city planning. Edition Axel Menges

Steinhardt N (1999) Chinese imperial city planning. University of Hawaii Press

Wheatley P (1975) The ancient Chinese city as a cosmological symbol. Athens Center of Ekistics, vol 39, no 232, pp 147–158. https://www.jstor.org/stable/43618477?seq=1

Xu Y (2019) Grids of Chinese ancient cities: Spatial planning tools for achieving social aims. Altralinea ed, Milano

Yang G (1999) A probe into the design ideas of Nanjing City Wall in Ming Dynasty (3):69–74

Yoon H (2008) Culture of Feng Shui in Korea: an exploration of East Asian geomancy. Lexington Books, London

Zhang Q (1985) The city planning of Nanjing in early period of Ming Dynasty. Journal of Nanjing Institute of Technology 3:113–123

Zhao R, Yao Y, Zhang Y (2018) A study on the architectural culture of drum tower in Ming Zhongdu. J Anhui Jianzhu Univ 26(5):39–46

Chapter 11
Archetypal Characters of Sacred Places and Cultural Landscape: Sansa, Korean Buddhist Mountain Monasteries Comparing Cistercian Architecture

Hee Sook Lee-Niinioja

Abstract A place expresses what is specific and local with a spirit to facilitate identity presence and actions. It emerges from the quality attributed to the place. When its spirit is recognizable, human's individual or shared perceptions of a specific landscape take place. The spirit of place bears the atmospheric quality of this specific landscape; the sense of place has four elements—(i) emotional: place attachment, (ii) cognitive: place identity, (iii) behavioural: place dependence, (iv) social: social bonding. Place identity can be explained through experience, memories and people's activity and is drawn up through reflections of these societal structures. Sansa is a collection of Buddhist Mountain monasteries distributed over the southern Korean peninsula (7–9 C), revealing universal characters through spatial arrangement and individual structures and objects. As a sanctuary, it sustained living centres of belief and practice, despite restriction, suppression or damage due to war and conflict. Sansa testifies to the historical development of Korean Buddhism and facilitated diverse Buddhist schools and popular beliefs. Its intangibility continues the traditions of coexisting contemplative, doctrinal studies of Seon Buddhism. This paper discusses the archetypal characters (landscape-people-architecture) that created the Sansa spiritual and cultural landscape and compares Cistercian sacred architectural settings in nature.

Keywords Sansa Mountain monasteries · Sense of places · Korean Buddhism · UNESCO World Heritage List of Korea · Cistercian architecture

11.1 Introduction

Within the World Heritage Convention, "Outstanding Universal Value" (OUV) denotes exceptional cultural and/or natural significance extending beyond national borders, which bears universal importance, locally and globally, as well as for present

H. S. Lee-Niinioja (✉)
Piritanaukio, 2B 18, 00150 Helsinki, Finland
e-mail: leeheesook@hotmail.com

R. P. B. Singh et al. (eds.), *Placemaking and Cultural Landscapes*,
Advances in Geographical and Environmental Sciences,
https://doi.org/10.1007/978-981-19-6274-5_11

and future generations of all humanity. A property must meet one or more UNESCO World Heritage six Criteria: (i) masterpiece, (ii) human values, (iii) testimony, (iv) typology, (v) land-use and (vi) associations.

Sansa (Fig. 11.1) was inscribed on the UNESCO World Heritage Lists in 2018 with Criterion (iii).

> Criterion (iii): Buddhism has a long history that has traversed a number of historical eras in the Korean Peninsula. The seven mountain monasteries—Tongdosa, Buseoksa, Bongjeongsa, Beopjusa, Magoksa, Seonamsa and Daeheungsa—offer a distinctively Korean instantiation of Buddhist monastic culture from the seventh century to the present day. These mountain monasteries are sacred places and provide an exceptional testimony to their long and continuing traditions of Buddhist spiritual practice.

The assimilation of Korean Buddhism is manifested through historic buildings, objects and documents, but its intangible aspects contain the continuing tradition of self-governing temple management, education, meditation and Seon Buddhism's doctrinal studies. At its culmination, the geographical characteristic of the Korean mountains led the Sansa temples to the supervision of five provinces and seven cities and counties.

The text for the Sansa World Heritage nomination related to OUV reads:

> (Integrity) Together the seven temples contain the elements necessary to express the Outstanding Universal Value of Korean Buddhist Mountain monasteries, including their mountain settings, well-preserved buildings for religious practice and daily living, worship halls and shrines, meditation areas, monastic academy spaces and dormitories for monks.

> (Authenticity) The authenticity … of the components for Buddhist spiritual practices and rituals … The architectural elements have been … maintained according to principles of repair and restoration, using traditional construction techniques … The religious traditions and functions of the Buddhist temples maintain a high degree of authenticity.

In connecting the Sansa World Heritage inscription texts with this paper's topic, it needs explanations of how Buddhism was first introduced externally into Korea, which subsequently developed syncretic Korean forms of Buddhism. They became represented within the Sansa in each local combination of landscape, people and architecture. The paper also draws some comparisons and contrasts to the situation in medieval Europe of the Cistercian Order. In other words, comparisons are pursued (i) between each temple of Sansa and (ii) between Sansa and the Cistercian sacred building.

11.2 Buddhism in Korea and Sansa Construction

Buddhism was introduced into Korea from China in the late fourth century CE. After the beginning of the seventh century, doctrinal Buddhism within Korea was characterized by four distinct sects. The Vinaya ('doctrine' or 'discipline') school prioritized interpretation and devotion to precepts. The Tiantai ('platform of the sky') school highlighted the specific teachings of the Lotus Sutra. The Huayan school

Fig. 11.1 Part of a poster for the Sansa inscription to UNESCO World Heritage (*Source* www.koreansasa.net, modified by Lee-Niinioja)

pursued Avatamsaka ('flower garland sutra') philosophy and the Bodhisattva path, while the Faxing school focused on the phenomenal appearance of the Dharma.

Although Unified Silla united the Three Kingdoms of Goguryeo, Baekje and Silla in the seventh century, Buddhism developed along diverse lines in religious philosophy. New schools were set up according to varying interpretations of scripture, doctrine and practice. The first mountain monasteries, such as Tongdosa (Vinaya school), Buseoksa and Bongjeongsa (Huayan school) and Beopjusa (Faxing school), had communities of monks with the laity. Following the rise of Chan Buddhism in China, a form spread to Korea, known as the Seon, and in response to that period's social changes, the Seon school reached throughout Korea in and after the ninth century. The Seon established Magoksa, Seonamsa and Daeheungsa.

One of the more critical individuals in elevating, interpreting and disseminating Buddhist thought within Korea was the distinguished monk Wonhyo (617–686). Through his commentaries on other Buddhist scriptures, he made the world of Buddhist thought available to Korea, and in turn, his work became known outside of Korea. Wonhyo played a pivotal role in bringing Buddhism to ordinary people. More temples were built in the major cities but also in rural provinces.

By the mid-seventh century, Buddhism in Korea had developed into a religion of the masses. Sansa was associated with sacred worship, public events and other spiritual practices as a group of multifunctional sites. What began as a trend of mountain sites had become the norm by the end of the seventh century. Adding to this complex, the Great Master Euisang (625–702) founded the Hwaeom (Avatamsaka) school and built Buseoksa. In the following century, Jinpyo led the Beopsang school in constructing Beopjusa.

The ninth century was a time of turmoil and change, during which Korea was receptive to the import and spread of the Seon (Chan) school, focusing on self-realization through meditation. Regional reformist movements founded the Seon sacred centres of Magoksa, Seonamsa and Daeheungsa for philosophical training and sacred places of worship.

Unified Silla (668) dissipated and brought on a later Three Kingdoms period. The Goryeo Dynasty (918–1392) (from whose name 'Korea' came) emerged to unify the whole country, adopting a centralized governing system and a society based on Confucian and Buddhist ideals. Goryeo Buddhism pursued the harmony of the Seon (meditative) and Gyo (doctrinal) schools. City temples facilitated centres of public religious practice where the state conducted official Buddhist ceremonies, and people from different classes gathered for everyday religious activity. By contrast, the mountain temples remained as venerated institutional retreats for spiritual practice and scriptural study.

The subsequent Joseon Dynasty (1392–1910) government preferred a Neo-Confucian path and suppressed Buddhism. Many schools died out, the Seon and Gyo schools being the exception. Due to the personal beliefs of members of the Royal Family, some Buddhist temples also survived (Fig. 11.2). Although the national role of Buddhism had been diminished, Buddhist monks rose to help to overcome the national crisis during the Japanese invasions of Korea (1592–1598).

The contribution of Buddhist adherents and institutions created a general revival of Buddhism within Korea. The restoration of buildings damaged by invasions gave a platform for developing Korean Buddhist architectural masterpieces. Temples played as the centres of widespread religious practice by performing rites and ceremonies attended by ordinary believers. In the wake of national grieving, Buddhist rituals to appease the souls of the dead became a widespread part of the country's recovery. Added to the syncretism, shrines assimilating native beliefs and commemorating the dead were in popular demand.

Throughout these periods, Sansa was a nursery for Korean Buddhism. It cradled an entire family and encompassed an accumulation of Sakyamuni, Maitreya and Amitabha Buddhist beliefs.

Fig. 11.2 Preparing Lotus Lantern Festival for the Buddha's birthday (lunar calendar: 8 April) at Jogyesa (1395), Seoul (*Source* Lee-Niinioja)

11.3 The Uniqueness of Sansa

Sansa is a sacred self-governing space where monks perform their daily lives and spiritual devotions. It also provides a communal space for public worship. The ideal of Sansa architecture is to blend with the natural landscape.

Typical settings for Sansa architecture are found amid forests, alongside streams and at the base of gentle mountain slopes. With an open structure, the buildings sit on low mounds reducing any interruption of the natural landscape. Three types of Sansa locations, classed by terrain and environment, are (i) valley bottom, (ii) slope and (iii) streamside. Sansa sites are typically located on sloping land where the viewpoint rises higher as the viewer moves from the front gate to the main hall, broadening the view along the way. The buildings cluster in defined interconnected areas with minimal change to the natural terrain. The fundamental spatial arrangement of the temples places major buildings around an earth-covered yard.

Fig. 11.3 Eungjindang shrine and temple complex around Seonamsa madang (*Source* Steve46814)

Today, the seven restored temples within the Sansa World Heritage area are reconstructions from the first half of the seventeenth century. As Sansa grew, its typical architecture changed. The configuration of a square-shaped yard (*madang*, Fig. 11.3), with the main Buddha Hall on one side, a pavilion serving as an entryway on the opposite side, and dormitories on either side of the main hall, became the standard form. The rites and assemblies for the souls of the war dead are held in the yard, where a hanging scroll painting of the Buddha is adorned. Wooden buildings of diverse forms embody religious values, while the open yard underscores the dignified beauty of the main Buddha Hall and the archetypal Korean Buddhism.

11.3.1 Tongdosa Temple

Tongdosa (Temple of the Mastery of Truth, Fig. 11.4) was built in 646 by Master Jajang, who first introduced the Vinaya school to Korea. The temple is located on a flat terrain by Yangsancheon Stream at the foot of Mt. Yeongchuksan. It has served as the centre of Vinaya Buddhism in Korea. The area embracing the Hall of the Great Hero and Diamond Precept Platform was constructed first. Subsequent construction prolonged the temple eastward following the stream, and by the seventeenth century, its scale and appearance took their present form.

Within the temple's overall plan, different buildings facilitate the varied activities of monks and lay believers. The Diamond Precept Platform enshrines relics (sarira) of the Buddha, and the Hall of the Great Hero focuses on the belief and worship of the historical Buddha. The Platform and the Hall are both contained with spaces for public religious practice. Buddhism's absorption of indigenous beliefs is incorporated into the Shrine of the Three Deities and buildings commemorating past patriarchs. Tongdosa is also known as the "Buddha Jewel Temple" and has served both as a spiritual place and a monastic teaching centre.

Fig. 11.4 Collection of four temples (from upper left: Tongdosa, Bongjeongsa, Beopjusa, Daeheungsa) (*Source* Steve46814, combined by Lee-Niinioja)

11.3.2 Buseoksa Temple

Buseoksa (Floating Rock Temple) was founded in 676 by Great Master Uisang. It has been the centre of worship for Amitabha. Located on a relatively steep sloping site at the foot of Mt. Bonghwangsan, it is not near a stream like the other temples. At the top of the site, the Hall of Patriarchs and the Hall of Infinite Life were built first, and additional buildings followed. Although the temple gained its present form in the ninth century, it had minor damage during the Japanese invasions. Spaces for religious activities include the Hall of Infinite Life, Ksitigarbha Hall and others. In the Shrine of the Three Deities, syncretic Buddhism assimilating indigenous beliefs is visible. Moreover, the Hall of Patriarchs commemorates Uisang, while the Shrine of Seonmyo tributes a young woman Seonmyo who assisted Uisang to establish the temple.

Buseoksa is devoted to the belief of Amitabha. The image of Amitabha Buddha is enshrined in the main hall, the Hall of Infinite Life, and sits facing east. The doctrine of Amitabha is reflected in the temple layout. The Hall of Infinite Life built in the thirteenth century is one of the oldest buildings in Korea, where the development of East Asian wooden architecture is apparent as distinctiveness. It assimilates southern and northern styles derived from China. In the open yard, a stone lantern enhances the sacredness of the main hall.

11.3.3 Bongjeongsa Temple

Bongjeongsa (Phoenix Dwelling Temple, Fig. 11.4) was constructed in 677 by Neungin, whose master was Great Master Uisang. Since that time, it has been the leading temple of coexisting Buddhist traditions Sakyamuni and Amitabha. It is unique in Korean Buddhism. The temple complex is located on a sloping site at the foot of Mt. Cheondeungsan by the side of Songyacheon Stream. The Paradise Hall was erected first towards the east of the complex, and later in the thirteenth century, the Hall of the Great Hero was added to become the central place. The temple avoided much damage during the Japanese invasions. Its current form was consolidated after constructing the Assembly of the Infinite Sea and the Pavilion of Eternity in the seventeenth century.

The monastery facilitates a full range of monastic and lay activities. Religious activities are performed in the Hall of the Great Hero and the Shrine of Three Deities, exemplifying Korean Buddhism's integrated characteristic space for public religious activity. Areas for more private practice include the Seon Meditation Hall, while the Avatamsaka Lecture Hall is for studying sutras. Daily life can be observed in the Old Golden Hall. Bongjeongsa is known for its cultivation of vegetable gardens and communal food self-sufficiency. The temple's authentic spirit continues a tradition of crop cultivation as a spiritual practice.

11.3.4 Beopjusa Temple

Beopjusa (Dharma Abode Temple, Fig. 11.4) was erected in the mid-eighth century by Vinaya Master Jinpyo and his disciple Yeongsim. Jinpyo was the founder of the Korean Beopsang school, characterized by repentance and belief in Maitreya (the future Buddha). Since the temple's post-invasion restoration, it has served as the centre for Maitreya. Situated on a broad, flat site at the foot of Mt. Songnisan, in the upstream part of Dalcheon Stream, the temple grew as it thrived. The Hall of Parijata, enshrining Maitreya, was the first structure built. The expansion was northward along the stream, and the temple gained its current scale and composition in the seventeenth century.

Outdoor services are held in a large open yard. The Hall of Eight Pictures, Korea's only extant preserved wooden pagoda, facilitates the full range of Buddhist activities. Religious spaces within the temple complex include buildings such as the Treasure Hall of the Great Hero and Hall of Eight Pictures. The Shrine of the Three Deities illustrates Korean Buddhist syncretism. Here, the Sarira Pavilion retains the true relics (sarira) of the historical Buddha. The spirit tablet of a concubine of King Yeongjo is enshrined within the Vow Shrine of Lady Seonhui (1765). It is sectioned off with enclosure walls with auspicious symbols and stars verifying the religious life of the royal court. The shrine has a three-portal front gate in the Confucian style.

11.3.5 Magoksa Temple

Magoksa (Hemp Valley Temple) was founded in the latter half of the ninth century when Seon school temples were more widespread, becoming a stronghold of Sakyamuni worship. Located on a flat site at the foot of Mt. Taehwasan by a bend of Magoksacheon Stream, the temple complex was expanded southward, reaching its present scale with two main areas on either side of the stream. In the northern area stands the Treasure Hall of the Great Hero, and in the southern, the Vulture Peak Hall. After damages from the Japanese invasions, the temple underwent a principal restoration acquiring its current spatial composition in the eighteenth century.

The monastery arranges a series of activities, providing religious activity space in buildings such as the Treasure Hall of the Great Hero and the Treasure Hall of Great Light. Magoksa was famed for its monk-artists, and the continuous tea ceremony has been observed to this day. The original form of the storehouse reflects the life of a traditional monk.

11.3.6 Seonamsa Temple

Seonamsa (Immortal's Rock Temple, Fig. 11.3) is a temple of the Seon school, built in the latter half of the ninth century. It exemplifies a Korean mountain temple made of multiple areas. Situated on a flat site at the foot of Mt. Jogyesan along the Seonamsacheon Stream, the temple was expanded northward to its present scale. The Hall of the Great Hero and two separate three-story stone pagodas were built first. Following that, the temple grounds were expanded further to the north. The complex was destroyed in the Japanese invasion and subsequently suffered more fires. Its current spatial composition dates from the nineteenth century.

Spaces for religious activities include the Hall of the Great Hero and Ksitigarbha Hall. A statue of Sakyamuni is featured in the Hall of the Great Hero, and Avalokitesvara Hall follows a T-shape floor plan commissioned by the Royal Family for their natal prayers to beget a prince. The Shrine of Mountain Spirit displays a syncretic architectural style. More private spiritual practices occur at the Seon training centre, the Hall of Bodhidharma, the House of Preaching on Meditation and the House of Sword Seekers for monk education. The Outhouse is the oldest extant form of a traditional toilet. The temple's fame for cultivating tea fields includes Seon's spiritual practices, such as the tea ceremony.

11.3.7 Daeheungsa Temple

Daeheungsa (Great Rise Temple, Fig. 11.4) focuses on Seon Buddhism from the latter half of the ninth century. It is a mountain riverside monastery with multiple discrete areas located on both sides of the Daeheungsacheon Stream at the bottom of Mt. Duryunsan. The first construction took place on the northern site around the Treasure Hall of the Great Hall. More buildings were added across the stream towards the south until the temple acquired its current composition in the nineteenth century. Monastery such as the Treasure Hall of the Great Hero facilitates all religious activities and performances. An example of architecture reflecting indigenous beliefs and Buddhism is the Shrine of Mountain Spirits.

The Confucian-oriented Joseon dynasty allowed the construction of Pyochungsa in honour of Great Master Seosan Daesa (1520–1604) and his disciple, who joined in defending the nation during the Japanese invasions. It was nation-protecting Buddhism and was a unique element at Daeheungsa. State officials held Confucian style rites at the shrine as well as an annual tribute to the Great Master Seosan. Pyochungsa testifies to the fusion of belief in the power of Buddhism and Confucianism together to protect the nation.

11.4 Comparison Between the Seven Sansa Temples

According to the comparison (Table 11.1), the seven Sansa temples began between the seventh and tenth centuries. Beyond the common underlying feature of primarily Korean Buddhist temples, their regions, spatial arrangements, affiliated schools and beliefs are dissimilar. Moreover, each temple has two Korean cultural property designations, either 'National Treasure' or 'Treasure'. This also applies to cultural heritage artefacts regardless of their numbers. A critical commonality among all of the temples is (i) a natural streamside setting, (ii) distinguished monks and (iii) masterpiece temple architecture. They comprise the archetypal characters of Sansa, integrating each other into a syncretic harmonious whole.

11.5 The Archetypal Characters in Formulating Sacred Sansa

Sansa is representative of the development and propagation of Mahayana Buddhism in East Asia. It reflects the integrated nature of Korean Buddhism through location, spatial composition and spaces for belief and spiritual practice. Each temple embodies a different type of monastery, but it has sustained monk communities (sangha), which have continued since their inception. These three archetypal characteristics are always present: riverside natural setting, community and exceptional architecture.

Table 11.1 Sansa, Buddhist Mountain monasteries in Korea

Temple/ Built	Tongdosa /646	Buseoksa /676	Bongjeongsa /677	Beopjusa /mid 8C
Region	Yangsan, Gyeongsang nam-do	Yeongju, Gyeongsang buk-do	Andong, Gyeongsang buk-do	Boeun, Chungcheong buk-do
Spatial arrangement	Multiple yards, valley bottom type	Single axis, slope type	Two axes, slope type	One yard, valley bottom type
School[a]	Vinaya	Avatamsaka	Avatamsaka	Dharma Character
Belief	Comprehensive[b]	Amitabha	Amitabha, Sakyamuni	Maitreya, Sakyamuni
Cultural property	NT: Daeungjeon, Geumgang Gydean	NT: Muryangsujeon T: Josadang	NT: Geungnakjeon, Daeungjeon	NT: Palsangjeon T: Daeungbojeon
Cultural heritage artefact	NT: 1 T: 15	NT: 5 T: 6	NT. 2 T: 6	NT: 3 T: 13
Temple/ Built this whole row should have a dark blue background like the beginning of this table.	Magoksa/ 2.half 9C this whole row should have a dark blue background like the beginning of this table	Seonamsa/ 2.half 9C this whole row should have a dark blue background like the beginning of this table	Daeheungsa/ 2.half 9C this whole row should have a dark blue background like the beginning of this table	this whole row should have a dark blue background like the beginning of this table
Region	Gongju, Chungcheong nam-do	Suncheon, Jeolla nam-do	Haenam, Jeolla nam-do	
Spatial arrangement	One yard, streamside type	Multiple yards, valley bottom type	Multiple yards, streamside type	
School[a]	Seon	Seon	Seon	
Belief	Sakyamuni, Avatamsaka	Sakyamuni	Sakyamuni, Nation-protecting	
Cultural property	T: Daegwangbojeon, Five-story Stone Pagoda	T: East and West Three-story Stone Pagodas, Daeungjeon	T: Three-story Stone Pagoda, Cheonbuljeon	
Cultural heritage artefact	T: 5	T: 14	NT: 1 T: 9	

Data source Part of Table 1 from https://whc.unesco.org/en/list/1562 (Sansa, Buddhist Mountain Monasteries in Korea) and rearranged by Lee-Niinioja

NT National Treasure. *T* Treasure

[a]The school of Buddhism at each temple was maintained (7-14C) but disappeared after the fifteenth century

[b]The beliefs practised at Tongdosa Temple include all indigenous Korean beliefs

(Landscape) Through the location and natural environment, Sansa articulates its religious values to harmonize with nature. The varied arrangements of buildings integrated with the terrain symbolize the principle of human coexistence with nature. Sansa generates a characteristic cultural heritage where human architecture coordinates with the natural landscape.

(People) The sustainability of the Buddhist community is primarily maintained by the daily life and spiritual practice of the resident monks. Extending the sustainability into the broader community are places for training, education and worship in religious activities and Buddhist ceremony performances by the monastic and lay community together, culminating in the participation and support of the Royal Family.

(Architecture) The buildings at each temple site manifest the core beliefs of its adherents. Tongdosa follows the Vinaya school of thought, the spirit embodied in the Diamond Precept Platform and the Hall of the Great Hero. Buseoksa's terraced layout reflects spiritual practice according to the Avatamsaka Sutra. In Bongjeongsa, the two-open-yard configuration for two main halls serves for both worships of Sakyamuni and Amitabha. With a statue of Maitreya, Beopjusa is a temple layout based on the Dharma Characteristics school. The Treasure Hall of the Great Hero at Magoksa and the Hall of the Great Hero at Seonamsa are structured to dedicate the belief of Sakyamuni. Pyochungsa Shrine at Daeheungsa reflects Buddhism as national protection.

11.6 The New Cistercian Order in the European Medieval Christian World

In 1098, Abbot Robert of Molesme Abbey and twenty-one of his monks founded a new monastery in the south of Dijon, France, to seek to live in stricter isolation following *The Rule of St. Benedict* (written in 516 by Benedict of Nursia). The new settlement's name was *Novum monasterium* (Latin for 'new monastery'), leaving the older tradition. A year later, Abbot Robert was ordered by the Pope to return to Molesme and was succeeded as abbot by Alberich. Soon, the rigour and reputation of the new monastery attracted the donations of local people and nobility. It led to the expansion of the abbey and its daughter institutions beyond Burgundy, the order becoming known by the original place name "Cistercium" (*Citeaux* in French).

11.6.1 Isolated Natural Settings Near Water

The Cistercians' reclamation of new land was not primarily their objective but rather a result of their pursuit of solitude. This philosophy encouraged them to select depopulated or forest areas near water. Cistercian monasticism was built upon the ethos of an older hermitage tradition that customarily stressed seclusion. The abbeys kept the turbulent world at a distance. Moreover, their locations, far from other habitations, brought significant consequences. Unlike other traditional monasteries of the period, they did not encourage the growth of settlements in their proximity; instead, they adopted a system of lay brothers (*Conversi*) for monastery labour (Leroux-Dhuys 1998).

The foundation of a new Cistercian abbey required religious devotion and twelve monks. Careful planning was also necessary to keep monks and their abbot away from severe weather during the first winter. It was ideal to have an aristocratic patron who could provide land, building materials and living essentials. Moreover, the patron had to arrange the clearing of the forest and the construction of the first buildings. Afterwards, the founder abbey would send an experienced monk to consult with the patron on the new monastery plans. Although the monks took responsibility for the operation, existing sources tell of construction work carried out by a professional building team. The monks and lay brothers restricted themselves to management and organization.

As a consequence of employing construction experts from diverse lands, there developed corresponding regional differences in Cistercian buildings. The Order's motherhouse handed down standard specifications and plans for church grounds and the layout of the monastic complex. However, these plans' execution varied according to the limitations of on-site materials, regional masonry techniques, building practices and architectural styles.

Between the first approval of a Cistercian monastery by Pope Calixtus II (1119) and Saint Bernard's death (1153), the Cistercian Order had built ten monasteries. Thirty years later, the Order had grown exponentially to 351 abbeys. Its political influence in Europe extended beyond the Order through the prestige of Saint Bernard, the voice of Cistercian monasticism.

11.6.2 Saint Bernard of Clairvaux (1090–1153)

Bernard of Clairvaux practised an unequalled religious and political power in the Western Christian Church for almost thirty years. As a man of God inspired by the principles of Gregorian reform, he was a charismatic abbot who counselled clerics, kings and princes intending to shelter Western Europe under a single overarching church. Miraculous legends within the Catholic hagiography about Saint Bernard obscure his pure virtues, but he should be undoubtedly credited for facilitating and guiding others. An early persuasive public act was his conversion of his brothers and closest relatives to enter the Citeaux Abbey (1098) with him.

Bernard's inspiration was to put his conduct under the flag of purity, invoked by the monastic reform movement and the Order of Citeaux. In his writing *On the Steps of Humility and Pride* (c. 1120), Bernard revealed the psychological tension between his wish to refuse the world through humility and an ambition to work in the world through charity. He called for a stricter return to *the Rule of St. Benedict*, whose spiritual motto composes "peace" (*pax*) and "pray and work" (*ora et labora*) besides the avoidance of luxury. The Cistercians' strict asceticism, emphasizing poverty, work and silence, brought a new type of church architecture combining austerity with the purity of line, recapitulated in the high standards of masonry.

11.6.3 Cistercian Architecture in Simplicity

Bernard of Clairvaux initiated a Cistercian building campaign when he laid the foundation stone of the abbey church and monastery at Clairvaux II in 1134. His philosophy of unity in architecture testified to relatively isolated and hidden Cistercian abbeys from the mainstream communities of his days. The Cistercian monasteries as an architectural type followed similar formal and functional schema across Europe. Approximately, 750 abbeys and buildings illustrate in their standard building and landscape schemes the pre-eminence of Bernard of Clairvaux's ideas—simplicity. They bore the imprint of controls over the work.

The Cistercians' main concern was to build a physical environment suitable to support their balance of monastic and communal life. The architectural programme interpreted *The Rule of St. Benedict* into spatial terms and was articulated further in the Citeaux text *Capitula* (IX, 6–7) about the abbey building. The monks daily lived out values of asceticism, poverty and pursuit of spiritual reflection, which manifested such values through the architecture. In other words, architecture needed to favour the growth of a monastic life closer to perfection.

The archetypal principles of Citeaux were distributed by St. Bernard in letters to every abbey of the Order. As embodied in their architecture, Cistercians preferred solutions over pretensions, simpler over superfluous. Cistercian buildings project a feeling of endurance, sustaining spirit over time. Accordingly, Bernard stirred controversy between Citeaux and Cluny over these principles. His famous writing *Apologia* (1125), sent to Abbot William of St. Thierry, stigmatized the Cluny monastic as overly concerned with architecture and ornamentation. Nothing, according to Bernard, should distract the eye and spirit from the contemplation of God. His thoughtful preaching word also placed the welfare of the poor in the care of the monasteries as being a higher priority than aesthetic obsession in the early twelfth century (Rudolph 1989).

> Oh vanity of vanities, yet no more vain than insane! The church is resplendent in her walls and wanting in her poor. She dresses her stones in gold and lets her sons go naked. The eyes of the rich are fed at the expense of the indigent. The curious find something to amuse them and the needy find nothing to sustain them. (St Bernard's *Apologia* translated by David Burr)

According to the Order's first chronicle, the *Exodium parvum* (pre-1119), Cistercian buildings prohibited any indications of arrogance and luxury. The General Chapter resolution (1123–1124) expressed the ban on sculptures and images except for painted crosses. The Cistercians felt that sculptures and paintings evoked trouble and distracted the meditations of the monks, threatening their austerity and contemplative solemnity. High stone towers, considered by the Cistercians to be the archetype of arrogance, were forbidden in 1157. The elimination of figural sculptures on buildings, wall paintings, altar tables with reliefs or images of saints and stained glass led to the Cistercians' underlining the architecture inside the monastery.

> None of our houses is to be built in cities, in castles or villages; but in places remote from the conversation of men. Let there be no towers of stone for bells, nor of wood of an immoderate height, which are unsuited to the simplicity of the order. (Brakspear 1910, p. 1)

The austere architecture of the Cistercian monks represents the physical form of moral and aesthetic ideals. It flourished at various times in the history of Western Christian religious communities.

11.6.4 Tintern Abbey: Sacred-Industrial-Tourist Place

Tintern Abbey was built in 1131 adjacent to Tintern village in Monmouthshire, on the Welsh bank of the Wye River (Fig. 11.5). It was the first Cistercian foundation in Wales and the second in Britain. The abbey fell into ruin in the sixteenth century, and its current remains are a mixture of building works covering 400 years (1131–1536). Departing slightly from the stricter older architecture, Tintern Abbey has the Decorated Gothic style, contemporary with its period of development. It has a cruciform plan with an aisled nave, two chapels in each transept and a square-ended aisled chancel.

The Dissolution of the Monasteries under King Henry VIII ended the monastic life at Tintern. The adjacent area became the industrial wireworks in 1568, one of the first of its kind in Britain. By the mid-eighteenth century, the Tintern Abbey ruins had become a fashionable visiting spot, set among the romantic, picturesque Wye Valley.

Tintern's intangible value includes the collective memory of its Cistercian followers and the tourists who have visited subsequently. It does not replace the tangible loss of the ruins, but it can be traceable in "Lines Composed a Few Miles above Tintern Abbey, On Revisiting the Banks of the Wye during a Tour", written by the English Romantic poet William Wordsworth (1770–1850) on 13 July 1798.

> Five years have past; five summers, with the length
> Of five long winters! and again I hear
> These waters, rolling from their mountain-springs
> With a soft inland murmur.- Once again
> Do I behold these steep and lofty cliffs,
> That on a wild secluded scene impress
> Thoughts of more deep seclusion; and connect
> The landscape with the quiet of the sky.

Fig. 11.5 A print of the River Wye landing at Tintern by Edward Dayes, 1799, Whitworth Art Gallery (*Source* https://eng21911.files.wordpress.com/2010/10/sce01096jp2-re3-final.jpg)

It wonders whether Wordsworth wrote this poem, reflecting the glory of the Cistercian Order that had invoked the Second Crusade (1146–1149) in Europe in the name of God.

11.7 Buddhist Sansa and Cistercian Sacred Building in the Monastery Setting

The comparison (Table 11.2) between Sansa and Cistercian buildings shows a few similarities and dissimilarities. Of the commonalities, the first is the sacredness of their architecture operated by monks and lay believers since their establishments. The second is the location of the sacred buildings, close to nature, where water finds for their daily meditations and practical activities. The difference points to their building constructions and architectural ornamentation. The Buddhist temples tend to be occupied by many Buddha statues and other objects, giving a full engagement between the building and the believers. On the contrary, Cistercian buildings endow a sense of emptiness surrounded by transparent glasses and massive Romanesque columns, requiring the monk's souls towards the main altar and the heaven, where God and his only Son Jesus reside.

Table 11.2 Comparison of Sansa and Cistercian

Type	Sansa Monastery	Cistercian Monastery
Religion	Buddhism (various Orders)	Catholic Cistercian Order
Prosperous period	7-9C	12-13C
Location in nature	Mountain, near water	Forest, near water
People	Monks and lay people	Monks and lay people
Architecture	Wooden temple	Stone church
Ornamentation	Full with Buddha figures and decorative motifs	A few Christian symbols, empty atmosphere
Open space	Madang for the monastic and laity	Cloister for the monastic
Entrance	One Pillar Gate	Portal with sculpture

Source Lee-Niinioja

11.8 Cultural Landscape and Sensing Spirits of Places

Although the term "landscape" can propose from an enclosed piece of land with its specific character to the whole nature vista on the horizon, the term's context needs consideration in translation. Each landscape comprises various historical sites, associations and natural and human environs and is subject to interpretations and understandings.

Tuan (1977) suggests how humans create "place" within a landscape by naming, endowing it with meaning and then blending events and attitudes into an experiential whole. Olwig (2002) argues that different ways of conceiving landscapes lie between physical and cultural geography. Moreover, the landscape is a way of seeing, underlining the interaction between human beings and nature over time. It involves cultural aspects of social behaviour and values besides a physical setting. Daniels and Cosgrove (1988, p. 1) claim:

> A landscape is a cultural image, a pictorial way of representing, structuring or symbolizing surroundings. This is not to say that landscapes are immaterial. They may be represented in a variety of materials and on many surfaces … on the ground.

Palang and Paal (2002) summarize: the landscape is (i) a reflection of the relationship between man and nature, (ii) a manifestation of dynamic natural and social events and (iii) an interface between the present, the past and the future. Related to culture, Jones (1991) characterizes the cultural landscape as a subjective, perceivable part of the landscape, composed of symbols, meanings and understandings. It is immaterial as it relies on context and culture. Anything perceived as a cultural landscape by one person might not be the same for another.

Moreover, Cosgrove (1984) argues that each socio-economic formation generates its landscape with symbols, magic, policy and history. Therefore, according to Johnston et al. (2000), describing a cultural landscape is to grasp it as a mental layer of human knowledge laid over the physical surface. A landscape is a memory that contains remnants of land uses and recollections of power relations in the past. It bears a set of narratives transmitted through generations that decide the identity of a landscape or a place.

In terms of sacred places, the spirit of place (as described in the ICOMOS Quebec Declaration 2008) includes extraordinary quality, ambience and unique character. (i) The spirit of place composes tangible and intangible components, contributing to making place and endowing spirit. Intangible cultural heritage gives a wealthier and fuller meaning to monuments, sites, landscapes, routes and collections of objects. (ii) The spirit of place is complex and multiform; thus, multidisciplinary research teams and traditional practitioners can preserve and transmit the spirit of place. (iii) The spirit of place is constantly reconstructed to meet the needs for change and continuity of communities shared by different groups.

Norberg-Schulz (1980) believes that the spirit of the place evokes a sense of belonging and well-being. Its erosion could diminish their quality of life due to its deeper meanings and emotional connotations to the inhabitants. Relph (2009) notes that the sense of place is often described in terms of the character attributed to it. Our shared or individual perceptions of a specific landscape are identified as the spirit of place. A place expresses what is specific and local with a spirit to provide the presence of identity and its actions. The spirit of the place has the atmospheric quality of the particular landscape, and a sense of place emerges from this quality attributed to it. When this quality identifies the spirit, human's individual or shared perceptions of a specific landscape emerge.

Four critical components as characteristics of a sense of place are proposed by Kyle et al. (2004). (i) Place attachment: the poignant element tells emotional connections to a setting. (ii) Place identity: the cognitive element contains the individuals' values, attitudes and beliefs about settings that permit them to assert their identity. (iii) Place dependence: the behavioural element refers to a functional reliance on a setting in facilitating the attainment of goals. (iv) Social bonding: a social element proposes that settings become meaningful through social relationships and shared experiences.

Place identity can be clarified through experience, memories and people's activities. Its structure is formulated through reflections on these social structures. Butina-Watson and Bentley (2007) hold the opinion that psychologists classify the structure of place identity to contain human identity, feelings, memories, perceptions and cognition as being exposed through particular social structures. Geographers take an alternative with landscapes customized by human involvement that creates cultural landscapes with human identity.

11.9 Concluding Remarks

This paper aims at finding the archetypal characters behind Sansa as a sacred place. In doing so, the notion of living heritage needs to be understood. Sansa houses both tangible and intangible cultural traditions maintained through the ages. It embodies a syncretic belief system of Korean Buddhism, including religious practices and worship of temple patriarchs and monastics who contributed to protecting the nation. At the temples, monks carry on the tradition of meditation practice and retreats. The temples have tea fields, kitchen gardens and communal works to sustain daily life.

Sansa architecture displays open spaces and temple precincts in harmony with the natural borders of the local environment. The spatial focal point of each is an open yard (*madang*) adjacent to the main Buddha Hall of the monastery. As an intermediary space linking buildings with various functions, the *madang* is a venue for large-scale outdoor rituals and accommodates the monks and public crowds in a living community. It highlights and refines the magnificence and beauty of the main Buddha halls, and its surrounding buildings testify to the openness of Korean Buddhism yet regional characteristics in the spatial layout.

Moreover, the One Pillar Gate at each temple entrance is a symbolic boundary between the sacred and the mundane worlds. The participant in the public ceremony passes through a series of gates, sensing the particular temple site's overall sacredness and the individual's path to enlightenment. Why so? The geometry of the temple arrangement is based on a natural order process from the temple entrance to the mountain. The architecture embodies the progress of the individual worshipper and the historical progression of Korean philosophy and architecture. On the other hand, the Cistercian Order's primary purpose is spiritual; its economic benefits follow a consistently applied discipline. It is no surprise that Cistercian monasteries reflect a balance of spirit and nature and a living heritage until today.

The Sansa and Cistercian monasteries inherited traditions with the zealous cooperation of communities of monks. Both Orders encountered local environments with a series of spiritual principles, from which practical compromises emerged a cumulative architectural tradition. It was also successive generations of faithful adherents who have transmitted and reinvented these traditions, connecting the past-the present-the future.

Behind all of these syncretic fusions, however, in the author's view, it is a harmonious cooperation of the three archetypal characters (landscape, people and architecture) that have transcended the sacredness of Sansa and Cistercian monastic buildings. The three characters are the absolute hidden contributors to endow a sense of identity and continuity across time and space.

The author's final thought recollects a critique through correspondence with a landscape architect (March 2021). "Aboriginal people in Canada ... have taught me that the spirit of place exists first, continuously, and last. All the rest is a changeable illusion". If he is correct, how is the hierarchical order of the three archetypal characters in sacred buildings? Nature or landscape is the ultimate to be counted as the first.

References

Brakspear H (1910) Tintern Abbey: Monmouthshire. HMSO, London

Butina-Watson G, Bentley I (2007) Identity by design. Architectural Press, Oxford

Cosgrove DE (1984) Social formation and symbolic landscape. University of Wisconsin Press, Madison

Daniels S, Cosgrove DE (eds) (1988) Iconography and landscape—the iconography of landscapes: essays on the symbolic representation, design and use of past environments. Cambridge University Press, Cambridge

Johnston RJ, Gregory D, Pratt G, Watts M (2000) The dictionary of human geography (fourth edition). Blackwell Publishers

Jones M (1991) The elusive reality of landscape. Concepts and approaches in landscape research. Nor Geogr Tidskr 45(4):153–169

Kyle G, Mowen A, Tarrant M (2004) Linking place preferences with place meaning: an examination of the relationship between place motivation and place attachment. J Environ Psychol 24:439–454

Leroux-Dhuys J-F (1998) Cistercian abbeys: history and architecture. h. f. ullmann Publishing GmbH, Postdam

Norberg-Schulz C (1980) Genius loci: towards a phenomenology of architecture. Rizzoli

Olwig K (2002) Landscape, nature and the body politic: From Britain's renaissance to America's new world. University of Wisconsin Press, Madison

Palang H, Paal P (2002) Places gained and lost. In: Sarapik V, Tüür K, Laanemets M (eds) Koht ja Paik II. Place and Location. Eesti Kunstiakadeemia, pp 93−111

Relph E (2009) A pragmatic sense of place. Environ Architectural Phenomenol 20(3):24–31

Rudolph C (1989) The scholarship on bernard of Clairvaux's *Apologia*. Offprint from Clteaux Fase 40(1–4):69–111

Tuan YF (1977) Space and place: the perspective of experience. University of Minnesota Press, Minneapolis

Web Sources

Bernard of Clairvaux: *Apology*. https://sourcebooks.fordham.edu/source/bernard1.asp. Accessed 6 Apr 2022

https://www.poetryfoundation.org/poems/45527/lines-composed-a-few-miles-above-tintern-abbey-on-revisiting-the-banks-of-the-wye-during-a-tour-july-13-1798. Accessed 17 Nov 2021

ICOMOS Quebec Declaration 2018 https://www.icomos.org/images/DOCUMENTS/Charters/GA16_Quebec_Declaration_Final_EN.pdf. Accessed on 1 Feb 2021

Sansa, Buddhist Mountain Monasteries in Korea. https://whc.unesco.org/en/list/1562. Accessed 1 Feb 2021

Chapter 12
Placemaking of the Barotse Cultural Landscape, Zambia

Chilangwa Chaiwa

Abstract In the purview of defining cultural landscapes by the World Heritage Committee, an attempt is made here to appraise and evaluate the effective management of the cultural landscape, illustrated with the Barotse Cultural Landscape (BCL) based on data obtained from the local community, traditional rulers and heritage professionals. The BCL is a floodplain traversed by man-made canals and rich with biodiversity and cultural resources and thus presents a rare example of combined works of nature and of man that have been well-preserved and sustainably utilised through a traditional management system by the Lozi people. The Lozis have occupied the BCL for over 400 years and have had a special and vibrant relationship with their environment, thus resultantly developing a highly centralised and elaborate political and administrative system that has ensured the sustainable management of resources. This has been achieved through councillors (*manduna*) who have power over the use of land and the natural resources. Since Zambia's independence in 1964, emphasis has been laid on the conservation of the BCL supported by associated laws and traditional practices. These have reinforced the formation and strengthening of various institutions for the conservation and protection of natural resources and cultural heritage.

Keywords Cultural landscape · *Manduna* · Traditional · Placemaking · Sustainable

12.1 Introduction

The concept of placemaking was first introduced in the 1960s but only fully developed three decades later in the 1990s. The concept came about in an effort to address the need of having communities whose development is not only focused on the physical

C. Chaiwa (✉)
National Heritage Conservation Commission, P.O. Box 320013, Plot 9304 Dedan Kimathi Road, Lusaka, Zambia
e-mail: chilangwa.chaiwa@icomos.org

R. P. B. Singh et al. (eds.), *Placemaking and Cultural Landscapes*,
Advances in Geographical and Environmental Sciences,
https://doi.org/10.1007/978-981-19-6274-5_12

aspect but also on the affability, accessibility, utilisation, complacency and identity of the place. All this with the aim of creating places that meet the needs of the people. Authors such as Jane Jacobs and William H. Whyte initiated ideas which contributed to the formation of the concept. Placemaking is not a one-off event but a process with a multifaceted approach to the development and management of public spaces. Community participation is vital to its success and therefore people are required to be at its centre.

Most communities have a connection to the place in which they and their forefathers have lived. The place is a part of who they are, it is a part of their identity and in order to enhance the connection they have with the place as well as to improve their well-being, decisions to do with the place should involve community participation. Too often professionals have made decisions for communities without involving the members of the community themselves without realising how cardinal their participation is to the success of any project or programme. Traditional knowledge systems have been used successfully in various parts of Africa and still used by some ethnic groups. Unfortunately, the success of some practices has been eroded by the introduction of technological systems, migration and many other factors. On the flip side, some cultures have hindered some developments as some developers have failed to appreciate the need to involve the communities who have in turn opposed developments some of which may have been beneficial to the community.

A good example of an ethnic group that has not discarded its knowledge systems and traditions is the Lozi. The Lozi people of the Western part of Zambia in Central Africa are a people closely connected to their roots. They have a rich culture which has withstood the test of time and globalisation by remaining authentic. They are a proud people who are very passionate about their origin and traditions. A large expanse of the area inhabited by the Lozi is the Barotse Cultural Landscape. A beautiful area home to various fauna and flora as well, a remarkable landscape with traditional structures built by the locals. The rich culture of the landscape has been preserved for centuries and contributes to the attraction of the place.

12.2 The Barotse Cultural Landscape

The Barotse Cultural Landscape (BCL, see Fig. 12.1) of the Western Province in Zambia is known for its rich culture and cultural features, which are both man-made and natural. It is mostly a floodplain (see Fig. 12.2) which is located in the Western part of Zambia in Central Africa and is an area of rich biodiversity and cultural resources some of which have evolved over the years. According to the World Heritage Committee, Cultural landscapes are cultural properties and represent the 'combined works of nature and of man' designated in Article 1 of the World Heritage Convention (Operational Guidelines for the Implementation of the World Heritage Convention). The landscape has evolved over the years under the management of the Lozi people.

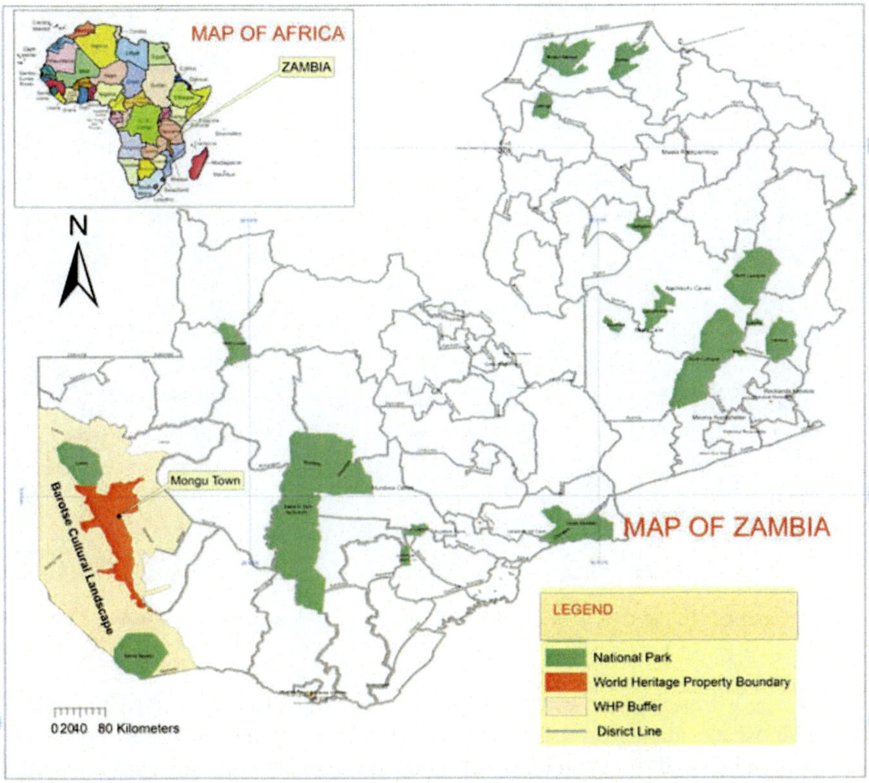

Fig. 12.1 Location map of Barotse Cultural Landscape World Heritage Property (*Credit* National Heritage Conservation Commission)

The BCL is one of the sites on Zambia's tentative list for United Nations Educational, Scientific and Cultural Organization (UNESCO) World Heritage status and is under consideration for inscription as a World Heritage site under the following criteria:

Criterion III. **Cultural Landscape that bears a unique and exceptional testimony to a living cultural tradition** or to a civilisation which is living or which has disappeared.

Criterion IV. Be an outstanding example of a type of building, architectural or technological ensemble or landscape which illustrates (a) significant stage(s) in human history and

Criterion VI. Be directly or tangibly associated with events or living traditions, with ideas, or with beliefs, with artistic and literary works of outstanding universal significance.

Fig. 12.2 Floodplain of the BCL (*Credit* National Heritage Conservation Commission)

The Lozi who have inhabited the BCL for over four centuries have lived in tandem with the landscape managing it in a sustainable manner under the control of councillors locally referred to as manduna (induna-singular) and headed by the Litunga (king). The manduna were and still are entrusted with power to manage different aspects of the landscape with any perpetrators of laid down regulations being made to account for them through the kutas which are the court system in place. It has also been a taboo to access certain resources such as fish in specified sacred areas and this is a practice still being observed.

The centralised political and administrative system which is termed as the Barotse Royal Establishment (BRE) has been instrumental in the effective and efficient management of the canals, fish, royal graves, sacred forests, game, sacred pools and sacred lakes. With the colonising of the nation in 1924 by the British and later being ruled by a republican government when Zambia gained its independence in 1964, the BRE has had to combine forces with these alternative authorities in the conservation and management of resources in Barotseland though various challenges have been faced affecting the effectiveness of the multifaceted approach to management.

12.3 Elements of the Barotse Cultural Landscape

The key elements of the Barotse Cultural Landscape are the royal burial sites, the Kuomboka Ceremony, historic canals, royal palaces, mounds, the Liuwa National Park and natural resources.

12.3.1 Royal Burial Sites

Twenty royal graves are currently found in the BCL, the location of each having been selected by the respective Litungas to be buried there (Zambia 2013). A caretaker known as Ng'omboti is selected by the Kuta to look after the royal grave and together with some selected villagers and their families, forms a village and lives around the grave once the Litunga has died. Whilst for many, the death of a person is the end of their role or usefulness to the community not so with the Lozis and their departed Litungas. The departed play an important role in the lives of the people and in the successful reign of the successive Litungas. The royal graves are of high spiritual value and this is reflected in the design of the burial grounds, modalities put in place for safeguarding of the graves and having a person charged specifically to look after the grave as well as act as mediator between the late kings and the people. The graves are also sacred sites where various rituals are undertaken.

Each grave is located on a mound, the tradition of not only burying but also living on mounds having been necessitated by the geography of the land which is predominantly a floodplain. The grave will be positioned at the centre and specific trees will be planted on the boundary (see Fig. 12.3). The grave itself will be enclosed by reeds which have spiked ends known as limbwata. As with other royal graves, they possess a significant feature in the form of two Y-shaped poles which mark the entrance into the inner chamber of the grave. The prominent royal graves to date are those of Mulambwa—son to Mwanawina I, Lubosi Lewanika (the uniter), Mwanawina I—son to Prince Mbanga and Mwanawina III—son to Lewanika, Sipopa—son to Sebeso, Ilute Yeta—son to Yeta III and Imwiko—son to Lewanika.

The deceased Litungas are believed to be providers of fertile land yielding good crop harvests. They are also believed to be a source of fruitful wombs for humans and animals and make it possible for fishermen to yield bountiful catch. For all these to be achieved, the Ng'omboti lays the role of a cult priest and has the responsibility of appeasing the spirits of the departed by conducting rituals. Important to note amongst the royal graves is the grave for the Queen mother Mbuyuwamwambwa from whom all the Lozi Kings originate. She was the first ruler and she established the Barotse Kingdom and therefore is held with high esteem by the Lozis. As per tradition, her grave is surrounded by a village which is known as the Makono Village.

Fig. 12.3 Lewanika I Royal Grave (*Credit* National Heritage Conservation Commission)

12.3.2 The Kuomboka Ceremony

Zambia has various traditional ceremonies most of which are celebrated annually. The ceremonies exhibit rituals, customs, oral history, spiritual and material culture. The Lozis of Barotseland hold the Kuomboka Ceremony (see Fig. 12.4) which brings together a number of ethnic groups. The Kuomboka which means to get out of water to dry ground is held when the Zambezi plains are flooded after the rains which occur from November to March (Zambia 2013). Even though villages in the plains are built on mounds, the flooding of the plains causes even the raised ground to be wet and as a consequence to be uncomfortable. The advent of the Kuomboka Ceremony came about due to the challenge presented by the flooding of the plains which affected the gardens and grazing. For the Lozi-speaking people, moving with their cattle to higher ground was a solution.

The transhumance which is preceded by the royal drums known as maoma which when played symbolises the commencement of the Kuomboka and summons the paddlers of the barges used to ferry people to higher ground. The main barge which is the one the Litunga travels on is known as the Nalikwanda. The Nalikwanda bears an elephant and the flag of the Litunga which are the official symbols of the kingship. The journey is made from Lealui to Limulunga and takes about seven hours. The Nalikwanda (for the people) is painted in black and white stripes which symbolises the people and spirituality, respectively. Its maintenance is done every year. All the ethnic groups along the Zambezi River which traverses the landscape contribute to its construction by bringing different parts. The Kwangwa supply different types

Fig. 12.4 The Kuomboka Ceremony (*Credit* zambianwatchdog)

of fibre, the Subiya supply the plants and the teak trees, the Mbunda supply the great masks and the Nkoya play procession the music. The transhumance is also a celebration characterised by the performance of dances and singing of songs. The Kufuluhela Ceremony although not publicised as much as the Kuomboka Ceremony is also very important to the Lozi as it also shows how they have managed to live positively in sync with the forces of nature. The Kufuluhela takes place when the water levels in the plains recede making it possible for the Litunga and his people to move back from Limulunga which is on higher ground to Lealui.

12.3.3 Historic Canals

The BCL is criss-crossed by a number of man-made canals which were constructed for irrigation, fisheries, navigation and drainage. The canals are divided into three main categories namely primary, secondary and tertiary canals. The Lozi used an organised system of co-operative labour known as the Makolo to construct the canals (Zambia 2013). The labour was also structured through organisational units called Lilalo which were headed by indunas. Sharpened wooden spades known as bishumo (singular—sishumo) were used for the construction and maintenance until 1887 when they started making iron spades known as li kalafu (singular—kalafu). However, over the years there has been a lapse in the effective maintenance of the canals causing their degradation. This has affected the effective functioning of the canals further hindering sustainable agriculture. Most of the canals have been affected by siltation and vegetation growth caused by the lack of periodic maintenance. This is further exacerbated by the use of manual labour which is not so effective.

The Government of the Republic of Zambia implemented the Pilot Program for Climate Resilience (PPCR) in 2013 for the Barotse and Kafue Basins. The programme was funded by the World Bank with NIRAS and BRL ingénierie as Consultants. A sub-project falling under this programme was the improved use of priority traditional canals. A Standards Norms and Procedures (SNP) manual for the management of all canals in the Barotse Basin was also prepared. The priority traditional canals selected for the project were Muoyowamo Canal, Lubitamei Canal, Fisheries Canal, Nebubela Canal, Nyombala Canal, Namitome Canal and Musiamo Canal.

12.3.4 Royal Palaces

There are three palace complexes in the BCL which are outlined below.

12.3.4.1 The Palace Complex at Lealui

This is the seat of power for the Litunga and the site from which the transhumance ceremony commences (Zambia 2013). Apart from the palace itself which is called the Kwandu (see Fig. 12.5), the complex is also made up of the house for the drummer (referred to as the limbetelo), the pavilion for visitors (Kashandi), the house for induction (Kamona), the private house for the king (Lilenge), the house for the queen (Nanda) and fences with pointed spikes (Imilombwe) which signify the presence of royalty. Built in 1890, the *Kwandu* is a grass-thatched frame building which has two rows of hardwood columns supporting the roof. The walls are painted white and the verandah floor is of a red traditional paint. In the village set-up, a house normally does not accommodate all the functions we find in a Western style house. Hence, even for the palace complex at Lealui, we have the various functions in different structures. For visitors at Lealui, there is the *kashandi* which was built in 1908. It is one single room enclosed by dwarf walls and surrounded by a raised verandah. The verandah has hardwood poles on its perimeter supporting a grass-thatched roof. The house of the Queen, the Nanda is unfortunately not the original one which was built in 1906 but was demolished. The current one is also unfortunately not a reconstruction but is of concrete walls with an iron roof.

12.3.4.2 Palace Complex at Limulunga

This is the seat of temporary power during the floods to which the Litunga moves.

Fig. 12.5 The Kwandu at Lealui (*Credit* National Heritage Conservation Commission)

Limulunga is on higher ground compared to Lealui and is the place to which the Litunga moves at the peak of the annual floods (Zambia 2013). The palace at Limulunga (see Fig. 12.6) was constructed by the British in the 1930s and becomes the seat of temporary power when the Litunga is there. Like Lealui, the Limulunga palace complex also has a Kashandi. Unlike the palace at Lealui, the one at Limulunga is not of a traditional/vernacular design and construction but rather European. This 'borrowed' Architecture as well as the British Royal attire the Litunga adorns when arriving at the palace symbolises the cultural change brought about by the interlinkages with foreign culture. The palace has French windows and double-pitched roofs with upper vents. It also has a large verandah. Both the Lealui and Limulunga Palace complexes have kutas to their east. The Kuta is part of the Lozi judicial system.

12.3.4.3 Palace Complex at Nalolo

The Nalolo is the seat of power for the female ruler and is comprised of the Kwandu (palace), Kusinga (the Litungas private abode during his visits), the limbetelo (drummers' house) and the Imilombwe (fence which is spiked). The kwandu was constructed from 1887 to 1889 although the palace complex was set up in the sixteenth century (Zambia 2013). The kwandu at Nalolo is the only palace which has double walls for extra protection.

Fig. 12.6 Limulunga Palace (*Credit* National Heritage Conservation Commission)

12.3.5 Mounds

It would not have been possible to undertake various activities in the BCL due to its being predominantly a floodplain. The people would not have been able to live there, bury their dead or cultivate their crops (except for those which grow in such conditions like rice) if the Lozis had not come up with a mechanism in which they could have their settlements raised above the floodplains (Zambia 2013). They came up with a process in which they piled up soil to form mounds on which they settled. This was a long process done over the centuries and maintenance works are still done by having more soil from the plains being heaped on the mounds to improve or increase their height above ground level. These man-made mounds are known as liuba. Natural mounds made by termites are also found in the floodplains and these are known as mazulu. Some of the mounds are now uninhabited with mango fruit trees being the only signs remaining of earlier settlements.

12.3.6 Liuwa National Park

The Liuwa National Park found in the Western Province of Zambia covers an area of 3,369 square kilometres. It originally served as hunting ground for the Litunga at that time, Lubosi Lewanika. Designated as a national park in 1972, the Liuwa National Park has both humans and animals inhabiting it and have been there for over a century. The Litunga installed people in the area in the 1880s when he had it designated as a protected area to be gamekeepers adhering to their traditional practices, taboos and regulations which are still in existence. When the park got designated as a national park, it became the responsibility of the Zambia Wildlife Authority (ZAWA) who are now known as the Department of National Parks and Wildlife falling under the Ministry of Tourism and Arts. Whilst management of the park went to the government in 1972, the BRE still maintained traditional rights. With challenges faced by ZAWA due to lack of capacity and resources, it was found necessary to enter into a partnership with Strichting African Parks Foundation (SAPF). This was done in 2004 and the agreement was signed by the BRE, SAPF and ZAWA. The agreement has helped to address issues of poaching, degraded infrastructure and low tourist numbers.

12.3.7 Natural Resources

Part of the BCL is a designated Ramsar site falling under the list of Wetlands of International Importance. The Convention on Wetlands is an intergovernmental treaty which was established in 1971 by UNESCO and currently has 171 contracting parties. The convention provides a framework for national action and international cooperation for the conservation and wise use of wetlands and their resources. Zambia's entry into force was on 28 December 1991 and it currently has 8 Ramsar sites. The BCL is home to aquatic vegetation, water birds, reptiles and amphibians and it is important for the Red lechwe (Kobus leche) and Sitatunga (Tragelaphus spekei). There are over 140 species of mammalian fauna found in the area typical of the southern savannas and the largest herd of migrating blue wildebeest in Zambia is found in the BCL. There is also a variety of bird life which is more abundant during the period when the landscape is inundated. The region also boasts of 34 reptile species and 14 amphibians. Amongst the forests found in the BCL, some are considered to be sacred. These include Moombo, Likonge, Mushituwambumu, Ikanungwe and Mutwiwambwa. Taboos and beliefs linked to the sacred forests restrict the locals from harvesting any forest resource unless authorised to do so by the Kuta. The sacred forests have been considered to be sources of construction materials for the royal barges and other wooden products needed for the Litunga and his people. The sacred forests were also the hunting ground for the Litunga.

12.3.8 *Other Elements of the Barotse Cultural Landscape*

Apart from the royal graves, there are also groves, lagoons, lakes and forests which are considered to be sacred. There are also coronation sites, royal barges and drums found in the landscape (Zambia 2013).

12.3.8.1 Sacred Forests

There are forests located in the Barotse Cultural Landscape which have taboos and beliefs associated to them from a link to the history of the Litunga and others who ruled the Lozi Kingdom. These forests were mainly considered as hunting ground for the Litunga and also as sources of construction materials for the royal barges and other wooden products required for the Litunga and his people.

12.3.8.2 Lagoons

There are a number of lagoons found in the landscape and each royal grave is deliberately located near a lagoon. The lagoons found near the royal graves serve as the abode of the barge and other objects used by the deceased Litunga. The items are sunk there after the demise of the Litunga. The Lozis believe that their kings do not die but merely leave their people and hence the need of the objects which were sunk near their graves. To avoid exposure of these items, dredging is prohibited. There are certain taboos linked to the lagoons which the Lozis believe in and follow. Hence, you find fishing and cultivation near the lagoons is prohibited during certain periods. Since it is believed that the Litungas do not die provision for their movements is catered for by connecting the royal lagoons to canals.

12.3.8.3 Nayuma Grove

Nayuma Grove is a sacred site found at Lealui near the canal where the Litungas' barges are kept. The grove is home to spiritual objects used in rituals which are conducted before the Kuomboka Ceremony. The ceremony takes place when the plains get flooded and the Litunga moves to the higher ground at Limulunga.

12.3.8.4 Sacred Lakes

The lakes which are termed as sacred in the landscape each have a link to royalty due to certain occurrences involving the Litungas. It is believed that Lake Makapaela developed from a well which was dug by Mange, the grandson of Mbuyuwamwambwa. Although Mange had been rejected by his diviners, he insisted on becoming chief and was made one in the forest area. It is further believed that on his way to establish his rule, he dug the well from which Makapaela developed. Various strange occurrences connected to Makapaela were witnessed by the community which they attributed to the spirit of Mange. It is understood that later Mange's spirit moved to Lake Nalulao in Chief Sikwanan'guyas area. Since then and even now, Lake Nalulao is attended by a caretaker referred to as Ndaleti. Lindaleti (plural for Ndaleti) control the times when fishing can be done as well as the methods to ensuring that the correct ones are followed. From all that is fished from the lake, it is mandatory to set aside a portion for the palace. Lake Ilute is another sacred lake where Litunga Ilute is believed to have bathed. Lake Likomokelo is also sacred as it is where Litunga Lewanika got amazed at the depth of the lake. The beliefs and taboos adopted by the Lozi with links to the lakes and lagoons are what have helped to preserve the fish species and the landscape as a whole up to this present day.

12.3.8.5 Coronation Site

Coronation sites are very important places in the Barotse Cultural Landscape with the most notable one being the one in the village of Makono as that is where the installation of all Litungas commences. The site for coronations has been used for generations and Makono is where the female ruler known as Mbuyuwamwambwa was buried. Mbuyuwamwambwa played a cardinal role in the history of the Lozi as she is the one who led them from the Kola Congo region to the Zambezi Plains. Special sacrifices are performed on the grave sites of Mbuyuwamwambwa and the first Litunga Mboo at Imwambo as part of the process of installation of a new king. This is done as the new king has to be presented and accepted by his ancestors.

The Makono village is unfortunately threatened by serious river bank erosion caused by the diversion of the channel. Over the years, the banks of the river have been eroding towards the village and has become more pronounced over the recent years. Any destruction to the village will mean a loss to the history and bedrock of the rich cultural heritage of the Lozi people. Other coronation sites are those for the installation of indunas and these are managed by a caretaker following a set of traditional norms which restrict access to the sites and prohibit other social activities to take place there.

12.3.8.6 Royal Barges

The Lozi have various royal barges which are kept at Siyubo village near the palace at Lealui and the Nayuma Sacred Grove. The barges serve various purposes and are constructed by experienced and specially skilled Induna who are also tasked to repair and keep the barges in good condition. The barges are used during the transhumance process (popularly known as the Kuomboka) when the plains get flooded. The royal barges include the Nalikwanda, the Notila, the Mbolyanga and the Natamikwa.

The largest barge which is the Nalikwanda is fifteen metres long and four metres wide. It can carry one hundred and fifty people and is used for carrying the Litunga during the Kuomboka Ceremony. The Nalikwanda also carries the Maoma war drums and the Mutango royal drum and its Mwatota. It also carries the Sinkoya royal drum as well as the Siluyana silimba or xylophone. No female is allowed on the Nalikwanda. The Notila is another royal barge used to transport the Litunga but is smaller than the Nalikwanda. It also carries the Siluyana and the Sinkoya royal drums.

The Mbolyanga is for the Litungas wife, Mooyo Imwambo and the Litungas mother (Mukwae Ngula or Makoshi) uses the Sabelele. The Natamwika is a royal canoe used for surveillance and is paddled by the Litungas body guards (mabuto). The Ngweshi which are the crimson-coloured royal spears are carried on the Natamikwa. The canoe is used by the Litunga when he disembarks from the Nalikwanda or Notila or when he wants to move from the Notila to the Nalikwanda or vice versa. The Mundende is also a royal canoe and is alternately used with the Natamikwa. The Matenda is a barge which carries royal property. The Namandimbwe is paddled by the kitchen staff and carries the kitchen utensils. The Nalikena is used by the Ngambela.

12.3.8.7 Drums

The royal drums which are of different types and sizes are kept at Nayuma Sacred Grove. They are used for different purposes which include the departure of the king from Lealui as he moves to Limulunga during the Kuomboka, the coronation and death of the Litunga and signalling the well-being and even the presence of the Litunga. Some of the notable drums used are the Moondo, Maoma, Mwenduko and Munwaa or Mutango. The Moondo which is also known as the talking drum signifies the assembling and dispersing of the Kuta. The Maoma summons the paddlers for the Kuomboka and its ‘reverse’ ceremony the Kufuluhela. When the Litunga is not spending the night at Lealui, the Mwenduko is played in the morning to make the people aware. The Mumwaa is played everyday at either Lealui or Limulunga depending on where the Litunga is and is used to signify his presence.

12.4 Management System

The Lozi is one of the main ethnic groups found in Zambia and inhabit the western part sharing borders with the North Western Province, the Central Province, the Southern Province, Namibia and Angola. The Litunga (king) is the supreme ruler who is highly revered by his people, a people with a strong belief in their traditional laws and taboos which has greatly contributed to the prudent management of the land, its people and resources. As with other tribes found in Zambia, the Lozi have also used traditional fables to transmit knowledge from one generation to another. The Lozis regard the Litunga as the 'Minya Mupu na Ng'ombe' or the owner of the land, cattle and all natural resources. He is also referred to as 'Mbumu wa Litunga' or king of the earth. The Lozis have over the years followed a centralised system of management (see Fig. 12.7) with the Litunga overseeing councillors or ministers locally referred to as manduna. The manduna are appointed by the Litunga to head Kuta (councils) which are the administrative, legislative and judicial bodies of the Barotse Cultural Landscape. Different manduna are in charge of the management of various aspects of the landscape including the use of land, consumption of resources and public works.

Due to a number of challenges, the traditional management system on its own has not been so effective and therefore partnering with government bodies and other external partners has been highly beneficial to the sustainability of the BCL. There are a number of Acts which apply to the BCL helping to protect and manage effectively important resources in the landscape. These are the National Heritage Conservation Commission Act, the Forests Act, the Zambia Wildlife Act, the Fisheries Act, the Lands Act, the Urban and Regional Planning Act, the Zambia Environmental Management Act and the Merchant Shipping Act. The management plan of the BCL produced as per UNESCO requirements will also contribute to strengthening the implementation of the various legislation in place.

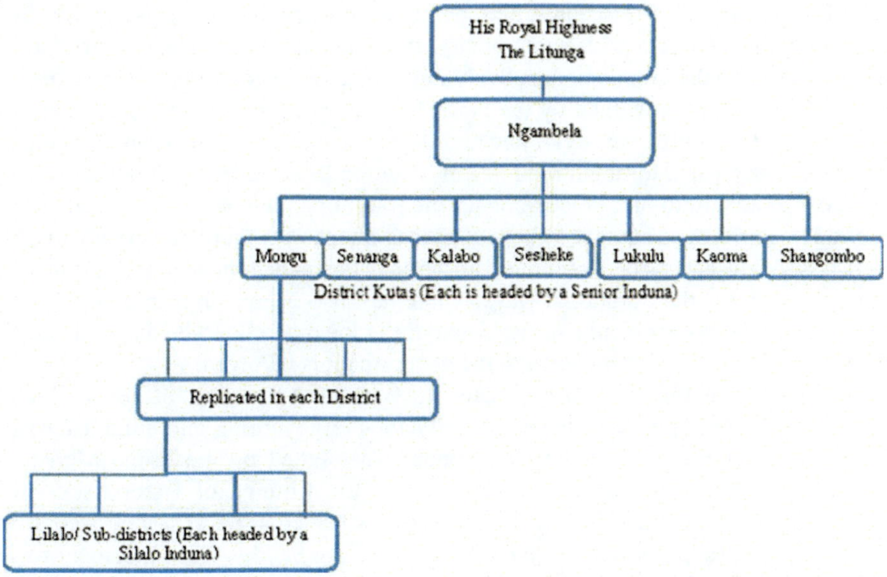

Fig. 12.7 Organisational structure of the Barotse Traditional Governance system

12.5 Factors Affecting the Management of the Barotse Cultural Landscape

Whilst arguments exist in favour of the effectiveness of traditional management systems, it must be noted that a number of challenges have surfaced over the years which necessitate the need to combine the traditional with more contemporary type of management system for success.

12.5.1 *Change with Time*

Various aspects of life change over time. There will always be factors which will influence change which may turn out to be positive or negative. Changes in the philosophy of life in the BCL has had a negative bearing on the maintenance practices. The practices have been a social demand which the community adhered to without querying. However, with the introduction of democracy and issues to do with human rights more and more people are questioning their mandatory involvement in traditional practices and feel that the traditional demand is abrogating their rights.

Prior to 1964, the year in which Zambia gained its independence from Britain, Barotseland was governed separately as a kingdom but it is now a province of the Republic of Zambia. Whilst there may be numerous benefits to be gained by the kingdom being part of a wider community, the wider politics of the country have been brought into the kingdom, negatively impacting on the traditional management system. Certain laws governed by the State, for example, those to do with fisheries or forests, will have different styles and approaches to managing the resources which the locals with their traditional methods may not be in agreement with. The locals were able to adhere to and practise under the traditional customs but now are also required to conform to another system of management with more players involved. There is also now an aspect of commercialisation in the management of the landscape which undermines the traditional ways. There has also been a shift in power from the people to the government causing some of the local people to be disgruntled or develop apathy to the affairs in which the government is involved.

A thorny issue for the Lozis is also the Barotse Agreement of 1964 which granted the Litunga principal local authority for the governing and administering of the affairs of Barotseland. The agreement was signed by the Prime Minister of Northern Rhodesia Kenneth David Kaunda, the Litunga of Barotseland Sir Mwanawina Lewanika the third and her Majesty's Principal Secretary of State for Commonwealth Relations and for the Colonies Duncan Sandys. The agreement was however cancelled by the Zambian Republican Government through the constitutional (Amendment) Act of October 1969 (Bull, Mutumba Mainga (2014) 'Reserved Area: Barotseland of the 1964 Agreement', *Zambia Social Science Journal*: Vol. 5: No. 1, Article 4.). This has been a contentious issue with various factions amongst the Lozi creating apathy to any involvement by the government in the management of affairs in Barotseland. The Government of the Republic of Zambia continues discussions with the Barotse Royal Establishment to come up with an amicable solution to resolve the issue.

Urbanisation which refers to the movement of people from the rural areas to the urban areas has been a common trend in Zambia from as far back as the colonial days before Zambia got its Independence. People have been drawn to the urban areas where most of the mining, manufacturing and service industries are located. Higher wages are paid when working in these industries as opposed to what can be earned in the rural areas and this thus gives people hope that by moving to the urban areas, they will be assured of better education for their children, better facilities, more business opportunities and generally a better quality of living. The Barotse Cultural Landscape has not been spared from its people moving away to urban areas and has had a negative impact on its management as there are those who due to their lineage would have been involved in the management but have opted to translocate to urban areas where there are more employment opportunities and social benefits. Most of the transfer of knowledge or skills to the next generations as with other ethnic groups found in Zambia has been by word of mouth and only to specific families, groupings or individuals. The movement of any of these beneficiaries to other areas has greatly disrupted the smooth functioning of life in these communities.

12.5.2 Adoption of a Foreign Religion

Enshrined in the traditional management system of the BCL are beliefs in the powers of the spirits which they believe aid them in the various affairs of the kingdom. Rituals are performed for certain practices to invoke the spirits. The advent of Christianity to the BCL has, however, caused some locals who have been converted to lose the impetus to continue in sustaining their traditional practices. In the eyes of Christianity, the traditional beliefs and rituals are viewed to be evil, barbaric or ungodly. Some people have failed to see how the traditional beliefs can coexist with Christianity meaning that their conversion to Christianity has automatically meant the discarding of the beliefs they and their forefathers have held on to and been guided by for generations. The same traditional beliefs are the ones which have helped them to manage the landscape in a sustainable manner and it is therefore unfortunate that some Christians do away with these beliefs.

12.5.3 Leadership in Barotseland

The leadership in place plays a key role in the management of the Barotse Cultural Landscape. Over the years, it has been found that the locals are more inspired by certain leaders more than others. Currently, it has been noted that some leaders do not observe certain traditions and there is some nepotism at play in the selection of some indunas. This has created some cultural dilution as some of those selected are not knowledgeable about their traditions and they require others to educate them when they are appointed as indunas.

12.6 Concluding Remarks

In as much as there are elements of the traditional management system which are highly beneficial to the protection and preservation of the Barotse Cultural Landscape, it has to be noted that on its own and with conditions which have negatively impacted on the affairs of the kingdom, it is no longer as effective. The way forward will require more thought to an integrated approach in the management of the BCL combining the traditional and conventional methods. Attention should be made to ensure that decisions are not made in the absence of the Barotse Royal Establishment and it should be noted that community participation will be key at all levels.

Government bodies and other key stakeholders who have various professionals under them tasked with development of the landscape in the various sectors should be mindful of the fact that the community has a big role to play in the success of any planned ventures even though some of the community members may not have the applicable academic qualifications but they are able to directly identify with the place and whose well-being is at stake. Their contribution at an early stage should therefore not be ignored.

For successful placemaking of the BCL, the diverse opinions and knowledge of the community, traditional leaders, government bodies and other stakeholders operating there should be integrated into a cohesive vision. This should in turn translate into a plan of works which should be implemented in a sustainable manner. The development of the plan should take into consideration applicable sustainable development goals which have been mainstreamed into Zambias Eighth National Development Plan running from 2022 to 2026.

References

Bull MM (2014) Reserved area: Barotseland of the 1964 agreement. Zambia Soc Sci J 5(1), Article 4

Zambia (2013) Nomination Dossier Barotse cultural landscape proposal for inscription on the UNESCO World Heritage List

Chapter 13
Pandemics, Situational Under-Tourism, and the Search for Sustainability

Daniel H. Olsen

Abstract With the continuing saga of the COVID-19 pandemic, several commentators have noted the rehabilitation of the earth's environment due in part to the substantial decrease in domestic and international travel. As such, there has been an increased call for a "new normal" in tourism, which includes limiting international travel. This chapter focuses on whether this new normal is feasible in the context of future tourism development. After discussing the concepts of epidemics and pandemics and the influence of the COVID-19 pandemic on the global tourism industry, attention is turned to examining what constitutes this new normal to argue that proponents of this stance overlook the influence of this new normal on places that suffer from chronic under-tourism, before concluding.

Keywords Over-tourism · Under-tourism · COVID-19 · Tourism development · Sustainability · Pandemics · Global tourism

13.1 Introduction

As the world continues to struggle with the continuing COVID-19 pandemic as well as its attendant variants, questions have arisen among academics and journalists regarding the long-term impacts of this pandemic in terms of quality of life and environmental sustainability. This has been particularly true in the context of tourism. While the lack of jet-setting visitors to far off lands has severely damaged economies that are particularly reliant on tourism dollars for economic sustainability, the shuttering of international and even domestic tourism has seemingly led to increasing environmental rehabilitation and recovery. As such, there have been several calls for a "new normal" in future tourism growth and management, focusing on things such as "human-centered tourism", ecological restoration and justice, racial reconciliation,

D. H. Olsen (✉)
Department of Geography, Brigham Young University, 690B KMBL, Provo, UT 84602, USA
e-mail: dholsen@byu.edu; triad@k-state.edu

R. P. B. Singh et al. (eds.), *Placemaking and Cultural Landscapes*,
Advances in Geographical and Environmental Sciences,
https://doi.org/10.1007/978-981-19-6274-5_13

and social inclusion at the expense of exploitative and unequal economic models of tourism development.

These calls for a new normal have also merged with concerns by tourism scholars, government officials, and grassroots or local organizations regarding "over-tourism." In both perceptual and real terms, over-tourism refers to an "excessive growth of visitors leading to overcrowding in areas where residents suffer the consequences of temporary and seasonal tourism peaks, which have enforced permanent changes to their lifestyles, access to amenities[,] and general well-being" (Milano et al. 2019: 1). However, questions arise regarding whether this call for a new normal will be heeded by both a tourism industry that is focused on short-term economic profit and government officials who rely on tourism for economic sustainability and support, particularly in destinations that suffer from what is termed here "chronic under-tourism."

The purpose of this chapter is to interrogate whether this new normal is feasible in the context of future tourism development. This chapter begins by clarifying the differences between epidemics and pandemics and highlights several instances of pandemics in both the ancient and modern world. The impacts of the COVID-19 pandemic on the tourism industry are then examined, after which the discussion shifts to discussing what constitutes this "new normal." Different types of under-tourism that this new normal does not consider are discussed, followed by questioning whether this new normal can be successfully implement in the immediate and/or distant future.

13.2 Epidemics vs. Pandemics

With the spread of the COVID-19 virus and its variants beginning in 2019 (which pandemic continues as this chapter is being written), the media has increasingly used the terms *epidemic* and *pandemic* in their reporting. However, historically, people have conflated these two terms even though they refer to different epidemiological events. According to Last (1988: 94), "An epidemic is the occurrence in a community or region of cases of an illness, specified health behavior, or other health-related events clearly in excess of normal expectancy." In contrast, a pandemic is an epidemic "distributed or occurring widely throughout a region, country, continent[,] or globally" (Morens et al. 2009: 1018) that usually affects a large proportion of a population. According to Cunningham (2008: 29), the term "epidemic" comes from the Greek word *demos*, meaning a disease that "falls on the mob," and refers to "any disease that kills many people, kills them quickly, [and] kills them in an unpleasant way" irrespective of sex and age. The term "pandemic" comes from the Greek word *pan*, meaning "everywhere," and "is meant to convey the scale and spread of the outbreak, not its greater severity or mortality" (ibid.). In this vein, Grennan (2019: 910) differentiates between the terms *endemic*, *outbreak*, *epidemic*, and *pandemic*:

- "An *endemic* condition is present at a fairly stable, predictable rate among a group of people";
- "An *outbreak* is when there is a sudden increase in the number of people with a condition greater than is expected";
- "An *epidemic* is an outbreak that spreads over a larger geographical area"; and
- "An epidemic that spreads globally is a *pandemic*."

While the definitions of epidemic and pandemic are closely related, what differentiates epidemics from pandemic is *scale*—an epidemic is regional in scale, while pandemics are global in scale.

Morens et al. (2009: 1019–1020) have noted several characteristics of pandemics, including:

- *Wide geographic range*: transregional, interregional, or global;
- *Disease movement*: the transmission of pathogens is traceable back to an origin;
- *High attack rates*: rate of geographic spread;
- *Minimal population immunity*: levels of immunity to a disease within a given population;
- *Novelty*: pandemics involve pathogens that are newly introduced into a population;
- *Infectiousness*: high infection rates;
- *Contagiousness*: multiple transmission modes; and
- *Severity*: high fatality rates.

Epidemics and pandemics are not new, having been a part of the human struggle for centuries, leading to the deaths of millions of people (see Table 13.1). Through much of world history, epidemics and pandemics were considered punishment by God or the gods for human misdeeds (Cunningham 2008)—and is still the case in some religious circles, where the HIV/AIDS pandemic is viewed as God's punishment for homosexual behavior (Snowden 2019). However, the underlying causes of these epidemiological events was not discovered until the nineteenth century (Kulikowski 2021). While there is little information about the causes of various epidemics and pandemics in the ancient era of world history due to gaps in the historical record, the earliest known small-scale epidemic was in 3000 BC in northern China, where in an unknown virus wiped out an entire village (now the Hamin Mangha archeological site) (Yonggang and Ping 2016). However, according to Hays (2005), the first of what he considers the best-documented epidemics and pandemics was a plague (possibly smallpox, typhoid fever, or Ebola) that affected the city of Athens, Greece, between 430–427 B.C., followed by instances of malaria in Rome, Italy. Hays then lists forty-eight other epidemics and pandemics up to the present day that have affected world history, which (beyond those listed on Table 13.1) include, among others, the Plague of London (1640–1644), Smallpox in London (1707–1709), the First Cholera Pandemic (1917–1824), the Sleeping Sickness in East/Central Africa (1900–1905), and Polomyelitis in the United States (1945–1955).

Table 13.1 A selective list of major epidemics and pandemics throughout world history

Event	Start–End, CA/AD	Location	Cause	Deaths
Antonine Plague	165–180	Roman Empire	Smallpox or measles	5 million
Plague of Justinian	541–549	Europe, West Asia	Bubonic plague	30–50 million
Black Death	1331–1353	Asia, Europe, North Africa	Bubonic plague	75–200 million
New World Smallpox Outbreak	1520 onwards	Americas	Smallpox	56 million
Italian Plague	1623–1632	Europe	*Yersinia pestis* bacteria	280,000
Great Plague of Seville	1647–1652	Spain	Bubonic plague	2 million
Russian Flu	1889–1890	Europe, North America	H2N2 (avian origin)	1 million
Spanish Flu	1918–1920	Worldwide	Influenza	100 million
Asian Flu	1957–1958	East Asia	Influenza A (H2N2)	1–4 million
Hong Kong Flu	1968–1969	East Asia	Influenza A (H3N2)	1 million
HIV/AIDS	1981–present	Worldwide	Human immune-deficiency virus	25–35 million
COVID-19 pandemic	2019–present	Worldwide	SARS-CoV-2	5–22 million

Sources Alfani and Murphy (2017), LePan (2020), Jedwab et al. (2021), Rosenwald (2021), https://en.wikipedia.org/wiki/List_of_epidemics

13.3 Pandemics and Tourism

Historically, there has been a strong connection between travel and the transmission of diseases, with human mobility being an important vector for the spread of disease. Concerns regarding the travel-related transmission of disease predate the modern era (Steffen 2017). For example, in 1865, a cholera outbreak in Europe that was traced back to Hajj participants in Saudi Arabia led to European governments moving to medicalize the Hajj, enforcing stringent European medical practices in the areas of sanitation, regimentation of medical practices, forced vaccination, and surveillance (Huber 2016; Olsen 2020). Historically, the diseased diffused through overland trade routes and via coastal ports—a combination of contagious and hierarchal forms of diffusion. However, with increases in transportation efficiencies and the democratization of travel after World War II, personal mobility increased and global travel patterns shifted, changing this hierarchal spread of disease from coastal ports to cities with airports. Increased contact between people in different parts of the world intensified the hierarchal expansion of epidemiological events, leading to growing concerns regarding the spread of various diseases, such as sexually transmitted diseases, respiratory tract infections, hepatitis A and B, and meningococcal disease that are generally associated with and facilitated by travel (Tatem et al. 2012; Olsen 2020). In spite of this, it was not until the 1970s that the collection of official statistics regarding human mobility and disease diffusion occurred and was assessed at various scales (Steffen 2017).

The rapid growth of the international tourism industry has led the World Health Organization to create a website specifically addressing international travel and health (https://www.who.int/ith/en/). Several academic journals that focus on travel and epidemiological research, such as the *Journal of Travel Medicine* and *Travel Medicine and Infectious Disease*, have been established to monitor and suggest preventative measures against the diffusion of diseases through travel. In addition, the International Society of Travel Medicine, which was founded in 1991 and is presently based in Atlanta, Georgia in the United States, focuses on promoting various travel health initiatives, including a Global Travel Clinic Directory to help tourists find medical providers when they travel (see www.istm.org).

Epidemics, pandemics, and other major disruptive events such as the September 11 terrorist attacks in 2001, the severe acute respiratory syndrome (SARS) outbreak in 2003, and the global economic downturn in 2008–2009 have the potential to disrupt the global economy, including international and domestic tourism industries. As Gössling et al. (2021) note, of the three aforementioned disruptive events, only SARS (−0.4%) and the global economic downturn (−4.0%) led to an actual decline in international tourism numbers. As such, it takes a severe pandemic or a long-term disruptive event, like the present-day COVID-19 pandemic, to several limit international tourism numbers. At the same time, when there is a disruption within the tourism system, it is typically developing countries or island states (e.g., The Caribbean) that face the greatest decline in tourism visitation (Goretti et al. 2021) because they have tourism-dependent economies that are highly sensitive to even minor disruptions in the tourism sector.

In the case of the COVID-19 pandemic, in March 2020, governments enacted almost draconian measures to prevent the spread of the virus. As a part of these measures, the global tourism industry was effectively shuttered, with most countries implementing travel bans (Seyfi et al. 2020) and ordering their citizens to people to stay at home in order to "flatten the curve." Because of these measures, international tourism numbers crashed, with a 74% (1 billion) average decrease in international tourists arrivals within various world regions (regional decreases in international tourist arrivals ranged from −68% in the Americas to −84% in Asia and the Pacific), leading to a 2 trillion USD decline in global tourism direct gross domestic product (UNWTO 2021a, b, c) and a loss of over 200 million tourism-related jobs (UN News 2020). While international tourism arrivals rebounded slightly during the third quarter of 2021, full international tourism recovery is expected to take several years (UNWTO 2021d; see Gössling et al. 2021; Seyfi et al. 2020; Sigala 2020).

13.4 New Normal

While governments and tourism business leaders are anxious to return to "business as usual," many academics and writers in the media have viewed the COVID-19 pandemic as a time to pause and rethink tourism development and management. According to some scholars, the COVID-19 pandemic and its resulting travel restrictions led to a process of "de-globalization" (Niewiadomski 2020), with the end result being "under-tourism" (Gowreesunkar and Thanh 2020)—in essence, an "anthropause" (McGreevy 2020). With the shuttering of the global tourism industry, as well as other sectors of the global economy, fossil fuel consumption, greenhouse emissions, waste production, and water usage decreased. As a result, and in conjunction with the sudden absence of tourists, there was a rapid natural ecological restoration in many tourist destinations, with short-term increases in the presence and activity of various bioindicators (i.e., flora and fauna) in these destinations (e.g., Crossley 2020; Rume and Islam 2020; Soto et al. 2021; Spalding et al. 2021). Examples of this restoration include the clearing of the water in the canals of Venice (e.g., Brunton 2020; Spalding et al. 2021) and the beaches around Acapulco, Mexico, Barcelona, Spain, and Salinas, Ecuador (Zambrano-Monserrate et al. 2020). However, there were also some ecological consequences to this lack of tourism to some destinations. One such example is the island of Stora Karlsö in Sweden, where the sharp decrease in tourists to the island due to the pandemic led to less white-tailed eagle chicks hatching. In this case, tourist activity was believed to scare away predators of the white-tailed eagles from the island, but with the disappearance of tourists, these predators returned (Machemer 2021).

The positive outcomes of the pandemic on the natural environment, as well as the uneven effects of the pandemic on regional economies, have led some scholars and social commentators to question whether "business as usual" within the tourism industry is the best thing long term. As Jamal and Higham (2021, p. 144) note, without serious consideration to the negative impacts of global forms of tourism, this return to business as usual will in essence just be "a return to neoliberal globalization." This recognition of the problems and tensions inherent in tourism development and management pre-COVID-19 has been spurred in part by "over-tourism" and concerns over environmental sustainability, quality of life, the deterioration of the visitor experience, and growing "anti-tourism"/ "tourism-phobia" sentiment in many parts of the world (see Capocchi et al. 2019; Bertocchi et al. 2020; Dodds and Butler 2019; Milano et al. 2019; Oklevik et al. 2019; Pechlaner et al. 2019). As such, many scholars have argued for a "new tourism normal" or "tourism reset," in which government and tourism officials take the tourism pause associated with the pandemic to study "ways in which tourism has supported neoliberal injustices and exploitation" (Higgins-Desbiolles 2020: 620) and think about how the tourism sector can be "reworked."

According to advocates of this perspective, this reworking of the tourism industry or the status quo would include examining how tourism can support vulnerable groups and ecosystems, the removal of structural inequalities, changing tourist behaviors, and a commitment to equity in all aspects of tourism development, including environmental reparation and ecological justice, climate action, biodiversity conservation, racial reconciliation and social inclusion, public health, and governance and finance (see Ateljevic 2020; Benjamin et al. 2020; Carr 2020; Cave and Dredge 2020; Higgins-Desbiolles 2020; Khmara and Kronenberg 2020; Niewiadomski 2020; Stankov et al. 2020). This reworking would also include reforming government tourism policy making (Lew et al. 2020). These changes to the tourism industry would require the development of less-exploitative tourism strategies in the areas of economics and the environment, along with "locally tailored solutions to redefine tourism based on local rights, interests and benefits" (Rastegar et al. 2021: 2), leading to a more "human-centered" tourism (Butowski 2021). From this perspective, tourism is viewed as a "sick industry" (Benjamin et al. 2020) and an unsustainable luxury (Hindley et al. 2022) that needs to be limited through changes in tourist preferences, the utilization of new technologies to limit business travel, and reorienting tourism-dependent economies to a more diversified economic portfolio (Babii and Nadeem 2021).

This post-pandemic new normal, however, would require governments and tourism companies to reject the "business as usual" economic model, focus on "degrowth" (Khmara and Kronenberg 2020; Butcher, in press), and limit in particular "international" travel, which is a major producer of greenhouse gases. This, of course, creates a strong tension between sustainability and competitiveness, which tension several scholars have tried to bridge (e.g., Angelkova et al. 2012; Seguí-Amortegui et al. 2019; Streimikiene et al. 2021), albeit unsuccessfully. This tension is in part aligned with the idea of "sustainable development," defined by the United Nations as "Development that meets the needs of the present without compromising the ability of future generations to meet their own needs" (The World Commission on Environment and Development 1987: 3). A careful reading of this definition suggests that "sustainable development" is an oxymoron, in that the phrase involves contrasting ideas that cannot easily be reconciled (Redclift 2005)—where "Continual population growth and economic development on a finite Earth are biophysically impossible" (Brown 2015: 1027). The term "sustainability" suggests the ability to maintain something at a certain level or rate over the long term, while the term "development" implies change, or taking something and manipulating it from its existing situation to a new, ideally superior, state. In many ways, how the term sustainable development is utilized depends on the weighting attached to "sustainability" (environment) and "development" (economics) (Redclift 2005) and the indicators used to measure these concepts (Njiro 2002).

13.5 Situational Under-Tourism

Within this vein, unfortunately, this focus on a "new normal" within the tourism industry with its focus on sustainability and limiting access to long-haul international travel does not consider different types of "under-tourism." For example, *situational under-tourism*, which is happening presently due to the COVID-19 pandemic, occurs when regional and world events, such as pandemics, natural disasters, and economic recessions, lead to a decrease in travel and tourism, with resultant high levels of unemployment and the closure of hospitality-related businesses. Until the event is over, countries will struggle coping with the loss of tourist revenue. This type of under-tourism differs from:

- *Chronic under-tourism*: where residents and officials in several world regions struggle with the lack of tourists, just hoping that tourists will show up (Koens et al. 2018; Tseane-Gumbi and Ani 2019; Wheeler 2019), and
- *Purposeful under-tourism*: where there are purposeful actions taken to slow and even stop tourism demand through strategies such as demarketing and degrowth (Clements 1989; Beeton and Benfield 2002; Medway et al. 2010; Tiwari et al. 2020).

On the surface, calls for a new normal would be beneficial to the resiliency of the tourism industry and equalize the economic, socio-cultural, and environmental benefits of tourism development. However, while the implementation of a degrowth mindset within the tourism industry through limiting international tourists would greatly benefit tourism destinations struggling with over-tourism—where tourism development and visitation degrades the quality of life of local residents—and would place a premium on quality over quantity, it would not necessarily benefit destinations where chronic under-tourism occurs. With many tourism destinations presently struggling with unemployment and underemployment due to the pandemic, the United Nations World Tourism Organization (UNWTO) has called upon national governments to accelerate the recovery of the tourism sector at all geographic scales. As such, any conversation regarding a new tourism industry normal must balance the sustainability/development divide and include discussions regarding destinations that suffer from chronic under-tourism. How can the tourism industry encourage tourists to visit less popular tourism destinations at various scales—encouraging tourist visitation from a city center to outlying neighborhoods or traveling to less-visited countries? In doing so, over-tourism at "must-see" tourism destinations will be lessened, new "must-see" tourism destinations will be developed, and the economics benefits of tourism development and visitation will trickle out to less-visited destinations.

13.6 Conclusion: The Search for Sustainability

This chapter focused on the effects of pandemics on the global tourism industry and how calls for a new normal, while noble, need to be tempered by an examination of chronic under-tourism. Calls for this new normal tend to give greater weighting to the "sustainability" side of the sustainability/development divide. While it is important to deal with questions related to equity, equality, quality, and access within the tourism industry to ensure its long-term viability, at its core tourism is economic in nature. Without a more balanced and integrative approaches regarding the interplay of sustainability and development in a new tourism normal context, this new normal may remain just a utopian idea. Indeed, without an equal balancing of sustainability and development within in this new normal, the result will be a non-zero-sum game, where those advocating for sustainability "win" and those advocating for development "lose." In addition, if this new normal means more regional and local tourism and less international tourism, what about island countries whose economies are oriented toward international tourism? What about countries and regions with countries that struggle with chronic under-tourism? As such, whatever this new normal looks like at the end of the COVID-19 pandemic, one must wonder what needs to occur to make this new normal a zero-sum game, where advocates on both sides of the sustainability/development equation win.

References

Alfani G, Murphy TE (2017) Plague and lethal epidemics in the pre-industrial world. J Econ Hist 77(1):314–343

Angelkova T, Koteski C, Jakovlev Z, Mitrevska E (2012) Sustainability and competitiveness of tourism. Procedia Soc Behav Sci 44:221–227

Ateljevic I (2020) Transforming the (tourism) world for good and (re)generating the potential 'new normal.' Tour Geogr 22(3):467–475

Babii A, Nadeem S (2021) Tourism in a post-pandemic world. International Monetary Fund. https://www.imf.org/en/News/Articles/2021/02/24/na022521-how-to-save-travel-and-tourism-in-a-post-pandemic-world. Accessed 18 Jan 2022

Beeton S, Benfield R (2002) Demand control: the case for demarketing as a visitor and environmental management tool. J Sustain Tour 10(6):497–513

Benjamin S, Dillette A, Alderman DH (2020) "We can't return to normal": Committing to tourism equity in the post-pandemic age. Tour Geogr 22(3):476–483

Bertocchi D, Camatti N, Giove S, van der Borg J (2020) Venice and overtourism: Simulating sustainable development scenarios through a tourism carrying capacity model. Sustainability 12:512. https://doi.org/10.3390/su12020512

Brown JH (2015) The oxymoron of sustainable development. Bioscience 65(10):1027–1029

Brunton J (2020) 'Nature is taking back Venice': wildlife returns to tourist-free city. The Guardian. https://www.theguardian.com/environment/2020/mar/20/nature-is-taking-back-venice-wildlife-returns-to-tourist-free-city. Accessed 18 Jan 2022

Butcher J (in press) COVID-19, tourism and the advocacy of degrowth. Tourism Recreation Research. https://www.tandfonline.com/doi/full/10.1080/02508281.2021.1953306

Butowski L (2021) Sustainable tourism: a human-centered approach. Sustainability 13(4):1835. https://doi.org/10.3390/su13041835

Capocchi A, Vallone C, Pierotti M, Amaduzzi A (2019) Overtourism: a literature review to assess implications and future perspectives. Sustainability 11(12):3303. https://doi.org/10.3390/su11123303

Carr A (2020) COVID-19, indigenous peoples and tourism: a view from New Zealand. Tour Geogr 22(3):491–502

Cave J, Dredge D (2020) Regenerative tourism needs diverse economic practices. Tour Geogr 22(3):503–513

Clements MA (1989) Selecting tourist traffic by demarketing. Tour Manage 10(2):89–94

Crossley É (2020) Ecological grief generates desire for environmental healing in tourism after COVID-19. Tour Geogr 22(3):536–546

Cunningham A (2008) Epidemics, pandemics, and the doomsday scenario. Hist Speak 9(7):29–31
Dodds R, Butler R (eds) (2019) Overtourism: issues, realities and solutions. De Gruyter, Oldenbourg, Germany
Goretti M, Leigh L, Babii A, Cevik S, Kaendera S, Muir D, Nadeem S, Salinas G (2021) Tourism in the post-pandemic world: economic challenges and opportunities for Asia-Pacific and the Western Hemisphere. International Monetary Fund, Washington, DC
Gössling S, Scott D, Hall CM (2021) Pandemics, tourism and global change: a rapid assessment of COVID-19. J Sustain Tour 29(1):1–20
Gowreesunkar VG, Vo Thanh T (2020) Between overtourism and under-tourism: impacts, implications, and probable solutions. In: Séraphin H, Gladkikh T and Vo Thanh T (eds) Overtourism: causes, implications and solutions. Palgrave Macmillan, Cham, The Netherlands, pp 45–68
Grennan D (2019) What is a Pandemic? JAMA 321(9):910
Hays JN (2005) Epidemics and pandemics: their impacts on human history. ABC-CILO, Santa Barbara, CA
Higgins-Desbiolles F (2020) Socialising tourism for social and ecological justice after COVID-19. Tour Geogr 22(3):610–623
Hindley C, Legrand W, Laeis GCM (2022) The pandemic of tourism: how tourism has become an unsustainable luxury. In Kotur AS, DixitEds SK (eds) The Emerald handbook of luxury management for hospitality and tourism. Emerald Publishing Limited, Bingley, pp 479–496
Huber V (2016) International bodies: the pilgrimage to Mecca and international health regulations. In: Tagliacozzo E, Toorawa SM (eds) The Hajj: Pilgrimage in Islam. Cambridge University Press, New York, pp 175–195
Jamal T, Higham J (2021) Justice and ethics: towards a new platform for tourism and sustainability. J Sustain Tour 29(2–3):143–157
Jedwab R, Khan AM, Russ J, Zaveri ED (2021) Epidemics, pandemics, and social conflict: lessons from the past and possible scenarios for COVID-19. World Dev 147:105629. https://doi.org/10.1016/j.worlddev.2021.105629
Khmarą Y, Kronenberg J (2020) Degrowth in the context of sustainability transitions: in search of a common ground. J Clean Prod 267:122072. https://doi.org/10.1016/j.jclepro.2020.122072
Koens K, Postma A, Papp B (2018) Is overtourism overused? Understanding the impact of tourism in a city context. Sustainability 10(12):4384. https://doi.org/10.3390/su10124384
Kulikowski CA (2021) Pandemics: historically slow "learning curve" leading to biomedical informatics and vaccine breakthroughs. Yearb Med Inform 30(1):290–301
Last JM (1988) A dictionary of epidemiology, 2nd edn. Oxford University Press, Oxford
LePan N (2020) A visual history of pandemics. World Economic Forum. https://www.weforum.org/agenda/2020/03/a-visual-history-of-pandemics. Accessed 11 Jan 2021
Lew AA, Cheer JM, Haywood M, Brouder P, Salazar NB (2020) Visions of travel and tourism after the global COVID-19 transformation of 2020. Tour Geogr 22(3):455–466
Machemer T (2021) Pandemic reveals ecological benefits of tourists in one seaside town. https://www.smithsonianmag.com/smart-news/pandemic-reveals-role-tourists-one-seaside-ecosystem-180976836/. Accessed 18 Jan 2022
McGreevy N (2020) Scientists propose a new name for nature in the time of COVID-19: the 'Anthropause'. https://www.smithsonianmag.com/smart-news/scientists-have-name-nature-time-covid-19-anthropause-180975224/. Accessed 18 Jan 2022
Medway D, Warnaby G, Dharni S (2010) Demarketing places: rationales and strategies. J Mark Manag 27(1–2):124–142
Milano C, Cheer JM, Novelli M (2019) Introduction. In: Milano C, Cheer JM, Novelli M (eds) Overtourism: excesses, discontents and measures in travel and tourism. CAB International, Wallingford, UK, pp 1–17
Morens DM, Folkers GK, Fauci AS (2009) What is a pandemic? J Infect Dis 200(7):1018–1021
Niewiadomski P (2020) COVID-19: from temporary de-globalisation to a re-discovery of tourism? Tour Geogr 22(3):651–656

Njiro E (2002) Introduction: sustainable: development an oxymoron? Agenda: Empowering Women for Gender Equity 52: 3–7

Oklevik O, Gössling S, Hall CM, Steen Jacobsen JK, Grøtte IP, McCabe S (2019) Overtourism, optimisation, and destination performance indicators: A case study of activities in Fjord Norway. J Sustain Tour 27(12):1804–1824

Olsen DH (2020) Disease and health risks at mass religious gatherings. In: Shinde KA, Olsen DH (eds) Religious tourism and the environment. CABI, Wallingford, UK, pp 116–132

Pechlaner H, Innerhofer E, Erschbamer G (2019) Overtourism: tourism management and solutions. Routledge, London and New York

Rastegar R, Higgins-Desbiolles F, Ruhanen L (2021) COVID-19 and a justice framework to guide tourism recovery. Ann Tourism Res 91:103161. https://doi.org/10.1016/j.annals.2021.103161

Redclift M (2005) Sustainable development (1987–2005): an oxymoron comes of age. Sustain Dev 13(4):212–227

Rosenwald MS (2021) History's deadliest pandemics, from ancient Rome to modern America. The Washington Post. https://www.washingtonpost.com/graphics/2020/local/retropolis/coronavirus-deadliest-pandemics/. Accessed 11 Jan 2022

Rume T, Islam SDU (2020) Environmental effects of COVID-19 pandemic and potential strategies of sustainability. Heliyon 6(9):e04965

Sigala M (2020) Tourism and COVID-19: impacts and implications for advancing and resetting industry and research. J Bus Res 117:312–321

Seguí-Amortegui L, Clemente-Almendros JA, Medina R, Grueso Gala M (2019) Sustainability and competitiveness in the tourism industry and tourist destinations: a bibliometric study. Sustainability 11(22):6351. https://doi.org/10.3390/su11226351

Seyfi S, Hall CM, Shabani B (2020) COVID-19 and international travel restrictions: the geopolitics of health and tourism. Tour Geogr. https://doi.org/10.1080/14616688.2020.1833972

Snowden, F.M. (2019). *Epidemics and Society: From the Black Death to the Present.* New Haven and London: Yale University Press.

Spalding M, Burke L, Fyall A (2021) COVID-19: implications for nature and tourism. Anatolia 32(1):26–127

Soto EH, Botero CM, Milanés CB, Rodríguez-Santiago A, Palacios-Moreno M, Díaz-Ferguson E, Velázquez YR, Abbehusen A, Guerra-Castro E, Simoes N, Muciño-Reyes M, Souza Filho JR (2021) How does the beach ecosystem change without tourists during COVID-19 lockdown? Biol Cons 255:108972. https://doi.org/10.1016/j.biocon.2021.108972

Stankov U, Filimonau V, Vujičić MD (2020) A mindful shift: an opportunity for mindfulness-driven tourism in a post-pandemic world. Tour Geogr 22(3):703–712

Steffen R (2017) Travel health-travel medicine. J Health Spec 5(1):42–44

Streimikiene D, Svagzdiene B, Jasinskas E, Simanavicius A (2021) Sustainable tourism development and competitiveness: the systematic literature review. Sustain Dev 29(1):259–271

Tatem AJ, Huang Z, Das A, Qi Q, Roth J, Qui Y (2012) Air travel and vectorborne disease movement. Parasitology 139(14):1816–1830

Tiwari P, Kainthola S, Chowdhary NR (2020) Demarketing: a marketing framework for overtourism. In Tiwari P, Kainthola S, Chowdhary NR (eds) Handbook of research on the impacts, challenges, and policy responses to overtourism. IGI Global, Hershey, pp 94–114

Tseane-Gumbi LA, Ani KJ (2019) Tourism and transformation in the South Africa's former homelands: Lessons and implications for Africa. Africa Renaissance 16(3):93–108

UN News (2020) COVID-19: impact could cause equivalent of 195 million job losses, Aays ILO Chief. Available at: https://news.un.org/en/story/2020/04/1061322. Accessed 17 June 2020

UNWTO (2021a) 2020: Worst year in tourism history with 1 billion fewer international arrivals. https://www.unwto.org/news/2020-worst-year-in-tourism-history-with-1-billion-fewer-international-arrivals. Accessed 2 Nov 2021

UNWTO (2021b) The economic contribution of tourism and the impact of COVID-19. https://www.e-unwto.org/doi/epdf/10.18111/9789284423200. Accessed 13 Jan 2022

UNWTO (2021c) Vaccines and reopen borders driving tourism's recovery. https://www.unwto.org/taxonomy/term/347. Accessed 13 Jan 2022

UNWTO (2021d) Global tourism sees upturn in Q3 but Recovery remains fragile. https://www.unwto.org/news/global-tourism-sees-upturn-in-q3-but-recovery-remains-fragile. Accessed 13 Jan 2022

Wheeler T (2019) A life best travelled. Big Issue Australia 589:18–19

World Commission on Environment and Development (1987) Our common future. Oxford University Press, New York

Yonggang Z, Ping J (2016) Investigation on the mass graves found in house foundations at the Hamin Mangha site, inner Mongolia: exploration and reflection on prehistoric catastrophic events. Archaeol Cult Relics 5:5

Zambrano-Monserrate MA, Ruano MA, Sanchez-Alcalde L (2020) Indirect effects of COVID-19 on the environment. Sci Total Environ 728:138813. https://doi.org/10.1016/j.scitotenv.2020.138813

Chapter 14
Post-COVID-19 Strategies: Cloisters as Urban Oasis to Reconnect Heritage and Communities

Carlo Francini, Alessia Montacchini, and Tatiana Rozochkina

Abstract This paper aims at illustrating the general contents and topics discussed within *Florence and the Cultural Inheritance of its Religious Heritage*, a project carried out by the Municipality of Florence in collaboration with the University of Florence and with the support of local stakeholders. The heritage of religious interest of the city is here considered from a World Heritage perspective, by following the conceptual framework of UNESCO's principles and programs. The present contribution also includes the presentation of the results of the "Survey on the impacts of COVID-19 in places of religious interest in Florence" and the possible post-COVID-19 recovery strategies identified during the project's meetings. In this respect, the paper focuses on the conceptualization of cloisters as a starting point for promoting a more general vision of "openness" of places of religious interest with mixed functions located in the metropolitan area of Florence. Cloisters are here conceived as greenspaces, open, shared and usable by the entire community in terms of hospitality and well-being enhancement. At the same time, the regeneration of cloisters will allow the community to rediscover the value of the heritage and to stimulate a renewed spirit of common responsibility toward these places.

Keywords Heritage of religious interest · Cloisters · Florence World Heritage · Post-COVID-19 strategies · Community

C. Francini (✉)
Florence World Heritage Office and Relations with UNESCO Office, Florence, Italy
e-mail: carlo.francini@comune.fi.it; mariacristina.giambruno@polimi.it

A. Montacchini
Department of Architecture (DIDA), University of Florence, Florence, Italy
e-mail: alessia.montacchini@unifi.it

T. Rozochkina
Department of History, Archeology, Geography, Fine and Performing Arts (SAGAS), University of Florence, Florence, Italy
e-mail: tatiana.rozochkina@stud.unifi.it

R. P. B. Singh et al. (eds.), *Placemaking and Cultural Landscapes*,
Advances in Geographical and Environmental Sciences,
https://doi.org/10.1007/978-981-19-6274-5_14

14.1 Introduction

We are currently witnessing an emergence of new challenges caused by the COVID-19 pandemic. In particular, the crisis has exposed the frailty of the urban management strategies in urban areas with massive and constant influx of tourists. The pandemic has underlined further all the complicacies and issues related to "overtourism" from an economic, social and cultural perspective, spanning from the residents' quality of life to the use of urban spaces.

Even in a renowned art city as Florence, the presence of millions of visitors per year has created over time many critical issues related to livability and residence in its historic center, which is a UNESCO World Heritage site since 1982. Moreover, mass tourism in addition to other phenomena—such as the reduction of residents and gentrification—has favored a strong loss of authenticity caused by the increasing homogenization of both activities and services oriented to meet, almost exclusively, the needs of tourists (De Luca et al. 2020). A city devoted to the "tourism mono-culture" has demonstrated, in these years of pandemic, all its structural weaknesses. In the medium-to-long term, the current crisis is likely to challenge the normative principles for the urban management and the heritage accessibility of the city, if the Sustainable Development Goals of the 2030 UN Agenda are to be achieved.

In this respect, in the last two years, the municipal administration of Florence has promoted a series of projects and initiatives in order to expand the cultural offer in a decentralized and widespread manner for both residents, city-users and visitors, with the aim of: (a) regenerating monumental complexes by adding new functions and services, (b) enhancing the rediscovery of lesser-known cultural "gems" of great cultural, artistic, environmental and social significance and (c) improving the accessibility of urban green areas as new points of aggregation (Comune di Firenze 2022).

As a contribution to these goals, the project *Florence and the Cultural Inheritance of its Religious Heritage*—promoted by the Municipality in collaboration with the University of Florence—has been focused on understanding the role of heritage of religious interest (here considered as any form of cultural asset with past or present religious or spiritual associations) and cloisters into the urban fabric, and has been trying to unfold their potential for urban wellness places for inhabitants and tourists. The intention is to increase the city's ability for positive adaptation to current and future shocks of social, environmental and economic nature, contributing, in this way, to the urban development of Florence as a sustainable and resilient city.

The conceptual framework of this paper is based on an analysis of the key outcomes of the thematic meetings annually held as part of the *Florence and the Cultural Inheritance of its Religious Heritage*, consisting of good practices shared by the local and national stakeholders and joint project proposals, conclusions and guiding principles for further management activities.

14.2 Florentine Heritage of Religious Interest and Cloisters

Florence is famous throughout the world for its stunning artistic and cultural heritage: a city that is "a unique social and urban achievement, the result of persistent and long-lasting creativity, which includes museums, churches, buildings and artworks of immeasurable worth" (UNESCO 2014). Florence, which has always been a commercial and cultural crossroads, has a long history of hospitality and interreligious and intercultural exchanges. Nowadays, the presence of the various religious communities that populate the city is visible in its different places of worship: besides the numerous Catholic churches of different orders, the city also boasts some Evangelical Baptist, Methodist and Waldensian churches, a Russian Orthodox church, an Israeli temple and an Islamic cultural center. The Fig. 14.1 shows the plurality of Catholic and non-Catholic worship places located both in the Historic Centre of Florence, UNESCO World Heritage site, and in its Buffer Zone. This area—which comprehends four different municipalities in the territory, meaning the municipalities of Florence, Bagno a Ripoli, Fiesole and Scandicci—provides an additional layer of protection to a World Heritage site and is intended to ensure the preservation of the immediate background, principal views and other structural and functional features of the site (UNESCO 2019).

The religious heritage of the city has broadly contributed to the inscription of its historic center into the *UNESCO World Heritage List* in 1982. The inscription of the site into the List not only ratifies the recognition of its importance, but also constitutes a responsibility toward a heritage that is no longer to be considered as belonging merely to the local and national reality, but becomes global, that is, of the entire humanity, involving past, present and future generations (UNESCO 1972). At the same time, this inscription is configured as a valuable opportunity for reflection and analysis of the opportunities for a development capable of involving local resources in a set of integrated actions of protection, conservation and cultural and economic enhancement. This precious recognition makes the site unique and of exceptional value worldwide and, therefore, the entire international community is obliged to participate in its safeguard (Francini 2013).

Like all the sites inscribed in the List, the Historic Centre of Florence has a Management Plan, a flexible, dynamic and practical tool useful for addressing proper protection and enhancement actions respecting the most important quality of a World Heritage site, its Outstanding Universal Value (OUV). The OUV is identified in the *Statement of Outstanding Universal Value* of the site and codified through a series of criteria that are set out in detail going to constitute the specific identity and value of the site itself according to the principles of the *Convention concerning the Protection of the World Cultural and Natural Heritage* of 1972 (Feilden and Jokilehto 1998).

Scrolling through the *Statement* of the Historic Centre of Florence, a vast list of palaces, churches, museums and monuments that characterize the site appears clearly; among these, there are numerous monumental complexes with individual buildings related to the cultural heritage of religious interest (localized in the map of the Fig. 14.2), in particular:

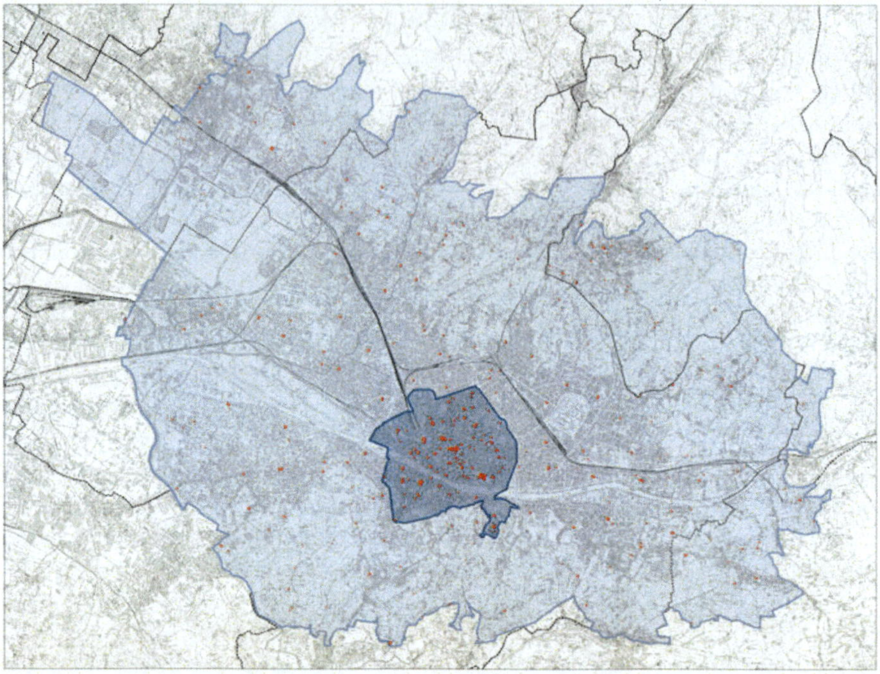

Fig. 14.1 Catholic and non-Catholic worship places (highlighted in red color) in the UNESCO World Heritage site "Historic Centre of Florence" (blue-colored) and its Buffer Zone (light blue-colored)

1. Cathedral of Santa Maria del Fiore with the Baptistery of San Giovanni and Giotto's Bell Tower
2. San Lorenzo Complex (with the Sacristy of San Lorenzo, the Medici Funeral Chapels and the Laurentian Library);
3. Santa Maria Novella Complex;
4. Santa Croce Complex with the Pazzi Chapel;
5. Basilica of Santo Spirito;
6. Church of San Miniato;
7. Convent of San Marco (UNESCO 2014).

Despite the multitude of Catholic and non-Catholic places of worship located in the Historic Centre of Florence World Heritage site and its Buffer Zone, the religious heritage of the city is made up of many other buildings that—mainly due to the secularization process of properties—are currently neglected or have changed their use and purpose over time. In fact, for what regards the Italian heritage landscape, a particular characteristic of the ensemble of religious heritage consists in the complexity of property relations: especially after the Italian unification in 1861, several waves of secularization of religious estates have resulted in a property puzzle, with some religious complexes being split between public, private and municipal properties.

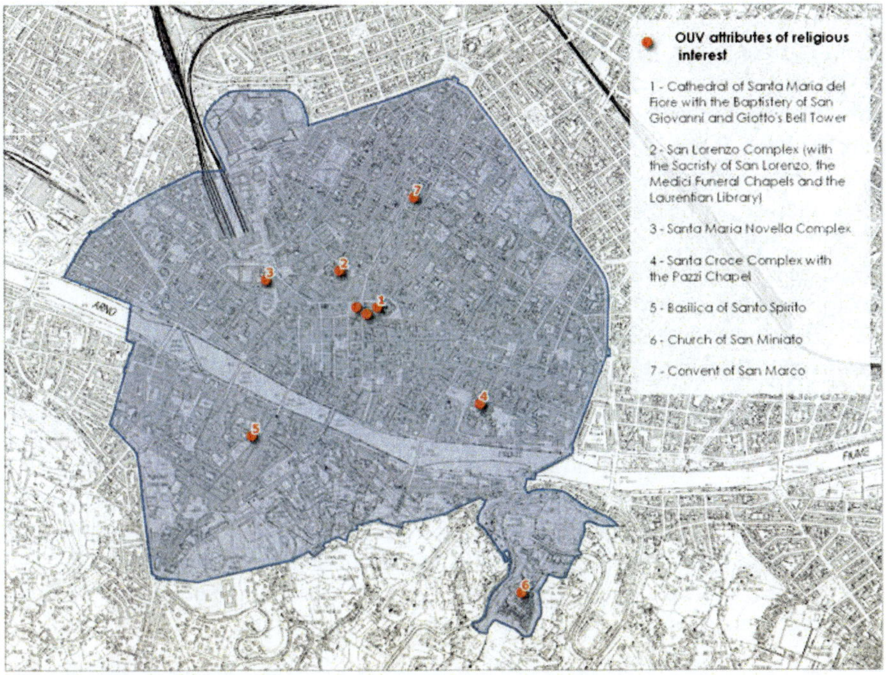

Fig. 14.2 Heritage of religious interest in the Historic Centre of Florence cited in the Statement of Outstanding Universal Value of the World Heritage site

Such intricate property relations, added to the shortage of financial resources and funding available for the maintenance of these complexes, have inevitably had consequences on the cohesiveness of the managerial strategies, significantly complicating the accessibility to these places and functional definition of their spaces (Colaianni 2012).

On the other hand, from a historical perspective, the secularization process in Italy—that in Tuscany, and therefore in Florence, began even before the Italian unification—was accompanied by the dissolution of monasteries and entire monastic orders. While some of these complexes would pass on to other religious orders preserving their original value for religious communities, other assets underwent a handover to a laical management and therefore were allocated for other intended uses, thereby losing their original religious connotation. Moreover, as a result of the significant decline in numbers of the clergy members, many of religious complexes would be left available to be later absorbed by the urbanization, acquiring new and diverse functions for secular uses.

This process is particularly felt in Florence that—as mentioned before—is a city that disposes of dense religious and monastic heritage. The succession of uses of these buildings, characterized by monumental architecture, has in some cases affected the memory of their original value, blurring the complete image of the complex historical evolution of the urban fabric. For instance, one pattern of the historical evolution of use of cloisters and adjacent structures that has been identified during the research and that has provided a common thread along which to link up different epochs to a religious heritage asset is the interrelated development of monastic complexes and health facilities. Indeed, the popular perception of historic hospitals in Florence today may fail to contextualize them in a long historical, social and religious evolution. The secularization of these structures led to the creation of modern hospitals that would now operate under the auspices of chairs of medicine. The former monastic herb gardens would develop into veritable botanic gardens where medicinal herbs used for treating disease were cultivated. One such example, still in operation to date, is the Hospital of Santa Maria Nuova in Florence, a complex which evolution is inseparable from the history of the adjacent Oblate monastic complex.

The re-functionalization of the ecclesiastical real estate and the consequent changes of the structure and the uses of the spaces is not only happening frequently in the Italian panorama, instead it is a common phenomenon currently going on in many other Western countries. In this transformative process, all changes should be recorded and documented in order to avoid the potential loss of those tangible and intangible values able to testify the asset's history. In fact, the original vocation, the evolution over time, the spirit of the place—namely its living, social and spiritual nature—are fundamental factors to be taken into consideration and acknowledged into the adaptive reuse of the buildings, with a view of creating a space that will also provide a narrative of the identity of the community of reference and the surrounding context (Oers and Haraguchi 2010).

14.3 The Framework of the UNESCO *Initiative on Heritage of Religious Interest*

A point of reference for the conservation and protection of cultural heritage, UNESCO plays a leading role in the development of guidelines for religious heritage conservation and management. Since 2016, UNESCO has coordinated a cycle of thematic consultations and workshops at the international level that brought together experts, stakeholders and representatives of different religious communities providing a unique platform for intercultural dialogue.

The necessity for a particular strategy for management of religious heritage has repeatedly been highlighted by the Advisory Bodies of the UNESCO World Heritage Centre, meaning ICCROM, ICOMOS and IUCN. This necessity stems both from the density of religious heritage as part of the general cultural heritage and from the numerosity of stakeholders, including clerical authorities, managerial bodies and the local community, that are involved in its management.

As a matter of fact, according to UNESCO, approximately 20 percent of the UNESCO World Heritage sites are places associated with the religiosity or spirituality of various cultures and traditions around the world: therefore, these sites constitute the largest category of properties on the UNESCO *World Heritage List.* For this category of properties, the Outstanding Universal Value—defined as "cultural and/or natural significance so exceptional as to transcend national boundaries and be of common importance to present and future generations of all humanity" (UNESCO 2019)—is given precisely by the strong, specific and unique spiritual significance to these places that these properties bear for one or more communities. It is because of this intrinsic value that the properties are eligible for inclusion on the *World Heritage List,* an inscription provided if they meet the Selection Criteria, the conditions of Integrity and Authenticity, and have an adequate System of Protection and Management.

Taking into consideration the high concentration and value of religious properties as a category of cultural heritage, and that many historic World Heritage cities have places of religious interest—with many of them also recognized as holy cities by various communities, the World Heritage Committee requested the UNESCO World Heritage Centre, in consultation with its Advisory Bodies, to facilitate the adequate protection of property of religious interest. Therefore, the Advisory Bodies have conducted a series of research studies, meetings and activities on the topic of heritage of religious interest and sacred sites. As a result, two recommendations, crucial not only for World Heritage sites but also for every site whose value is connected to spirituality, were elaborated.

The first one was *Conservation of Living Religious Heritage*, a reference volume containing the proceedings of the "Forum on the conservation of Living Religious Heritage" organized by ICCROM in 2003. During the Forum, various challenges were addressed, such as the pressing issue of conservation of sacred movable objects and places of heritage value, which must be reconciled with their function as part of a living religion. The second important publication was the 2005 ICOMOS General Assembly resolution, which called for "the establishment of an International Thematic Program for Religious Heritage" (§32. Religious Heritage).

Subsequently, with the kick-start of the *Initiative on Heritage of Religious Interest*, UNESCO has been playing a leading role in the development of overarching guidelines on the topic, and since 2016 has coordinated a cycle of thematic consultations and workshops at global level, during which experts, stakeholders and representatives of different religious communities come together in a spirit of dialogue, sharing and rapprochement of different cultures.

In the last years, the *Florence World Heritage and relations with UNESCO* Office of the Municipality of Florence has been seeking to promote a dialogue between the numerous stakeholders involved in the conservation and the management of the religious heritage in the Historic Centre of Florence—World Heritage site. The project *Florence and the cultural inheritance of its religious heritage* was conceived within the framework of UNESCO *Initiative on Heritage of Religious Interest* in order to promote its general principles. The establishment of the project was intended to promote awareness among the general public, so as to enhance the feeling of common responsibility toward the heritage of religious interest, both still in use and which function has changed in time.

14.4 The Florentine Experience

Taking into consideration the extent and the variety of the religious heritage of Florence, and in continuity with the normative action of UNESCO, the Municipality of Florence—through the Florence World Heritage and relations with UNESCO Office and with the scientific support of HeRe_Lab-Heritage and Research of the University of Florence, has elaborated a tailored, site-specific strategy of enhancement, conservation and management of religious heritage embodied in its project *Florence and the Cultural Inheritance of its Religious Heritage.* The launch of the project in 2017 coincided with the 35th anniversary of the inclusion of the Historic Centre of Florence in the UNESCO World Heritage List (December 17, 1982). On this occasion emerged the need to place greater attention on the heritage of religious interest present in the World Heritage site and its Buffer Zone.

The ultimate goal of the project is to deepen, together with different stakeholders, how much this heritage is or can be fundamental for a reading and an interpretation of the OUV, opening an intercultural front related to the intangible dimension of the heritage of religious interest. The involvement of those who are called to manage this reality was essential (a) to better understand the current criticalities and opportunities, (b) to open the dialogue on topics that range from examining the various conservation and maintenance actions undertaken over time to understanding how the forms of management and enhancement (in its broadest sense) have strengthened or weakened the OUV, as well as (c) to create a network—that had not been yet developed—among stakeholders (Comune di Firenze 2018).

The project was integrated into the 2018 *Monitoring of the Management Plan of the Historic Centre of Florence World Heritage site* and appears in the Action Plan within the updated 2022 Management Plan. In fact, it has been selected and included into the Plan given its strategical support to the enhancement of site's Outstanding Universal Value, as well as to the achievement of the Sustainable Development Goals of the UN 2030 Agenda (specifically, Goal n.8, 10, 11, 13 and 16). The Table 14.1 briefly explains the project's expected results and the efficacy indicators that will be employed by the Office to monitor and test the proper implementation of targeted actions and the progress of activities.

Cloisters are a significant characteristic architectural feature of the religious heritage of the city and particular attention has been lately drawn to their role respecting the Florentine territory. Cloisters can be either part of monastic foundations or adjacent to churches, even if numerous structures have lost their original religious function over time. The choice of this topic for the last project activities has been dictated by two crucial challenges faced by the Historic Centre of Florence.

The first issue regards the difficult accessibility, which arises from the fact that many fabrics that embody cloisters not publicly accessible since they (a) belong to a living religious community, (b) are allocated to a private or cultural use or (c) are underused/neglected spaces in need of strong restoration works. In particular, the intent of the initiative launched by the Municipality was to identify and possibly

Table 14.1 List of expected results and efficacy indicators for the monitoring of the project Florence and the Cultural Inheritance of its Religious Heritage

Expected results	Efficacy indicators of the project
• The creation of concrete connections with the subjects involved in the project	1. Percentage growth of local, national and international participants in the project's annual meetings
• The joint identification with the subjects involved of short-, medium- and long-term strategies for the sustainable development of the city's heritage of religious interest. The identified strategies will be included within the updating of the Management Plan of the World Heritage site;	2. Number of *Heritage Impact Assessment* conducted for neglected religious complexes under development intervention
• The drafting of a document summarizing the topics addressed. This document will highlight the peculiarities that emerged during the various meeting with the subjects involved and will help to identify local best practices for future confrontations with other national and international realities	3. Creation of a complete database on places of religious interest
• The realization of an in-depth study, such as the elaboration and the diffusion of plans and general methodologies aimed at optimizing the management of the heritage of religious interest at metropolitan level, with a focus on the World Heritage site;	4. Number of publications and activities to disseminate the results of the project (e.g., an atlas of the places of religious interest in the city and their state of preservation and use)
• The development of a digital inventory aimed at supporting the conservation and enhancement of the heritage of religious interest	

find a tactic to recover and "open" those cloisters where the religious function is not exclusive, and that have until now been difficult, or impossible, to access.

The second challenge lays into the extraordinary conditions caused by COVID-19 pandemic. Its consequences have drastically affected the World Heritage site area, bluntly exposing the problems of gentrification and depopulation of the city center, with its network of services designed, to a great extent, for the accommodation of the touristic market demands. These conditions drew the public attention to the problem of the lack of urban green spaces and peaceful accessible gathering spaces in an enclosed environment within the built-up city center, causing the Municipality to step up its efforts to make it a better living and *livable* place for the residents and city-users. In this respect, the cloisters have been conceptualized as key places to ward off the negative tourism-related effects and partly compensate for the heavy toll that COVID-19 pandemic has taken on urban experiences.

A goal-oriented conference has been held by the Municipality as part of the project in January 2021. A preparatory meeting took place in September 2020 and was aimed at identifying the main challenges and critical issues for the religious heritage brought about by the pandemic. The participants were invited to take an active part in the discussion by sharing briefly their experiences on the measures taken by each party during the lockdown period. The exchange of good practices with regard to urban spaces accessibility prompted the discussion topics for the upcoming conference, defining, at the same time, potential strategic actions for addressing the emergency situation created by COVID-19 in the city's places of worship.

Consequently, from November 2020 to January 2021, the Municipality carried out a "Survey on the impacts of COVID-19 in places of religious interest in Florence" involving different local stakeholders. The survey was administered throughout the entire Florentine territory considering both the historic center of the city as a UNESCO World Heritage site and its Buffer Zone. The primary objective was to collect more detailed information on the measures taken by the different managing bodies to deal with the impact of the COVID-19 crisis in the places of religious interest, as well as to understand the criticalities that were mostly affecting this typology of heritage.

The conference of January 2021, titled "Post Covid-19 Strategies: Rehabilitating Cloisters into Urban Oases as a means of reviving the past", was organized in view of the 2022 update of the Management Plan of the Historic Centre of Florence. This event was a crucial moment to analyze further the survey's results and to integrate the project with new initiatives, strategies and solutions so as to face the identified pressing issues. Starting from the main common issues identified, ideas were gathered to outline possible guidelines for the enhancement of cloisters related to the heritage of religious interest of the city.

With regard to cloisters, two main issues, corresponding to the discussed challenges that loom large in the Post-COVID-19 period, were addressed. The first one was related to the topic of the "public openness", in which stress was laid upon the fact that the appearance of cloisters as green, open, public areas in many cases proves to be a deceptive one. The reason for it lays in the formerly discussed intricate property relations.

Indeed, most of these "green oases" in the Historic Centre of Florence are often not publicly accessible. The opening of cloisters—such as those of Santa Croce, Santissima Annunziata, Santa Maria del Carmine, Santo Spirito and San Lorenzo—was thus seen as a strategic action, that would not only result in enhancing these places by putting forward ambitious proposals of re-functionalizing the often little-known or forgotten areas, but will also offer a place of well-being for residents, students and tourists, without interfering with the religious functions. In this case, where complexes of religious interest are managed by different subjects and communities, the improvement of cloisters' accessibility can possibly be achieved by harmonizing the communication between the stakeholders involved in management of these structures, reaching consensual agreements that would accommodate the interest of the entire community without corrupting the religious and spiritual use of the place.

The second topic of the discussion concerned the interrelation of history and heritage as embodied in cloisters. Tracing the links back into the history of these public spaces, several research proposals focused on the enhancement of places of religious interest have been put forward. Besides, the discusses need of historical and social contextualization of these architectural elements flowed into a new cultural thematic route offer.

The arguments and opinions of the participants involved provided the Municipality with a vision for the topic, and formed the ground-plan for future developments of the project. In particular, the conference resulted in a detailed mid-term agenda proposal entirely aimed at enhancement of cloisters, minor squares and religious heritage in the Historic Centre of Florence that was articulated into concrete actions gathered into three main key concepts:

(A) Hospitality

- Creation of an organic enhancement system: inclusion of cloisters into alternative urban and peri-urban itineraries and introduction into tourist circuits;
- Establishment of concrete connections with others subjects involved in the management and protection of the heritage of religious interest at the local level for network management;
- Establishment of diocesan (or inter-diocesan) offices for ecclesiastical cultural assets and places of worship, i.e., identification of personnel responsible for valorization (which the Italian Bishops' Conference began to require in 1992).

(B) Re-functionalization of neglected places of religious interest

- Use of the Heritage Impact Assessment methodology (developed by ICOMOS in 2011) for assessing possible impacts due to the adaptive reuse;
- Understanding the percentage of underused/neglected buildings and cloisters in the territory; .
- Monitoring the future or ongoing recovery work progress.

(C) Conservation

- Development and updating of preservation guidelines;
- Digital cataloging to support preservation.

14.5 Future Perspectives: Need of a Census and a Geo-referencing of Cloisters

The *Florence and the Cultural Heritage of Religious Interest* project—being considered as one of the strategic projects for the conservation and enhancement of the Outstanding Universal Value of the Historic Centre of Florence World Heritage site—the Municipality aims to carry out activities aimed at promoting protection, use and enhancement of the heritage of religious interest in the Florentine territory in its overall and unitary vision.

During the five-year period of reference, the intention is to establish, in collaboration with a plurality of subjects both internal and external to the municipal administration, a flexible set of operational rules, project ideas and actions aimed at achieving the project's objectives. To this effect, the project intends to organize further meetings with stakeholders (both public and private), a cooperation method that proved to be an effective form of monitoring the progress achieved, as well as a useful data collection tool for identifying new challenges and consequent new directions for the project.

In addition to the involvement of the many stakeholders on the ground, the agenda will place particular emphasis on the integration of the local community, often the first to signal the need for change and to perceive its consequences. By reserving a central role for the community, the project intends to promote not only participatory and collaborative planning but also to contribute to a rediscovery of the value of religious heritage to stimulate a newfound spirit of shared responsibility toward these places.

One of the pressing issues to be addressed in the coming years remains the harmonization of the complex system of actors involved in the management and valorization of the religious heritage. The constellation of stakeholders (owners, managing bodies and responsible actors) coming from different backgrounds represents an asset as it brings together approaches and methodologies proper to each field, reflecting the complex and polysemic character of the religious heritage. Nevertheless, this factor tends to complicate the management of both the heritage of religious interest as a whole and of individual church buildings.

With respect to the issue of public space, as previously pointed out, the Municipality shall carry out a project with the view to render underuse and neglected cloisters of the historic center accessible to the population. As a preliminary step, a research within historical urban development processes is being developed. For the purposes of the project, the study dwells on urban transformation processes and the evolution of the social value of Florentine cloisters and minor squares over time. Concurrent to the research is the work on a database of assets of religious interest situated both in the World Heritage site and its Buffer Zone. This assessment of the density of the religious heritage, which is necessary to plan sustainable enhancement policies for individual assets and systems, meets the broader, national objective.

A complete database of religious heritage both of public, private property and belonging to ecclesiastic authorities is a long-awaited goal for Italian heritage management. Unlike Italian clerical estate, which integral catalogue is publicly accessible, the data on religious heritage of other property types are still based on partial samples and patchy census campaigns carried out with different objectives and outcomes, which make their results obtuse and incongruous with an organic reading of the real consistency of these assets on the national territory (Rizzo and Rizzo 2015).

Playing its part in the national objective, the Municipality of Florence with the support of the University of Florence thus intents to elaborate a structured database of these complexes differentiating between the types of property relations, value of the single features, reporting, at the same time, presence of cloisters and, as related to heritage items which spiritual function was discontinued, the current use of the building. Presented as a digital thematic map, the database will be integrated into the *Florence Heritage Data System* project, another strategic project included in the

Action Plan of the 2022 Management Plan for the Historic Centre of Florence. The development of this Geographic Information System, integrated with the Municipality of Florence's Territorial Information System—Mapstore2, will be able to provide valuable information on the Historic Centre of Florence, the density of its religious heritage resources and its state of preservation and use.

A first set of cloisters, almost 40, located in the Historic Centre has been already identified as part of a preliminary census within the project. Given the multitude and variety of cloisters in the Florentine territory, this database is currently being integrated. In Fig. 14.3, it is possible to observe the location of some of the most significant cloisters that have been yet classified into the database.

In addition to cataloging the presence of cloisters in the World Heritage site and its Buffer Zone, those cloisters that are underused/neglected and that therefore can undertake a path toward a "3r" strategy (recovery, re-functionalization and reopening) will be highlighted: in fact, these peculiar cloister can be easily re-adapted, recovered and given back to the population since they do not have an exclusive use of religious type, and thus there are more possibilities to apply new projects that aim at reconnecting heritage with the community.

Next to it comes the realization of an urban planning project that will put particular emphasis on the involvement of the entire community. This step involves

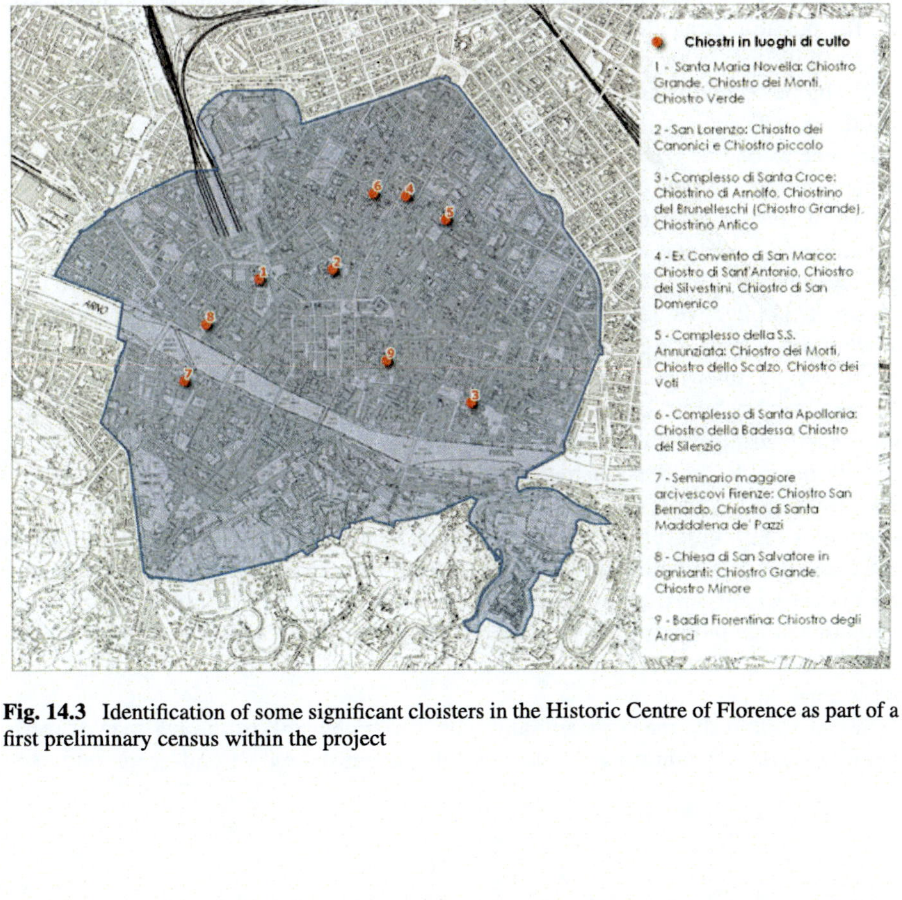

Fig. 14.3 Identification of some significant cloisters in the Historic Centre of Florence as part of a first preliminary census within the project

the identification of cloisters which use could be enhanced in collaboration with stakeholders.

The identification of target areas will be followed by assessment of the main critical issues related to these assets. As a result, integration of the offer of cultural itineraries with reference to the heritage of religious interest in the World Heritage site is expected to be achieved, an initiative that seeks to promote greater awareness of the material and immaterial qualities of the assets linked to spirituality and the historical value of the identified places.

14.6 Concluding Remarks

The COVID-19 pandemic emphasized preexistent structural management issues—particularly linked to the effect of overtourism—in the Historic Centre of Florence, World Heritage site. In this critical situation, the lack of urban green spaces and the need for more peaceful accessible gathering spaces in an enclosed environment within the built-up city center was underlined.

The events organized within the context of project *Florence and the Cultural Heritage of Religious Interest* provided an ideal platform for discussing the impacts of the COVID-19 epidemiological crisis in places of religious interest in Florence and the possible strategies for making urban cloisters accessible to citizens, city-users and visitors: a special focus will be dedicated to those cloisters that have lost their religious function or are not used exclusively by religious communities for not interfering with their worship needs.

In conclusion, the present paper proposes a twofold interpretation of cloisters that would reintroduce them as key urban features actively contributing to the articulation of the historic urban landscape. On the one hand, cloisters are interpreted as "urban oasis": safe and quiet spaces for social relations, for cultural activities and for slow tourism. Their accessibility and usability would represent an answer to the accentuated collective need to have open spaces available and usable by the entire population for improving their well-being, and to boost the livability of the historic center. On the other hand, the recovery, the re-functionalization and the reopening of underused/neglected cloisters will foster a reconnection and a reconciliation of history and heritage with the community.

With regard to the Sustainable Development Goals of the Agenda 2030—especially linked to the goals n. 8, 10, 11, 13 and 16—cloisters were thus conceptualized as features that could embody the close interrelation between community, urban fabric and the surrounding landscape at metropolitan level. By rendering the cloisters on the whole territory accessible and promoting social activities, this part of Florentine urban fabric fulfills its potential as a *fil rouge* for community involvement initiatives and strengthening the sense of common responsibility for the built heritage of Florence.

In fact, within the management and recovery framework of the heritage of religious heritage, communities should not only be considered as end-users, but also as custodians and active contributors (Council of Europe 2005). In this way, by following the principles of the Convention on the Value of Cultural Heritage for Society—best known as Faro Convention, adopted by the Council of Europe in 2005 and ratified by the Italian Government only in September 2020—the reopening and the adaptive reuse of the spaces will simultaneously become less "institutionalized" and more "socialized". A new collective consciousness is likely to arise by urging the need of a shared responsibility between communities and authorities with respect to heritage.

References

Colaianni N (2012) La tutela dei beni culturali di interesse religioso tra Costituzione e convenzioni con le confessioni religiose. Stato, Chiese e pluralismo confessionale, n. 21/2012. Available at: https://riviste.unimi.it/index.php/statoechiese/article/view/2204

Comune di Firenze (2018) Monitoring Management Plan of the Historic Centre of Florence – UNESCO World Heritage site. Second update. Florence, Comune di Firenze

Comune di Firenze (2022) The Management Plan of the Historic Centre of Florence–UNESCO World Heritage site. Florence, Comune di Firenze

Council of Europe (2005) Convention on the value of cultural heritage for society. Available at: https://www.coe.int/en/web/culture-and-heritage/faro-convention

De Luca G, Francini C, Liberatore G, Shirvani Dastgerdi A (2020) Sustainable cultural heritage planning and management of overtourism in art cities: Lessons from Atlas World Heritage 12(9): 3929. Available at: https://doi.org/10.3390/su12093929

Feilden BM, Jokilehto J (1998) Management guidelines for World Cultural Heritage Sites. Roma, ICCROM. Available at: https://www.iccrom.org/it/publication/management-guidelines-world-cultural-heritage-sites

Francini C (2013) Dicembre 1982: Il Centro Storico di Firenze è Patrimonio dell'Umanità. Storia di Firenze. Il portale per la storia della città 16 novembre 2013, pubblicazione online https://www.storiadifirenze.org/wp-content/uploads/2013/11/16-francini-unesco.pdf

Rizzo RG, Rizzo LS (2015) Religious heritage in Italy: websites and geolocalization. A new appraisal. Bollettino dell'Associazione Italiana di Cartografia, EUT Edizioni Università di Trieste, c(154): 121–136. Available at: https://www.beweb.chiesacattolica.it/UI/attachments/3_Patrimonio_religioso_Italia_siti_web_e_geolocalizzazione_AIC154.pdf

UNESCO (1972) Convention concerning the protection of the World Cultural and Natural Heritage. Paris, UNESCO. Available at: https://whc.unesco.org/archive/convention-en.pdf

UNESCO (2014) Retrospective Statement of Outstanding Universal Value of the Historic Centre of Florence. Paris, UNESCO. Available at: https://www.firenzepatrimoniomondiale.it/wp-content/uploads/2015/11/Retrospettive-Eccezionale-Valore-Universale.pdf

UNESCO (2019) Operational Guidelines for the Implementation of the World Heritage Convention. Paris, UNESCO World Heritage Centre. Available at: https://whc.unesco.org/en/guidelines/

Van Oers R, Haraguchi S (2010) Managing Historic Cities. World Heritage Papers No.27, Paris, UNESCO World Heritage Centre, Available at: https://whc.unesco.org/document/105467

Part III
Cultural Landscapes: Emergence and Scenario

Chapter 15
Archaeological Sites in Northern Japan: Interfacing Landscape and Sacred Rituals

Olimpia Niglio

Abstract The history of the ancient origin of Japan is extraordinary. Every country must know and preserve its history, and Japan has always demonstrated a great capacity for the research of ancient origins. The complex history of this country has also cancelled many ancient sites and sacred places, but in these last years, the archaeological studies allowed to redeem important and fundamental information and heritage on these sites. In Japanese history, the Jomon culture is very important to also understand the actual traditions that continue the contemporary society and especially at the Ainu Community. This paper aims to propose a reflection on 17 archaeological sites between Hokkaido and the part to the North of Honshu. A series of settlements, burial area, ritual, and ceremonial sites of which stone circles, among other sites, are the remains of a unique sedentary, hunter-fisher-gatherer society developed spanning a very long period of time from 13,000 to 400 BCE. This heritage is extraordinary to understand the organization of ancient Japanese society and above all to know the rituals and sacred sites of this community that have left important traces for Japan's history. This paper will present some reflections on this sacred historical context, thanks to the support also of special images.

Keywords Jomon culture · Hokkaido · Archaeological sites · Cultural landscape heritage · Images

15.1 Introduction

On July 27, 2021, the Jomon Prehistoric Sites in Northern Japan were inscribed on the UNESCO World Cultural Heritage List. This important inheritance is located between Hokkaido and in three prefectures of Honshu: Aomori, Iwate, and Akita.

O. Niglio (✉)
Department of Engineering and Architecture, University of Pavia, Pavia, Italy
e-mail: olimpia.niglio@unipv.it

Faculty of Engineering and Design, Hosei University, Tokyo, Japan

R. P. B. Singh et al. (eds.), *Placemaking and Cultural Landscapes*,
Advances in Geographical and Environmental Sciences,
https://doi.org/10.1007/978-981-19-6274-5_15

The project research was realized under the guidance of the Agency for Cultural Affairs, Government of Japan that for many years has worked on it with the support the academic institutions and specialized scholars. All relevant local governments, including the prefectural governments of Hokkaido, Aomori, Iwate and Akita, have developed the Comprehensive Preservation and Management Plan, setting out the basic policies for conserving of this important Japanese heritage in its entirety.

The Jomon Prehistoric Sites in Northern Japan consist of 17 interesting archaeological sites: Odai Yamamoto, Kakinoshima, Kitakogane, Tagoyano, Futatsumori, Sannai Maruyama, Ofune, Goshono, Irie, Komakino Stone Circle, Isedotai Stone Circles, Oyu Stone Circles, Kiusu Earthwork Burial Circles, Omori Katsuyama Stone Circle, Takasago Burial Site, Kamegaoka Burial Site, and Korekawa Site. These sites are characterized by a series of settlements, burial areas, and ritual and ceremonial sites with stone circles that are remains of a unique sedentary, hunter-fisher-gatherer society developed spanning a very long period of time from 13,000 to 400 BCE.

Northern Japan, where the nominated property is located, has varied landforms such as mountains, hills, plains, and lowlands as well as inner bays, lakes, and rivers with an abundant flow of water. This area was blessed with forest and aquatic resources. Cool-temperate deciduous broadleaved forests ("Boreal Beech Forest") provide abundant forest resources such as chestnuts and walnuts, etc. The intersection of warm and cold currents creates ideal fishing conditions off the coast. Also, migratory fish such as salmon and trout swim upstream in the river. Under such environmental conditions, people living in Northern Japan were able to secure food in a stable manner, and they initiated a sedentary way of life around 15,000 years ago as indicated by the use of pottery. Over the subsequent 10,000 years, they continued hunter-fisher-gatherer lifeways without changing to an agrarian culture, adapting to environmental changes such as climate warming and cooling and the corresponding marine transgression and regression (Kosugi et al. 2009).

Already in the very early stage of sedentary life, they developed a complex spiritual culture. They made graves and also created ritual deposits, artificial earthen mounds, and stone circles that were used for rituals and ceremonies, probably functioning as places for ancestor and nature worship and confirming a social bond across the generations and the settlements.

Thus, the Jomon Prehistoric Sites in Northern Japan have Outstanding Universal Value as a testimony to the pre-agricultural lifeways and complex spiritual culture of prehistoric people, attesting to the emergence, development, and maturity of sedentism of a globally rare sedentary hunter-fisher-gatherer society in Northeast Asia which developed over a very long period of time.

This paper intends to analyse this specific cultural heritage in the North of Japan for sharing some knowledge very important to understand the Japanese history and the development of this community in the extreme Eastern area on the Pacific Ocean. In this paper, some sites of Jomon Prehistoric Culture will be analysed in detail (Underhill and Habu 2006) (Fig. 15.1).

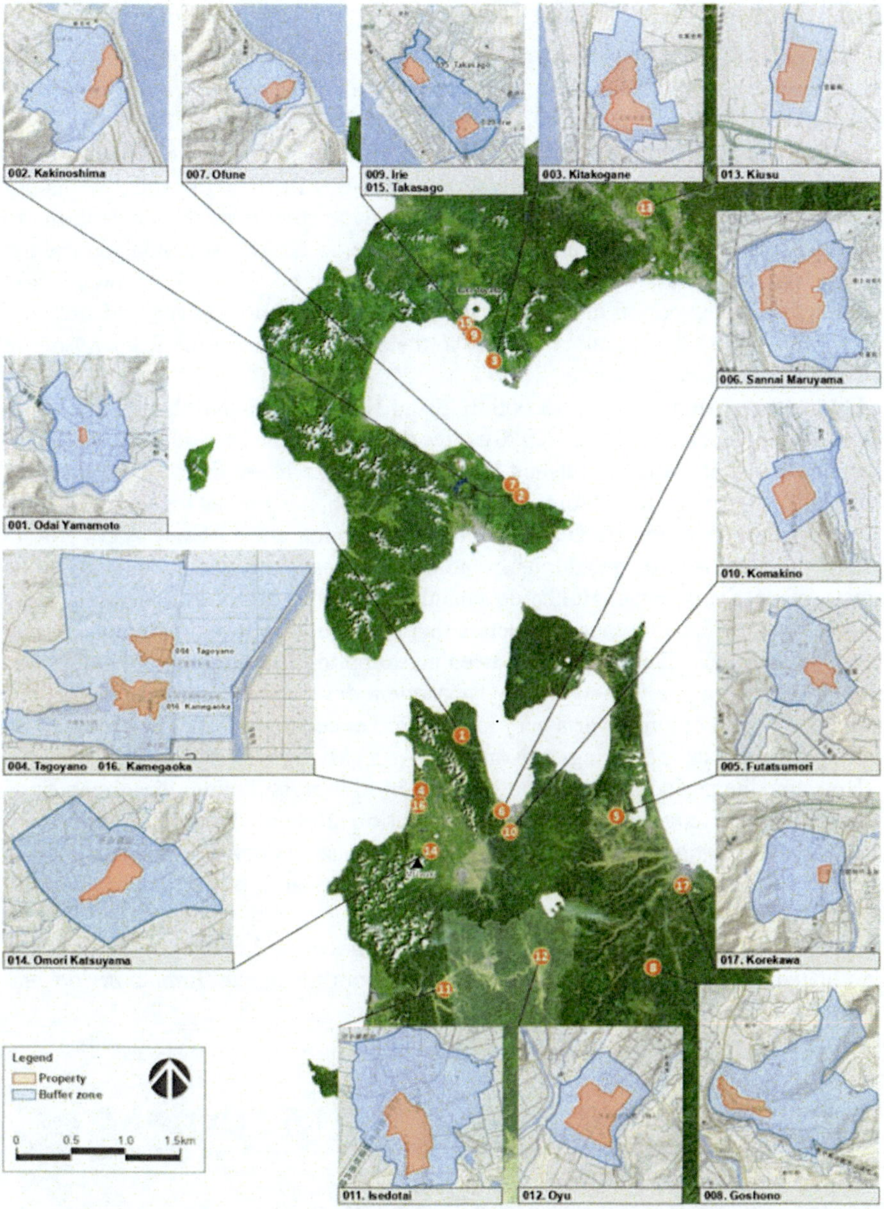

Fig. 15.1 Main Jomon Prehistorical Sites in the North of Japan, between Hokkaido and Honshu: (1) Odai Yamamoto, (2) Kakinoshima, (3) Kitakogane, (4) Tagoyano, (5) Futatsumori, (6) Sannai Maruyama, (7) Ofune, (8) Goshono, (9) Irie, (10) Komakino Stone Circle, (11) Isedotai Stone Circles, (12) Oyu Stone Circles, (13) Kiusu Earthwork Burial Circles, (14) Omori Katsuyama Stone Circle, (15) Takasago Burial Site, (16) Kamegaoka Burial Site, (17) Korekawa Site (*Source* Dossier Nomination, Agency for Cultural Affairs, Government of Japan)

15.2 Jomon Culture: The History

The Japanese history attests to a pre-agricultural lifestyle led by prehistoric people in the North of Japan, which started around 13,000 BCE and continued up to around 400 BCE. This time is categorized as the Jomon period in the history of Japan but it's interesting to understand as this period develops in dialogue this the history with other cultures. It is contemporaneous to a wide span from the Palaeolithic to Neolithic periods, and even partly the Bronze Age, in Northeast China and the Far East Russia of Northeast Asia, from the Palaeolithic to the Spring and Autumn period and Warring States period in China to the south of the Yellow River, and from the Palaeolithic period through the Iron Age to the establishment of the Roman Empire in Europe (Imamura 1996) (Fig. 15.2).

During this period, between 13,000 BCE and 1000 BCE the North of Japan, as the global climate became warmer rapidly after the Palaeolithic period, a rice-farming agricultural society was established along the Yangtze River of China against the backdrop of evergreen laurel forest. In Northeast Asia, it is known that hunter-fisher-gatherer societies were established in Northeast China and in Far Eastern Russia against the backdrop of deciduous broadleaved forests and deciduous coniferous forests, respectively. From Hokkaido Island to Okinawa Island in the south, there appeared hunter-fisher-gatherer societies that adapted to specific geographical and climatic conditions. This society has been an exceptional testimony to the development of sedentism from emergence to subsequent development and maturity and of complex spiritual culture over a very long period exceeding 10,000 years, based on hunting, fishing, and gathering, in Northeast Asia (Mizoguchi 2002).

However, the periodization of the Jomon Period is all still under discussion since archaeological discoveries in continuous evolution allow to characterize with more and more information this specific period of Japanese history. For this reason, it is essential to compare it with important studies and research projects carried out by Japanese scholars.

Professor Taniguchi Yasuhiro, a Japanese scholar at Kokugakuin University, reevaluated transition processes from the Palaeolithic to the Jomon period and

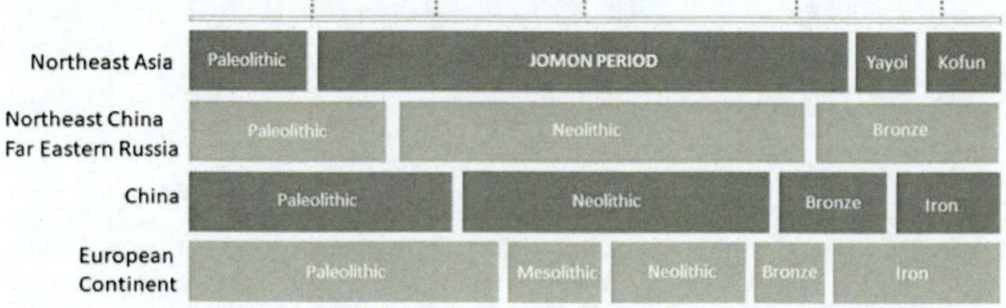

Fig. 15.2 Jomon period in comparison with the world history (*Source* Olimpia Niglio 2021)

concluded that his concept of "the Jomon period" could be applied only after the beginning of Early Jomon. He stressed that this should be the framework of the Jomon culture and the Jomon period. (Taniguchi 2010). However, analysing very well this topic it's very important to start from the article edited by Tanifuji Yasuhiko, scholar at Gunma Archaeological Research Foundation that has proposed an interesting list of the main publications and projects research on Jomon Prehistorical Sites. Indeed, this publication introduces many scientific references and the different opinions from Japanese scholars involved in this specific topic (Tanifuji 2013).

Other important references were archived in the Bulletin of the National Museum of Japanese History that, between 2009 and 2013, has edited many articles on this complex period of Japanese history; these articles allowed us to understand very well this long Jomon period and to know the cultural development in relationship with the local communities that lived in different sites from the north to the south of Japan.

This classification in different historical periods is fundamental to analyse and detail the different organizational methods of the settlements founded by the Jomon community in different places in northern Japan. About periodization, the theory of Professor Keiji Imamura is very important. In fact, Imamura does an excellent job of synthesizing the most recent results of Japanese prehistoric archaeology with a particular focus on the studies of Jomon culture. (Imamura 1996; Habu 2004). Meanwhile, despite his innovative discoveries, unfortunately his scientific research contains only a discussion on the theoretical and methodological implications of the studies of the Jomon culture in the context of world archaeology. The isolation of studies of the Jomon culture in the context of world archaeology stems from the fact that the results of archaeology are published mainly in Japan and only in Japanese. Even before the 70s, when the amount of information available was really very small, presenting the results of Jomon archaeology in non-Japanese languages was a very difficult task.

Meanwhile, in recent years numerous documents have been published, especially on the investigations of archaeological excavations, and this does not facilitate the synthesis of the results on Jomon archaeology also in other languages. At the same time, large differences in theoretical and methodological approaches do not facilitate intercultural dialogue between Japanese culture and other archaeological traditions in the world. (Habu 2004). On the one hand, many Japanese archaeologists, who were trained in the scientific tradition that considers archaeology as history, believe that Japanese prehistoric cultures, including Jomon, are historically unique. As a result, they believe that sharing with other prehistoric cultures does not contribute to improving knowledge of the Jomon culture data. In truth, many non-Japanese archaeologists interested in Jomon's archaeology are frustrated by the overemphasis on the types of pottery created by Japanese researchers, as well as their orientation towards historical-cultural and/or empiricist research. This frustration is evident in the writings published by North American and British archaeologists especially in scientific journals.

After clearing up this complex issue about the historical periodization, the main purpose of this contribution now is to show some archaeological sites of Jomon Culture in the north of Japan with a specific focus on the sacrality and their ancient rituals. Indeed, these rituals have contributed to the transformation of the lands and also for organizing the urban structure of the settlements. On these specific topics, we focus our observations now.

15.3 Jomon Archaeological Sites

The research by Professor Yunko Habu proposes an interesting methodological approach to understanding the Jomon settlements. He suggests two different theoretical perspectives: the first perspective is derived from ecological anthropology, which can be briefly defined as "the study of cultural behavior in its natural and social environment, in terms of its relationship to this environment". The second perspective that underlies the discussion in his research is the recognition that no archaeological practice is free from the social contexts in which it is conducted (Habu 2004). Following this interesting vision, it was possible to approach the Jomon culture, above all for a foreign scholar that knows Japan's history very well but always needs clarifications in relation to the specific conditions of these ancient sites.

The analysis of these interesting ancient sites also introduces another important topic: the different cultural approach between Japanese scholars and Occidental scholars that studied but today continue to deep these issues about the Jomon culture. Indeed a topic is the interpretation's method.

> The issues dealt with in Jomon archaeology include various controversial topics such as the origins of pottery and the development of hunter-gatherer cultural complexity [...] However, given the complex social and academic milieus that surround both Japanese [...] detaching the results of Jomon archaeology from their social and academic context would hinder accurate understanding of the current status of Jomon studies. Discussions on various controversial issues should be evaluated without dismissing the social and academic contexts in which Jomon data have been collected, presented, and interpreted. (Habu 2004, 18)

Enhancing the ecological anthropology and archaeological practice proposed by Professor Yunko Habu, the analysis of the sites reflects the different methodology for preserving these ancient settlements.

Japanese archaeologists were trained in the tradition of "archaeology as history", believe that every single archaeological site is unique and therefore should be protected as much as possible. Under the land development policy of the Japanese government, however, the ideal of site preservation is typically substituted by systematic rescue excavation: while the site itself would be lost, at least information contained in the site is documented in the excavation record (Kudo 2012). This practice is common in Japan and today many governmental institutions realize museum to preserve the memory but not the site. In the metropolitan areas and at the big cities, this practice is frequent and so many ancient sites were cancelled, preserving only some findings considered important for the memory of the site.

Differently, in many rural areas, the solutions were different because the pushing for building real estate has not been strong and incisive as well as in the big cities. In this specific perspective, it's important to analyse some Jomon archaeological sites.

Surely, in these sites, a common element was the discoveries of pottery's objects that allowed us to understand ancient uses and daily practices, as well as the importance of hunting and fishing for survival.

15.3.1 Odai Yamamoto Site

Odai Yamomoto is located in Sotogahama Town in the central part of the Tsugaru Peninsula in Aomori Prefecture, in the north of Japan, on a river terrace at an altitude of 26 m on the left bank of the Kanita River. It is near a river that salmon and trout, which are important food sources, seasonally ascend and also in a place where high-quality stone materials that are suitable for stone tools necessary for daily activities are available.

This settlement attests to the use of pottery for cooking food and boiling water and specific ancient technique for the construction of this location. The emergence of pottery, which is not suitable for a mobile lifestyle, is eloquent evidence of the inception of sedentism, or a sedentary lifestyle in a fixed location. The dwelling site was an oval area measuring 26 m North–South and 20 m East–West, and contained places where pottery was used, and stone tools were crafted. From this, it is thought that there were some land-use rules or regulations. However, rituals and ceremonies were not yet differentiated. The radiocarbon dating of the carbonized material on the unearthed pot shards shows that they date back to around 13,000 BCE, the earliest in Northeast Asia.

The unearthed stone tools show the same characteristics as the artefacts of the preceding late Palaeolithic period, but at the same time, stone arrowheads have also been unearthed, which suggests that people started to use bows and arrows. This site dates from the period when global warming occurred, and people's lifestyle was drastically changing from nomadic lifeway to sedentism (ca. 13,000 BCE). People moved from the mountain area to the plains and selected the location of the sedentary life to secure food stably and maintain their living. It is an important archaeological site attesting to the inception of sedentism and people's livelihood near the river.

15.3.2 Sannai Maruyama Site

Always in Aomori Prefecture, there is the Sannai Maruyama Site. It's located in the central part of Aomori Prefecture, on a river terrace at an altitude of 20 m on the right bank of the Okidate River flowing into Mutsu Bay. It faces on the inland bay and river mouth, with deciduous broadleaved forests abundant in forest resources expanding in the hinterland. The settlement site dates from the latter half of the development stage of sedentism. It was a long-lived hub settlement, comprising various facilities. It covered the whole terrace. The northern part was the residential area, consisting

Fig. 15.3 Aomori Prefecture. Sannai Maruyama Site. Archaeological remains of a large pit dwelling (*Source* Olimpia Niglio 2021)

of pit dwellings and a large pillar-supported structure; the eastern part was the burial area. There was a clear distinction between these areas.

In the burial area, graves were aligned in rows. Several large artificial earthen mounds that are considered to have been the places to perform rituals and ceremonies were constructed over a long period of time. A large number of clay figurines and ritual tools have been unearthed, indicating that nature worship and ancestor worship were practiced continually. In addition, as a result of archaeological excavations, hunting tools, fishing tools (hooks and harpoons), crafting tools, bones of various fishes, and seeds of chestnut and walnut have been unearthed. This shows that people in this settlement managed to use natural resources (Okada 2003).

This archaeological site is a large hub settlement, mainly from the second half of the development stage of sedentism (ca. 3,000 BCE–2,200 BCE). It is an important archaeological site attesting to the livelihood around the inland bay, large hub settlements, and the diversity of rituals and ceremonies (Fig. 15.3).

15.3.3 Komakino Stone Circle

Near the Sannai Maruyama Site is located Komakino Stone Circle, an interesting Jomon settlement on a tongue-shaped plateau between two rivers at an altitude of 80–160 m, with the western slopes of the Hakkoda Mountains extending to the southeast. In the hinterland, deciduous broadleaved forests abundant in forest resources expand. This stone circle dates from the first half of the maturation stage of sedentism.

Since it is not only a cemetery but also a ritual place, it shows a high degree of spirituality. Because no other stone circles have been found in the surrounding area, it is considered to have been a ritual place accompanied by a cemetery that was used and maintained jointly by several settlements over a wide range.

The stone circle consists of three rings measuring, respectively, 2.5, 29, and 35 m in diameter. In addition, it is surrounded by another set of stones placed in a ring measuring 4 m in diameter and a row of stones partly forming the 4th ring. The total area is as large as 55 m in diameter. For the outer and middle rings of the stone circle, flat stones are aligned alternately in longitudinal and latitudinal directions, as if in a stone wall, showing a distinctive alignment pattern. Before the stone circle was made, the ground had been levelled in advance by cutting the higher land and filling the lower land with the cut earth. Besides, the stone circle, pit graves, and burial jars have also been discovered (Fig. 15.4).

In addition, mainly from the stone circles' burial area, and dumping grounds, not only pot shards and stone tools, but also ritual artefacts have been unearthed, such as clay figurines, miniature pots, animal-shaped clay objects, bell-shaped clay objects, triangular stone plates, and stone discs. In particular, more than 400 triangular stone plates have been excavated. They were probably used in rituals and ceremonies at the stone circle. This site dates from the first half of the maturation stage of sedentism (ca. 2,000 BCE). It is a ritual archaeological site, centering on a stone circle. It is an important archaeological site that attests to the livelihood as well as rituals and ceremonies in the hill area.

Fig. 15.4 Aomori Prefecture. Komakino Stone Circle, sacred site (*Source* Olimpia Niglio 2021)

15.3.4 Oyu Stone Circles

In Akita Prefecture, near the Kazuno city, we find an important Jomon site, very famous for interesting sacred places that it preserves. This site dates from the first half of the maturation stage of sedentism (ca. 2,000 BCE–1,500 BCE). It is an archaeological site of a burial area, centering on stone circles that attests to the livelihoods in the inland area as well as rituals and ceremonies (Fig. 15.5).

The settlement is located in the northeastern part of Akita Prefecture, on a plateau at an altitude of approximately 180 m on the left bank of the Oyu River, a tributary of the Yoneshiro River. It sits close to the rivers where salmon and trout ascend and can be caught as food source, and deciduous broadleaved forests abundant in forest resources expand in the hinterland. These stone circles date from the first half of the maturation stage of sedentism. They are not only a burial area, but also a ritual and ceremonial space. Because no other stone circles have been found in the surrounding area, it is thought that several settlements over a wide region jointly used and maintained this ritual place.

There are two stone circles: the larger one with a diameter of 52 m is called Manza Stone Circle and the smaller one with a diameter of 44 m is called Nona-kado Stone Circle. Both consist of two rings made of river stones, which are aligned in various manners. The shapes of these two stone circles are similar in many ways.

Fig. 15.5 Akita Prefecture. Oyu Stone Circles, a sacred place of Jomon culture (*Source* Olimpia Niglio 2021)

Particularly, the stones at the centre of these stone circles and the so-called sundial stonework appear to lie on the same straight axis. Therefore, it is highly probable that these stone circles were built in association with each other.

It has been found that, in the surrounding area of the stone circles, stone alignments that are thought to have been burial areas are widely distributed and that the archaeological remains of pillar-supported structures, storage pits, and pit graves remain concentrically, centering on each stone circle. A large number of clay figurines and other ritual tools have also been unearthed. These clearly show that this place was a burial area and a ritual place. Also, since hunting and fishing tools, processing tools, and nuts such as chestnut and walnut have been unearthed, it is understood that people practiced hunting and gathering.

15.3.5 Kitakogane Site

In Hokkaido Island, there are six important Jomon Sites: Kakinoshima, Kitakogane, Ofune, Irie, Kiusu Earthwork Burial Circles, and Takasago Burial Site. Since 2021, the sites are in the World Heritage list. In this island, the Jomon culture is very important also to understand the development of Ainu community (Kitahara 2018).

Surely, a site important to understand the formation of the first settlement on this island is Kitakogane site that is in Date City on the eastern bank of Uchiura Bay in the southwestern part of Hokkaido, on a hill at an altitude of 10–20 m. It faces Uchiura Bay, abundant in aquatic resources, and is backed by deciduous broadleaved forests abundant in forest resources.

This is an archaeological site dating from the first half of the development of sedentism. The residential area, burial area, and shell mound were sited closely on a hill. In the lowland, there remain spring and archaeological remains of a watering-place. Near the residential area, a shell mound and a burial area were formed in an integral manner. Because there are various types of archaeological remains, this site is considered to have been an important settlement to know the development of this area. The shell mounds and the residential area were on the top of the inland hill, and the shells found in the shell mounds were mainly those of common orient clam, which inhabit the warm environment. This is interesting information to know also the moving of these communities from the mountains to the lowland near the sea. Indeed, in the settlements near the sea, a big number of the shells of common orient clam were found.

The way that environmental adaptations were made can be seen in the change in the location of shell mounds and the residential area as well as in the composition of shellfish species found in the shell mounds. These settlements near the sea were conditioned also by climate changes. Shells unearthed from the shell mounds (common orient clam, oyster, and scallop), bones of fish (tuna and flounder), and bones of marine animals (fur seal and whale) show the characteristics of livelihood of this region that was based on fishing.

From the spring and the archaeological remains of a watering place, a great number of deliberately broken stone mortars and stone dishes have been unearthed. It is thought that this place had a ritual function associated with the disposal of stone implements. In addition, animal bones were intentionally aligned in the shell mounds, which also shows a high degree of spirituality. This site dates from the first half of the development stage of sedentism (ca. 5,000 BCE–3,500 BCE). It is an archaeological site of a settlement accompanied by shell mounds. It is an important archaeological site attesting to the livelihood in the coastal area, people's adaptation to climate change, facilities that give a clue to the characteristics of a settlement with sophisticated spiritual cultures.

15.3.6 Irie Site

This site is located in Toyako Town in the southwestern part of Hokkaido, on a terrace overlooking Uchiura Bay at an altitude of about 20 m. It faces Uchiura Bay abundant in aquatic resources, and deciduous broadleaved forests abundant in forest resources expand in the hinterland. The archaeological site dates from the first half of the maturation stage of sedentism (ca. 1,800 BCE). It is an urban settlement distributed in the surrounding area of the ritual centre and cemetery and supported the latter. It is an important archaeological site that attests to the livelihood and spiritual lifestyle in the coastal area, mainly dependent upon aquatic resources.

This site is located in a contemporary planning area, and for this reason, for many years it was interested from real estate speculations.

This settlement site dates from the first half of the maturation stage of sedentism, and extended near the periphery of the terrace, comprising the residential area with pit dwellings and the burial area with pit graves. Shell mounds were formed in the periphery of the terrace and on the slopes of the terrace. The size of the settlements became smaller, as compared to those in the preceding stage. The distribution of settlements became more dispersive. From the shell mounds, shells (Japanese cockle and mussel), bones of fish (herring, rockfish, sea perch, and tuna), bones of mammals (Hokkaido deer and dolphin), and other animals have been unearthed in large amounts. The fact that various fishing tools, fish bones, and shells have been unearthed shows that fishing was actively practiced. In addition, special bone-made ornaments have been discovered indicating that there were the characteristics of ritual place and showing a high degree of spirituality.

From the burial area, the human bones have been found without pit graves. Among them, one adult skeleton showed that the person was infected with polio (infantile paralysis) and managed to live long with paralysed limbs probably with the help of other people (Fig. 15.6).

Fig. 15.6 Hokkaido, Otaru, City Museum. Jomon pottery (*Source* Olimpia Niglio 2021)

15.4 Sacred Stone Circles

The particularity of Jomon settlements is sacred places. The first communities (13,000 BCE) already built big stone circles, placing gravels in a ring or a circle. They were ritual places accompanied by communal cemeteries.

Stone circles vary in terms of location and the way that stones are aligned. Some are a single stone circle and others are multiple stone circles existing in one place.

Some stone circles comprise a single ring and others comprise concentric multiple rings. In some cases, stone circles indicate special relations to the orientations of the sun at summer solstice, winter solstice, spring equinox, and autumnal equinox, or surrounding landforms and landmarks such as mountains. Also, some did not last long, and others did. Each stone circle is quite distinctive. In some cases, they were found accompanied with pit dwellings, storage, facilities, and pillar-supported facilities. But these facilities are considered to have been built for the purpose of maintaining and managing stone circles, instead of for residential purposes. Stone circles are usually built outside settlements. As settlements became smaller and sparse during the same period, it is thought that people built stone circles also to confirm a bond among them by working together for construction and rituals (Kaner 2011).

In these sites, in the north of Japan, the sedentary communities had developed these sacred places in dialogue also with agriculture. In the world, it's very common the relationship between astrology, sacred places, and agriculture. This dialogue has been very strong in Jomon culture, and for this reason in these sacred places, seeds and fossilized pollens of domesticated plants dating from the Jomon period have been unearthed as a result of archaeological excavations (Habu 2014) (Fig. 15.7).

However, no archaeological remains of paddy fields, farmlands, and so forth have been found. The domesticated plants were limited to gourds and legumes; neither rice nor wheat has been found. Therefore, this property is not considered to be of an agrarian culture. The agrarian culture in the Japanese archipelago was basically the cultivation of rice as a single crop, which entailed significant landforms change and land development to make paddy fields, farmlands, water channels, and levees. Special tools for agriculture were used to manage and cultivate rice and wheat mainly. Around 1,000 BCE, rice cultivation was introduced from the Asian continent to the northern part of Kyushu Island of the Japanese archipelago through the Korean Peninsula. It spread to the northern part of Tohoku over the next 700 years but did not reach Hokkaido. The period of agriculture that spread in the Japanese archipelago is called the Yayoi period. With the beginning of the Yayoi period (300 BC), the period

Fig. 15.7 Akita Prefecture. Sacred stone circles (*Source* Olimpia Niglio 2021)

of this property based on hunting, fishing, and gathering for livelihood came to an end (Insoll 2011).

In these sites, during the archaeological excavations, many potteries were found and thanks these discoveries the archaeologists understood more also about artistic techniques and delay traditions.

Clay figurines, locally known as "dogu", are the most universal ritual artefacts that have been unearthed from the archaeological sites of the Jomon period. It is not clear exactly for what purpose they were used. However, because most of them represent women, some researchers maintain that they were related to delivery or childbirth and were used to pray for fertility or rebirth. In the early phase, they were in the shape of simple tablets. Around 3,000 BCE, the details of the human body came to be represented, such as the face, arms, stomach, and breasts. Furthermore, around 2,000 BCE, various expressions appeared, such as standing and seated postures. The expressions related to pregnancy and childbirth became particularly distinctive. Because of this presence of pottery and figurines, Japanese and Korean archaeologists have traditionally called the Jomon and Chulmun periods the "Neolithic" (Jordan & Zvelebil 2009).

During the long period of Jomon culture (from 13,000 to 400 BCE), it's interesting to verify the changes of the community's organization, and also the development of the settlements in different ways. A sedentary way of life continued to be formed in the Yayoi period. However, it had different characteristics from those of the sedentary way of life of a hunter-gatherer culture, as follows. Settlements gradually came to be concentrated in the plains that were suitable for rice cultivation. The locations of the settlements became standardized, and the diversity in settlement locations of the hunter-fisher gatherer culture was lost. Also, settlements came to be clearly delineated with large moats and so forth. The different concept of settlement area was born from that of a society based on hunting, fishing, and gathering.

In the Yayoi period, metal tools such as bronze bells, which were the communal property of the settlement, were used in rituals and ceremonies to pray for stable agricultural production. They differed completely from the ritual tools of the Jomon period, which were small clay or stone objects to pray for the good catch and stability of hunting, fishing, and gathering. Also, in the Yayoi period, the productivity of agriculture improved, and the difference in productivity gave rise to economic disparity as well as social hierarchy among different settlements. Economically advantageous settlements annexed and controlled the surrounding settlements and became larger. Within the settlement, powerful individuals appeared and took the leading position. These individuals began to preside over rituals and ceremonies. Shrines were built in association with the worship of individuals, and special tombs to bury individuals began to be built. Then, rituals and ceremonies for individuals came to be newly performed. These styles of rituals and ceremonies in the Yayoi period were significantly different from those of the hunter-fisher-gatherer culture in which, it is thought, members of the settlement were involved in rituals on equal terms.

All these changes were analysed by Professor Junko Habu that affirmed:

> […] Factors such as population increase, climate change, long-distance trade, craft specialization, and social inequality have been suggested as conditions, causes, and consequences of the development of early sedentism. It is unlikely that a single factor was the sole cause that facilitated the phenomenon. Rather, a combination of multiple factors that were closely linked with local and regional environments and sociopolitical systems has resulted in the unique historical trajectories of subsistence-settlement practice in each region. (Habu 2014, 726)

These observations are very appropriate because they clearly anticipate the social organization that characterized Japan throughout the Meiji period and how this system, although veiled in other forms, continues to exist. This is also a fascinating factor in the history of the Japanese people.

15.5 Concluding Remarks

In Japan, the Agency for Cultural Affairs has designed an interesting plan for the protection and the valorization of these Jomon sites, under the Law for the Protection of Cultural Properties (1950 and 2004). Every year, a week from November 1st to 7th is denominated Cultural Properties Protection Week. During this week, many important cultural sites are opened and visited in cooperation with the local governments and municipal boards of education.

In 2021 the Japanese Government has implemented also other important topics as the UN Agenda 2030 and climate change in dialogue with COP26 in Rome and Glasgow. After the nomination of the seventeen Jomon sites in the north of Japan (July 27, 2021), it has been interesting to verify like the knowledge of this ancient culture had a growing interest also in the young generations. This history allowed many schools to reflect on many issues very important during the pandemic: the change of the social organization and the climate change.

Indeed, the issues have characterized the long history of Jomon communities, and to read again the challenges that these people faced it's fundamental for understanding the development of the Japanese culture.

Many Japanese and foreign scholars have analysed the different changes of the social organizations of the Jomon communities in relationship with the ecological and climatic needs. These ecological approaches have favorited a deep knowledge of the epochal changes of this Jomon civilization that known nature's laws very well (Kirch and Hunt 1997; Kawahata et al. 2009; Habu and Hall 2013).

So, today, this historical knowledge defines an important base of reflection also for analysing the actual changes of a social organization more complex, but which needs are interconnected to the nature. Jomon culture allows us to deepen the dialogue between human needs and cultural landscape, and also to appreciate the contemporary design paradigms that find important references in this ancient culture. Beyond these 17 nominated sites, in Japan, there are many other important Jomon sites which transformations are examples that we must valorize.

Analysing again the Jomon history in dialogue with the issues of climate change and sustainability, it helps us to understand the cyclicity of life and the contemporary of the past. Indeed, the old experiences and knowledge are important supports to face the present and to build the future.

References

Habu Y (2004) Ancient Jomon of Japan. Cambridge University Press, Cambridge

Habu Y (2014) Early Sedentism in East Asia: From Late Palaeolithic to Early Agricultural Societies in Insular East Asia. In: Renfrew C, Bahti P (eds) Cambridge World Prehistorv. Cambridge University Press, Cambridge, pp 724–741

Habu J, Hall ME (2013) Climate change, human impacts on the landscape, and subsistence specialization: Historical ecology and changes in Jomon hunter-gatherer lifeways. In: Thompson V, Waggoner J (eds) Archaeology and the historical. Ecology of small scale economies. University Press of Florida, Gainesville, pp 65–78

Imamura K (1996) Prehistoric Japan: New perspectives on insular East Asia. University of Hawaii Press, Honolulu

Insoll T (ed) (2011) The Oxford handbook of the archaeology of ritual and religion. Oxford University Press, Oxford

Jordan P, Zvelebil M (2009) Ceramics before farming. Left Coast Press CA, Walnut creek

Kaner S (2011) The archaeology of religion and ritual in the prehistoric Japanese archipelago. In: Insoll T (ed) The Oxford handbook of the archaeology of ritual and religion. Oxford University Press, Oxford, pp 457–469

Kawahata H, Yamamoto H, Ohkstshi K, Yokovama V, Kimoto K, Ohshima H, Matsuzaki H (2009) Changes of environments and human activity at the Sannai-Maruyama ruins in Japan during the Mid-Holocene Hypsithermal climatic interval. Quatern Sci Rev 28:964–974

Kirch PV, Hunt IL (1997) Historical ecolagy in the Pacific Islands: Prehistoric environmental and landscape change. Yale University Press CT, New Haven

Kitahara J (2018) An introduction to Ainu studies, edited by Center for Ainu and Indigenous Studies, Hokkaido University, Sapporo

Kosugi Y, Taniguchi Y, Nishida Y, Mizunoe K, Yano K (eds) (2009) An International Perspective on Jōmon Archaeology, in Jōmon jidai no kōkogaku (Archaeology of the Jōmon Period) 12 (1). Doseisha, Tokyo

Kudo Y (2012) Environment and culture history of the upper paleolithic and the Jomon Period. Shinsensha, Tokyo

Mizoguchi K (2002) An archaeological history of Japan: 30,000 BC to AD 700. University of Pennsylvania Press, Philadelphia

Okada Y (2003) Jomon culture of northeastern Japan and the Sannai Maruyama Site. In: Habu J, Savelle JM, Koyama S, Hongo H (eds) Hunter-Gatherers of the North Pacc Rim. Senri Ethnological Studies vol 63. National Museum of Ethnology, Osaka, pp 173–186

Tanifuji Y (2013) Jomon Period, Japanese. Journal of Archaeology 1:60–63

Taniguchi Y (2010) The beginning of the Jomon Period: Rethinking the incipient Jomon. In The Archaeology of the Jomon Period, vol 1, pp 79–97

Underhill AP, Habu J (2006) Early communities in East Asia. In: Stark M (ed) Archaeology of Asia. Blaclcwell MA, Malden, pp 121–148

Chapter 16
(Dis)Continuous Spiritual Significance and Ritual Use of Çatalhöyük and Ephesus Archaeological Landscapes: A Comparative Historical Overview

Zeynep Aktüre

Abstract As of 2022, most of Turkey's World Heritage properties are or include archaeological sites or landscapes featuring places of religion and ritual of past and often present significance and use. This chapter focuses on two of them—Çatalhöyük and Ephesus—to highlight their similarities and differences in terms of the continuities in the discontinuity and discontinuities in the continuity of their spiritual significance and ritual use, which have transformed them into associative cultural landscapes through placemaking practices. This was through public archaeology at Çatalhöyük where New Age "Goddess communities" were numerically the most important among special interest groups by the restart of scientific research in the 1990s; and through spatial segregation of local ritual users, spiritual pilgrims, and cultural tourists of diverse ethnicity at a number of religious sites scattered in the landscape, following dramatic changes in the user profile before the House of Mary and Isa Bey Mosque have become the most densely-used components for ritual use at Ephesus. The concluding comparative evaluation of these strategies reveal their success in avoiding the risk of tension and conflict, as a sound basis for establishing mutual respect of values and constructive dialogue towards building a shared sustainable future.

Keywords Çatalhöyük · Ephesus · Associative spiritual values · Ritual use · Goddess · Mother Mary

Z. Aktüre (✉)
Independent Scholar, Izmir, Turkey
e-mail: zeynepakture@iyte.edu.tr

R. P. B. Singh et al. (eds.), *Placemaking and Cultural Landscapes*,
Advances in Geographical and Environmental Sciences,
https://doi.org/10.1007/978-981-19-6274-5_16

16.1 Introduction: World Heritage of Spiritual Significance as a Source of Conflict or Dialogue?

As of 2022, armed conflict and war is listed at the top of the major threats that could have negative effects on the values of properties inscribed on the UNESCO World Heritage List (WHL) which may result in their inclusion in the List of World Heritage in Danger. The Danger List currently contains fifty-two of the 1154 properties on the WHL, with archaeological landscapes destroyed by fundamentalist groups on the grounds of religious observances constituting a particular category. Immediately remembered would be the live broadcasted destruction of Palmyra in 2015, among five other archaeological properties on the Danger List from Syria since 2013, five more from Libya since 2016, and three from Iraq. For some stakeholders, the live broadcast of the first Friday prayer in 2020 which inaugurated the Hagia Sophia in Istanbul as a Grand Mosque after eighty-nine years of museum use was equally disturbing. This annoyance lies behind identifying the recent function change as a potential threat for the attributes that reflect the monument's Outstanding Universal Value (OUV) as a component in "The Historic Areas of Istanbul", which is inscribed on the WHL since 1985. The resulting tension between the major stakeholders caused harsh disputes among the representative Member States in the 44th World Heritage Committee Session held in Fuzhou (2021).

These examples highlight a potential in cultural heritage of outstanding religious significance (including that of dead religions) to become an arena of conflict by virtue of being an attribute on whose basis ethnic identity is defined and sustained. Religious monuments and landscapes of multi-layered use may be examples of both continuity in discontinuity and of discontinuity in continuity of spiritual significance and ritual use. The former includes ongoing religious significance after a historic structural change, as Hagia Sophia's for both Christian and Muslim communities after its museumification in 1934. The latter corresponds to fundamental changes in observances despite continuity in ritual use, as in the current practice of veiling figural imagery in Hagia Sophia during Islamic prayer times.

This chapter focuses on two World Heritage archaeological landscapes in Turkey—Çatalhöyük and Ephesus—to highlight differences in the types of challenges stemming from such (dis)continuities in their spiritual significance and ritual use. Archaeological sites and landscapes constitute a majority among the World Heritage properties in Turkey. In addition to being well-researched examples, these two have been selected due to the fundamental differences in their attributes of OUV while popular consideration as "Mother/Goddess" sites facilitates their comparison.

Fig. 16.1 Çatalhöyük and environs from the North (© Orhan Durgut, 1997)

16.2 Tolerance Without Acceptance: "The Goddess" at Çatalhöyük

Çatalhöyük was inscribed on the WHL in 2012 as a vast well-preserved Neolithic settlement documenting early adaptation of humans to sedentary life and agriculture in its distinctive adobe living units without door or window openings, attached in a layout without streets. In addition to sheer settlement size in two forking mounds from which the name "*çatal*" comes (Fig. 16.1), symbolic and artistic sophistication in wall-paintings and reliefs, distinguishing "shrines" from ordinary earthen dwelling units, was among key attributes in Çatalhöyük's portrayal by James Mellaart, its first excavator in 1961–1965, as the earliest-known town with ten thousand inhabitants (Mellaart 1964, 1967). A crucial claim for this identification was proof of organized religion found in figurines interpreted as representing (a) Neolithic Goddess(es), still the best-known symbol(s) for Çatalhöyük. Thirty years later, Mellaart's interpretation was challenged by the Çatalhöyük Research Project (ÇRP 1993–2018) under the direction of Ian Hodder. The project outcomes suggested a simpler egalitarian community in transition to settled life but yet without indications of urban complexity. Mellaart's "Goddess shrines" are now interpreted as intending to protect and communicate with dead ancestors buried in dwelling units (Hodder 2006, Farid 2014). Neither of the interpretations supported a widely-embraced proposal that "the first cities arose as ceremonial centres—places of holy ritual" which were voluntarily supported by peasants (e.g. Lynch 1989, 73). Although religious and spiritual significance and use was thus never suggested as a primary function, Çatalhöyük was transformed into a living sacred landscape over the decades between the two campaigns, at least for a minor group of users.

In the 1960s' and 1970s' climate of radical countercultural political opposition in the West, New Age practitioners had started to frequent Catholic Marian pilgrimage shrines (Turner and Turner 1978, 21–22 cited in Zwissler 2011, 330). Lauren Zwissler (2011, 330) describes these practitioners as self-identified "Goddess spirituality" pilgrims who visited those shrines also for "a different Mary"—Mary Magdalene who represented those aspects of the "Divine Feminine" (or the "Goddess") that were historically rejected by the Catholic hierarchy in the canonical image of the Mother of Jesus. A historical basis for this wholistic idea(l) of the "Divine Feminine" was established by Marija Gimbutas (1989, 1991, 1999) in a trilogy constructing the world of the "Great Goddess of Life, Death, and Regeneration" out of Palaeolithic and Neolithic finds. This is a peaceful "Old" Europe before seizure of power and control of resources around 3500 BC by Indo-European-speaking pastoralists from the north Black Sea steppes—"a patrifocal, mobile, warlike, ideologically sky-oriented society that was 'indifferent to art'" (Tringham and Conkey 1998, 23, 39). Archaeological and art historical research referencing "Mother Earth/Goddess" sites and sanctuaries have continued ever since (e.g. Meaden 2012; Polat 2019).

In Gimbutas' grand narrative, wall-paintings of vultures attacking headless bodies in Çatalhöyük "shrines" were taken as evidence for the practice of removing the head immediately after death and leaving the body out for excarnation by birds, among key symbols of death (Gimbutas 1990). Human skulls mounted under bull heads, supposedly representing uteruses for regeneration, were interpreted as the seats of the soul while handprints below other bull heads and numerous "Mother Goddess" figurines stood for "the Goddess herself" and her power's presence. Alongside other remains from Çatalhöyük, the best-known of these figurines is exhibited in the Museum of Anatolian Civilizations in Turkey's capital Ankara, some 300 km away from the site (see Fig. 16.2 below).

This has made this museum a highlight in Goddess Tours in Anatolia since the adoption of Çatalhöyük as a quintessential "Goddess site" among a huge range of others in Europe and elsewhere (Rountree 2002, 476). Modern followers of Goddess spirituality visited these places for "inspiration, healing, learning, and celebration of the divine feminine" (Rountree 2007, 9). Lydia Ruyle (2005b) was among organizers of the earliest of those tours in 1990 when her friend Asphodel Long performed a ritual at the spot where a "Goddess birthing figure" was found and later reproduced in the Ankara museum. For the performers and participants, this may have been a re-enactment of a millennia-long tradition rather than an invention of a New (Age) one. Yet Hodder's comments reveal that the locals evaluated such unaccounted uses of the site by outsiders as strange and inappropriate, as also did the wary Turkish government officials (Rountree 2007, 19):

> A traditional society in which women are covered and expected to be deferential is likely to look askance at New Age feminists, naked Goddesses, and groups dancing and chanting on the mound. Many in the local community are wary of newcomers and outsiders. At some moments it has been important to take a stand and not to participate in Goddess events so as not to confront local feelings. (Hodder 2000, 11)

So, instead of a success story in establishing mutual respect of values and constructive dialogue, the immediate response to the outsiders' confrontation of the locals through their spiritual use of the site was the latter's rejection, even of a centre set up by "Goddess community" members to help rejuvenate local crafts by women with partial funding from UNESCO (Hodder 2003). Nevertheless, Goddess visitors were still among the specifically addressed special interest groups in the site by the year 2000, possibly because they were numerically the most important and among the actual or potential funders for the project (Hodder 2000). Five years later, another donation (this time to ÇRP) was accepted for suggested use to clean up an orchard, earlier planted by the research team, as a welcoming place for "Goddess pilgrims" to meditate and do ritual (Ruyle 2005), possibly in compensation for the ongoing prohibition of physical access to actual remains for ritual or any other purpose due to their extreme fragility.

Usually characterized by a complex fusion of spiritual and political feminism, neo-paganism, and environmentalism, "Goddess spirituality movement" is an umbrella term designating a variety of mostly Western women and some men ranging "from hardliner feminists, to eco-feminists, to Goddess worshippers, to women simply interested in the role of women in early times" (Hodder 2000, 4). Coming mostly from the USA, and others from Canada, Europe, and Australasia (Rountree 2007, 18), those visiting Çatalhöyük lacked a consensus on the site's significance, as reflected in their alternative websites (Hodder 2000, 4). They had diverse motivations and expectations in visiting the site, in a range from spiritual experience in a believed place of prehistoric Goddess worship to ritual-making and political interest in past gender relations; going to such extremes as seeing archaeological excavation as a form of raping the Earth Mother and sending Hodder death threats to stop it (Rountree 2007, 16, 18). Their agenda seems to have had its share in the eventual re-consideration of gender roles in neighbouring Küçükköy where local village men gradually accepted ÇRP's employment of village women because of the financial benefit to families (Rountree 2007). As a successful outcome of ÇRP's keen focus on public archaeology, this has been an invaluable contribution in achieving gender equality and empowerment of all women and girls, among the United Nation's Sustainable Development Goals (UN-SDG).

Men were also employed as seasonal diggers, despite the local belief that mounds in Çatalhöyük's whereabouts contain spirits seen moving from mound to mound at some nights (Hodder 2000). Recorded and analysed by David Shankland (1996), these beliefs reveal further diversity in different stakeholders' perception of the Çatalhöyük landscape. Although ÇRP had set out with multivocality among its primary objectives, ÇRP researchers' "official interpretative authority" dominated in time over this diversity in site interpretation and presentation (Rountree 2007, 7; 2015, 164). Goddess feminists' "apparently dissenting interpretation" was reduced to the contents of two portable panels Kathryn Rountree prepared upon request of Hodder in 2003 to present the site from the perspective of Goddess visitors alongside one of local women from Küçükköy (Rountree 2007, 11, 19). This demonstrates that Çatalhöyük's becoming a pilgrimage destination and living sacred place for an interest group did not result in its management as a sacred landscape per se, in the sense exemplified in some WHS of spiritual significance. A striking illustration of this may be found in the use of an image usually interpreted as the Mother Goddess as a unisex sign for the site toilets, which is something very unlikely to happen with "images of male deities of mainstream religions" (Rountree 2007: 17). As to information management, tracing the "Goddess community" now requires a deeper dig into the archive reports (e.g. Andersson 2003; Rountree 2003) shared through the ÇRP website which once included a discussion with a member of the community (Louise and Hodder 1998) and links to Goddess pages (Hodder 1999). A general search with the keyword "goddess" now yields a single link to a page on "Life at Çatalhöyük" with the "official" interpretation:

> Today, many people believe that a mother goddess was worshipped at Çatalhöyük based on the discovery of numerous female figurines. However recent reinterpretations of the data have suggested that male and animal figurines were just as common as their female counterparts, if not more so. Equality in all senses appears to have been of primary importance for the people of Çatalhöyük. (http://www.catalhoyuk.com/site/life)

Now organized rather in the spirit of a thematic cultural route, Goddess Tours of Turkey (by Archaeologous, Awaken the Goddess, Ebruli Tourism, Talisman Tours, among others) continue to stop by Çatalhöyük. Ruyle and Rountree's first contact and guide, Reşit Ergener, appear equally influential in putting Çatalhöyük on those tours as the academic work of Gimbutas, Long, and others. With degrees in economics from Yale and Oxford, Ergener was a founder of Turkish Friends of Çatalhöyük in 1995, and of Anatours specializing on Goddess-oriented tours in Turkey to promote sustainable cultural tourism at the site, with the aid also of two books he authored (Ergener 1988, 2020). In this economic perspective, their being middle-class-educated travellers (Hodder 1999; Rountree 2002; Cusack 2009, 359) may have given the "Goddess community" visibility at Çatalhöyük in the 1990s and a pretext for their toleration by the wary local communities and authorities despite the above-mentioned clash of worldviews concretizing in their respective interpretations and demanded uses of the site. These may have otherwise resulted in conflict and tension. Nevertheless, ritual use of the mound by Goddess pilgrims proved to be physically and socially unsustainable in time despite its ongoing economic feasibility. Scholarly criticism since the early 1990s of the essentialist world of the "Great Goddess of Life, Death, and Regeneration" as constructed by Gimbutas and others may also have had its share in this decline (e.g. Tringham and Conkey 1998).

Universalizing accounts like Gimbutas' claim continuity in a symbolic repertoire centring on diverse anthropomorphic figurines displaying enormous variability in form, decoration, and degree of abstraction but treated homogeneously and interpreted as "Mother Goddesses" without satisfactorily accounting "why the basic symbolic matrix ... remained essentially intact" (Tringham and Conkey 1998, 24, 26–27). Another example may be a series of figurative images starting with the Çatalhöyük "Mother Goddess" (5750 BC) and ending in a Phrygian "Kybele" from Ancyra (6th c. BC) with Hittite Kubabas from Boğazköy (Hattusha) and Karkamış (9th–6th c. BC) in-between. Thus engaged is another WHS in Turkey, "Hattusha: the Hittite Capital" that was inscribed in 1986. This series was once the last display before entering the "Artemis Hall" in Efes Museum (Selçuk, Izmir) where the next step in the continuum—Artemis Ephesia—was displayed and presented in guidebooks as an intermediary step towards the Mother of God (Aktüre 2009). Though no less open to scrutiny than Gimbutas' world of the "Great Goddess", this proposal is still maintained in an improved version after Efes Museum's renovation in 2014 (Fig. 16.2). Narrowness of this particular Anatolian identification of "The Goddess" reveals when compared to the global network of places visited by the "Goddess community".

16.3 Acceptance Without Tolerance: "The Mother" at Ephesus

A Mother Goddess Tours advertisement explains the series as the story of the Anatolian woman appearing "as a Goddess since the Neolithic age, progressing to Hattia, Hittite Kubaba, Phryg[ian] Cybele, turning to Carians' Hekate, then Aphrodite and raised to Artemis, finally to Virgin Mary" (Ebruli Tourism). Inscribed on the WHL in 2015, the four-component "Ephesus" serial World Heritage property offers a physical setting for this narrative, starting from the prehistoric "Çukuriçi Mound" producing "Goddess" figurines, moving in time and space to the "Ancient City of Ephesus" wherein the Artemis Ephesia statues displayed in Efes Museum were unearthed, and then to "Ayasuluk Hill, Artemision, and Medieval Settlement with the Basilica of St. John and Isa Bey Mosque", and "House of Virgin Mary" (locally known as Panaya Kapulu). The sequence parallels geographical and historical changes that resulted in settlement relocations, the final being in the fifteenth century when the Orthodox Greek population moved to the mountain-village Çirkince/Kırkınca (modern Şirince) 7 km to the east of modern Selçuk town centre. In this expansive landscape, Pagan, Christian, and Islamic monuments from different periods have constituted important nodes for different ethnic groups though time.

Although Ephesus has been an important stop in Goddess Tours, there is no record of open-air rituals (of the type Ruyle reports for Çatalhöyük) at any of the inscribed components. This is confirmed by Cengiz Topal (January 15, 2022), site manager of "Ephesus" WHS, and director of Efes Museum; and also by Dr. Mustafa

Fig. 16.2 "Cybele/Mother Goddess" panel in Efes Museum and one of the Artemis Ephesia statues exhibited in the next hall. (© Zeynep Aktüre, 2015; Niki Gail © OAW/OeAI)

Büyükkolancı (January 24, 2022), a 45-year devotee of research at Ephesus first as a museum expert then as director of Ayasuluk Hill and St. John Monument project. Rare "Goddess" events included Ruyle's "9,000 Years of the Goddess in Anatolia" (1995) banner exhibition in the Library of Celsus, from which ten works based on Mellaart's drawings were donated to the newly-started ÇRP a decade before the above-mentioned visit to the site, in the period of the 2005 Goddess Conference in Efes Museum (Ruyle 2005a). By then, restoration of the Library of Celsus in 1970–1978 (Strocka 1979) had already made it *the* symbol for Ephesus and the hottest spot in the modern "tourist ritual" (Aktüre 2017).

Intermediary steps in this eventual shift of focus away from the long-ruined Temple of Artemis, which once was a major religious and pilgrimage centre in the Mediterranean, importantly included a period of prominence for Christianity. This is attested in several traditions including Ephesus' being a refuge for St. Mary in the company of St. John up to her bodily Assumption. St. Mary was given the title "God-bearer" (Theotokos) by the Third Ecumenical Council of 431 that met in the Market Basilica near the Roman period harbour of Ephesus which was by then converted into the three-aisled Basilica of St. Mary (Foss 1979, 52, 97). Destroyed by fire in the seventh century and replaced by a smaller cross-domed church that was abandoned in the Middle Ages, this early church is one of the two spots where small Christian groups currently perform preauthorized private ceremonies during their visits, in a green oasis off the crowded tourist track.

The other spot is the Justinian Basilica of St. John whose remains tower over the Artemision from an artificial terrace presumably constructed over the Apostle's burial along the south slope of Ayasuluk hill where the late antique and medieval settlement was located (Fig. 16.3). Tombs of the Seven Sleepers and St. Timothy, a disciple of St. Paul whose body was removed to the Constantinian Church of the Holy Apostles in Constantinople in 356, were among other attractions Ephesus had

Fig. 16.3 Major religious monuments around Ayasuluk Hill at Ephesus: Temple of Artemis in the foreground; İsa Bey Mosque on the left, below the Seljuk Castle, and the Basilica of St. John on the right (© Zeynep Aktüre, 2013)

to offer for pilgrims from the fifth through the end of the fifteenth century. They often came for a miraculous dust (*manna*) reportedly rising on a certain day from St. John's tomb. A document dating to 1336 or 1341 records Seljuk Turks, then in control of the area including the township now named after them, charging a penny for admission to the tomb and part of the building as a market for silk, wool, wheat, and other products. (Foss 1979, 33, 43, 147).

This was before Isa Bey Mosque was constructed, possibly by an architect from Damascus (Ürüm 2014, 296), in visible distance from the Artemision and St. John's, as one of the oldest and most impressive architectural monuments from the period of Aydinid principality. Since its repair and reopening in 1975 (T.C. İzmir Valiliği İl Kültür ve Turizm Müdürlüğü 2012, 175) as one of the two official living religious places in "Ephesus" WHS, the mosque has started to receive a special group of international devotees who believe it to have been modelled on the Great Umayyad Mosque of Damascus. This mosque is considered the fourth holiest place in Islam as the burial place of John the Baptist's head and of Prophet Muhammed's grandson as well as the return place of Jesus.

The second regularly functioning religious place is the House of Mary on Bülbüldağı (Mt Nightingale, ancient Mt Koressos) some 9 km to the south of the Ancient City. The spot's discovery in 1891 is popularly connected to a 1818 vision of a stigmatic nun, Anne Catherine Emmerich (Deutsch 1965, 1–7; Gallagher 2016). Yet, the Catholic Lazarists of Smyrna (Izmir), who are accredited with the discovery noted that the "monastery" (not the "house") they discovered was already in use.

They reported that the Orthodox Greek community of four thousand living in Çirkince/Kırkınca was celebrating mass on the feast of Mary's Dormition or Assumption on August 15, at a primitive altar constructed in the ruined building (Gallagher 2016). For their weekly Sunday and other celebratory prayers, Orthodox Greeks were using a small church constructed over St. John's on Ayasuluk, as reported by John Turtle Wood, the discoverer of Artemision, for the 1869–1870 season (Wood 1887, 164). So, there was continuity in the Orthodox Christian use of St. John's though in a less monumental setting. Excavation of the basilica had to wait for 1921–1922 when Smyrna and its environs were under the control of Greece, with some remains including part of the bones in the burial chamber presumably ending up in Greece by the end of the conflict (Ürüm 2014, 294–295).

Wood also reported that a group of some fifty Catholics residing in Izmir used to come to Ephesus specifically to celebrate certain apostolical feasts, meeting and praying at an altar in the "Great Mosque" by Ayasuluk (Wood 1887, 164). The identification is confirmed in an undated postcard, published by P.L. Dermond, of Isa Bey captioned as "258. Smyrne. –Eglise de St-Jean à Ephèse". One of the masses reported by Wood on April 24, 1872 was led by Archbishop Spaccapietra of the Catholic Church in Smyrna who had a seat at the First Vatican Council of 1869–1870 (Wood 1887, 205–206). Wood described the archbishop's sustaining the others' belief that the monument was initially a church while he himself knew of its construction as a mosque by the Turks (Wood 1887,164).

What enabled such appropriation of a major Muslim religious monument for Catholic rituals with the necessary permission from the Ottoman authorities should be the smallness of the Turkish community resident in the area. Those employed by Wood in the 1873–1874 season were going to mosques in nearby Scala Nova (modern Kuşadası) for the first morning prayer of the Ramadan feast as there were no resident imams at Ayasuluk for regular prayers, and only one of the small mosques was occasionally used for the purpose (Wood 1887, 242). This reveals that in a period when Wood's excavations were starting to be frequented more often by foreign visitors arriving through the port of Scala Nova, different ethnic communities of Selçuk and Smyrna had appropriated religious buildings located in different parts of the Ephesian landscape for their major celebrations. Such segregation in space may have lowered the tension expressed in fights between Wood's Greek and Turkish workers (e.g. Wood 1887, 229–232).

Further religious variety is implied by Wood's note on a sermon organized by Pastor Cook in English on 31 January 1873 as the last Protestant sermon on Ayasuluk (Wood 1887, 252). He additionally mentioned Armenians celebrating a holy day at Ephesus, climbing Mt Koressos at night in crowded groups (as is still done during the Assumption) and staying there in large tents set up for them by railway station attendants (Wood 1887, 164–165). They reportedly believed certain large stone blocks on the mountain to have belonged to a church, conducting their mass there. The celebrated feast (not specified by Wood) may well be the Assumption, which is one of the five major Tabernacle Feasts of the Armenian Apostolic Church (Daghavar) celebrated for nine days starting from the nearest Sunday to August 15.

The mention of railway station attendants raises the possibility that the group came from Smyrna. This was four decades before the drastic impacts of World War I on cultural landscapes in Anatolia, including deportation of the Orthodox Greeks of Çirkince/Kırkınca and Armenians of Smyrna and environs. Their tragic departure marks an abrupt discontinuity in the continual Christian uses of various religious sites scattered in the Ephesian landscape, with or without intentional destructions of the type witnessed in the Middle Eastern archaeological sites mentioned above. This further complicates attempts to understand the dynamics of ritual use at the House of Mary during this period of turmoil.

While August 15 is celebrated as one of the most important Marian feasts by all Christians, Orthodox Greeks commemorate St. Mary's Dormition, associating it with Jerusalem where they also locate her final home. Never establishing an ecclesiastic presence at the House of Mary (Gallagher 2016), Turkish-speaking Orthodox Greeks of Izmir currently celebrate August 15 at the Church of St. Dimitrios in Şirince (Selçuk Ticaret Odası). This is an indication of continual spatial segregation in the performance of major religious rituals and raises a question as to whether the Christian community reported by its Lazarist explorers as celebrating mass on August 15 at the House of Mary may have been the Armenians from Smyrna instead of Orthodox Greeks from Çirkince/Kırkınca. In either case, the Lazarists had replaced the users by their discovery of the House of Mary through its legal ownership and declaration as a place of pilgrimage for the Catholic Church after the Archbishop of Izmir allowed religious ceremonies there in 1892. A detailed account of this establishment was authored by Rinaldo Marmara in 2021 based on Vatican and other archives. These took place between the first pilgrimage visit by Pope Leo XIII in 1896 and the outbreak of World War I in 1914, to resume after the World War II in 1948 (Senior 1953, 11), and without much local participation except a few Çirkince/Kırkınca residents (Gallagher 2016). The House of Mary stayed under the stewardship of the Lazarist community based in Izmir until its last owner bequeathed it in 1952 to an association founded in 1951 by the then Catholic Archbishop of Izmir (Gallagher 2016), following dogmatization of the Assumption in 1950 by Pope Pius XII as one of the four Marian dogmas of the Catholic Church (Senior 1953, 12).

This was also when the Turkish government started constructing a road from Selçuk to the House of Mary (Senior 1953: 13) in extension of the Ancient City's official opening for visitation in 1951. Restoration at the House of Mary started in the same period, having hosted visitors in 1898–1900 under a jerry-built roof that was collapsed in the period 1913–1930 (St. Virgin Mary's House 2022). Restored historic remains at the site include an excavated dried cistern (*impluvium*) that is often presented by tour guides as an ancient baptismal font along the path leading to the three-partitioned small main building. Declaration as a pilgrimage site by Pope Ioannes XXIII in 1961 (Türker 2016, 159) was followed by papal visits of Paul VI (1967), John Paul II (1979), Benedictus XVI (2006), and Francis (2014), among historic events that made "Ephesus" one of the most visited archaeological sites in Turkey.

16.4 Discussion: (Dis)Continuities in Comparative Perspective

With 1.85 million visitors in 2019, Ephesus ranked fifth after Hagia Sophia while Çatalhöyük's annual 31 thousand was in the daily visitor range suggested by Hagia Sophia's annual 3.73 million (T.C. Kültür ve Turizm Bakanlığı). Thus, Çatalhöyük and Hagia Sophia are at the two extremes of the visitor number range among WHS in Turkey, with the Ancient City of Ephesus nearing the upper end and the House of Mary marking the middle with some 700–800 thousand annual visitors (Selçuk Ticaret Odası 2016). In line with the UN-SDG of protecting, restoring, and promoting sustainable use of terrestrial ecosystems and sustainably managing forests, the House of Mary is currently managed as a cultural component in a nature park extending over Bülbüldağı. This sustains spiritual traditions associated with the natural landscape features such as grottos, springs, and isolated trees where St. Mary was venerated for many centuries at pilgrimage sites (Meinardus 1996, 93–94). Yet, Ephesus seems to have maintained its spiritual importance rather through continual reappropriation of religious monuments and sites to meet communal needs of diverse ethnic groups through placemaking. In the past century, this has been achieved with support from archaeological and local religious institutions that discovered and/or (re)invented bonds between particular spots and monuments in the landscape and sacred objects or persona. These bonds enable a consideration of Ephesus as an associative landscape that has been attractive also for cultural tourists of diverse and dynamic profile.

Tourism statistics in Turkey fail to provide data on visitors' motives but numerous pilgrims who come to see and perform their worship in Christian sacred buildings are known from Ephesus (Ladstätter et al. 2016: 422). At Çatalhöyük, on the other hand, museumification through construction of protective shelters has strengthened perception as an archaeological site rather than an associative landscape. So, despite their being equally important highlights of contemporary Goddess Tours, a comparative analysis of Çatalhöyük and Ephesus would reveal their difference in the types of associations with past and present religious and spiritual values and uses, as well as a variety in those values and uses themselves, and the valuers and users.

This gives an idea about the colourfulness of the spiritual landscapes of Anatolia when viewed in the micro scale. As a supporting example from another region, centuries after his insistent opposition (that St. Mary was the bearer of the human Christ against the Marian doctrine adopted in the Third Ecumenical Council of Ephesus) had led to his discharge from the office and excommunication, eighty–ninety thousand followers of the Constantinopolitan Patriarch Nestorius were still living in Southeast Anatolia as landed and wandering tribal subjects of the Ottoman Empire, under their own patriarch (Mar Shimun) appointed by the metropolitan bishop seated at a complex locally known as the Dark Church and characterized by the lack of figural decoration except a single cross (Erdost 2016, 58, 60). Known also as Assyro-Chaldiens, they played an active part in a period of turmoil and armed conflict over national sovereignty in the area well into the 1930s (Erdost 2016).

Extant publications on the "Goddess communities" visiting Çatalhöyük reveal an equally colourful range of motivations with which individuals have joined Goddess Tours or made individual visits. Historic and current records of religious ceremonies at Ephesus demonstrate similar liturgical diversity in time and space, highlighting the dynamic nature of spiritual and ritual uses of archaeological landscapes. Some visiting ethnic groups (including Armenians of Smyrna) were replaced in time by cultural tourists. They arrive in masses to become (intentionally or not) pilgrims by participating in established rites while the rites become secularized through their participation (Gallagher 2016). A visitor survey to distinguish spiritual tourists from secular and religious pilgrims, as categorized by Carole Cusack (2016, 228), would greatly expand data on this issue (see also: Zwissler 2011).

Instead of the major architectural monuments that were among the most important of their own periods (such as St. Mary's where the Third Ecumenical Council met or the medieval pilgrimage destination of St. John's), the current destination of religious mass tourism especially during the Assumption is the House of Mary, which is a highly renovated small domestic building that was intentionally hidden in a mountainscape. Historically important religious buildings of monumental scale including the Artemision alongside St. Mary's and St. John's, on the other hand, are maintained under state ownership and in a largely ruined state, and used only by small groups for preauthorized private ceremonies in addition to personal acts of devotion during touristic visits. No large-scale communal ceremonies are currently allowed in any of those monuments to maintain their "museumified" neutrality. All these activities of different types and scales create and maintain powerful cultural and spiritual bonds with various nodes, providing very special opportunities to interact with the cultural landscape for different types of Ephesus visitors. This observation would support the argument of those committed to the primacy of "place" in their research on religion after the so-called spatial turn (e.g. Knott 2010).

The current experience is very different at Çatalhöyük. The setting is rather "museumlike" in the sense Theodor Adorno (1983, 175) expounds the German word "*museal*" as describing "objects to which the observer no longer has a vital relationship, and which are in the process of dying. They owe their preservation more to historical respect than to the needs of the present". The reference, in the introduction of this paper, to Hagia Sophia's change of status from museum back to a living religious monument is intended as an analogy to grasp the difference in the type (and not necessarily scale) of relationship with museumified archaeological sites like Çatalhöyük (Fig. 16.4) and living spiritual landscapes like Ephesus with tangible and intangible attributes (Fig. 16.5). As Hodder shared with Pia Andersson (2003), who was then researching the relationship between archaeology and new religiosity, already by 2002 Çatalhöyük was no longer visited by busloads of Goddess pilgrims. This is the reason why this present paper had to rely on already existing research and publications on modern ritual uses of the site.

These reveal significant similarities *and* differences between associative spiritual values and valuers of Çatalhöyük and Ephesus. In both cases, the groups producing rivalling narratives on spiritual significance are marginal when compared to the overall visitor and stakeholder profile—at Çatalhöyük, archaeologists and the

Fig. 16.4 South Shelter at Çatalhöyük (© Çatalhöyük Research Project, CC BY-NC-SA 2.0)

"Goddess community"; at Ephesus, various Christian and Muslim communities but, importantly, not the Turkish State contrary to Hagia Sophia. In both cases, there are crucial differences between the interpretations and uses intended by different interest groups, and those of the keepers of the sites under the auspices of the Turkish State—i.e. the archaeological research team at Çatalhöyük and Catholic Lazarists at the House of Mary—have become dominant. Nevertheless, the dominating groups are voluntarily allowing other voices heard, and at the House of Mary this includes exhibiting, on indoor and outdoor signboards, the directives of the Directorate of Religious Affairs now operating under the Presidency of Turkey, for proper intent and conduct according to the Sunni Islamic tradition. Last but not the least, the dominant discourses among the interest groups support (and is supported by) the most dominant activity that is cultural tourism, again under the auspices of the Turkish State which considers faith tourism as a major economic opportunity (Türker 2016) though without detailed consideration of the particular type(s) of spiritual experience(s) to be offered by each site designated for the purpose.

In this respect, the contemporary "museumlike" experience Çatalhöyük offers is very different from drinking (presumably healing) water from a fountain below the House of Mary and attaching a wish tag on an adjacent wall alongside many other (mostly Christian and Muslim) visitors in continuation of pagan rituals. Even for those who doubt its healing powers, this is a rare opportunity to enjoy natural clean water, whose sustainable management currently has priority on the UN-SGD agenda. Healing and fulfilment of wishes are observed as two uses still common to all religious and spiritual sites that usually make them a destination for all believers in miracles regardless of their religious affiliations (also Türker 2016, 153). In the Goddess pilgrims she analysed, Rountree (2002, 486) has observed a similarly strong motive of self-healing from the "wounds of patriarchy" as a basic component of the healing of society, by consciously engaging bodily with the journeyed sacred place through performing symbolic acts and conducting rituals. These practices transform physical spaces into a "threshold"—"a liminal place and moment 'in and out of time,' where the pilgrim directly experiences the sacred through miraculous healing or transformation" (Rountree 2002, 485). Although not emphasized in the current management strategy, the House of Mary has a thick collection of miraculous healing

Fig. 16.5 House of Mary at Ephesus. The house at top left, with an open-air ceremonial space in front, the wall of wishes and fountain on the lowest terrace (© Zeynep Aktüre, 2011)

stories that suggest possibility of ensuring healthy lives and well-being by all at all ages through mindful and respectful use of the Nature's resources, among the UN-SDGs. These stories have had their share in the site's conversion "from an obscure place of local pilgrimage to an international, interreligious shrine attracting popes as well as Turkish officials, in addition to hundreds of thousand annual visitors" (Gallagher 2016).

16.5 Concluding Remarks

Collective use for healing and wish-making is often used as a symbol for interreligious peace bridged by Mother Mary (e.g. Meinardus 1996, 101), as remarked in the open-air mass Pope Benedict XVI celebrated at the House of Mary during his Ephesus visit in 2006 (Gallagher 2016). As such, Ephesus would support Dionigi Albera's observation that, while the three monotheistic religions tend to exclude each other from their major monumental centres in the Mediterranean [through purification, privatization, and profanation (Knott 2010), as in Hagia Sophia], continuity and synchronic use is more commonly experienced in cultic sites associated with prophets and saints who are commonly recognized as holy figures by distinct traditions (Albera

2008; also Pozo 2020: 309–310). The most common figure is St Mary by virtue of having a special place in Islam's holy book, Qur'an, as the mother of Prophet Jesus (Ponzo 2020), which embraces also the Nestorian creed. St. Mary additionally is the only woman whose name is openly mentioned in the Qur'an in recognition of her innate purity. Such mainstream acceptance "The Mother" has enjoyed at Ephesus has been unparalleled by "The Goddess" at Çatalhöyük beyond tolerance in view of socio-economic expectations. Yet, their mutual respect for "The Mother" and co-presence alongside cultural tourists at the House of Mary especially for healing practices does not necessarily mean that believers from the three monotheistic religions and their internal divisions tolerate each other's dogmatic differences in mutual acceptance. Those between Catholic and Orthodox Christian dogmas on St. Mary's final resting place and the corresponding Assumption vs Dormition dogmas currently determine the values the two communities associate with different locations of the Ephesian landscape and their ritual uses to maintain peaceful segregation.

This chapter has been formulated with the conviction that ritual continuities in discontinuous religious use (as at Isa Bey under different religions), and ritual discontinuities in continuous religious use (as in the House of Mary albeit by different sects with irreconcilable dogmatic differences) may help in healing the scars of wars, ethnic conflict, and other calamities that have caused the discontinuities as would (re)invention of (new) meanings and uses such as "Goddess spirituality" and other New Age paradigms. The concluding comparative analysis of Çatalhöyük and Ephesus would hopefully contribute in the growing academic literature aiming to demonstrate the potential contribution of such archaeological landscapes in the enjoyment of a shared peaceful life as part of an inclusive society among UN-SDGs, through direct engagement with and experience of their natural and cultural attributes.

References

Adorno TW (1983) Valéry Proust Museum. In: Prisms. The MIT Press, Cambridge, Massachusetts, pp 173–186

Aktüre Z (2009) Reading into the mysteries of *Artemis Ephesia*. In: Chaplin S, Stara A (eds), Curating architecture and the city. London and New York: Routledge, pp 145–163

Aktüre Z (2017) Ephesus as ritual space: from Roman civic to modern tourist processions. In Foletti I, Palladino A (eds), Ritualizing the city: collective performances as aspects of urban construction, from Constantine to Mao. Viella and Masaryk University, Roma and Brno, pp 47–71

Albera D (2008) Situations de coexistence interreligieuse en Méditerranée: esquisse d'une analyse comparative. Chronos 18:129–149

Andersson P (2003) Holy place or wor… and working place: the challenges of multivocality in the meeting of science and religion at Çatalhöyük today. In: Çatalhöyük 2003 archive report. http://www.catalhoyuk.com/archive_reports/2003/ar03_19.html. Accessed 14 Jan 2022

Archaeologous. Sacred goddess tours in Turkey by Archaeologous. https://www.archaeologous.com/blog/237/sacred-goddess-tours-turkey-archaeologous/. Accessed 12 Jan 2022

Asphodel P. Long. http://www.asphodel-long.com/index.html. Accessed 12 Jan 2022

Awaken the Goddess. Goddess tour to Turkey, April 2022. https://www.goddessawakeningretreats.com/goddess-tour-to-turkey-april-2021. Accessed 12 Jan 2022

Cusack CM (2009) The return of the goddess: mythology, witchcraft and feminist spirituality. In: Pizza M, Lewis J (eds) Handbook of contemporary paganism. Brill, Leiden and Boston, pp 335–362

Cusack CM (2016) Medieval pilgrims and modern tourists: Walsingham (England) and Meryem Ana (Turkey). Fieldwork Relig 11(2):217–234

CNN Türk. Çatalhöyük'te 9 bin yıl önce hem yerleşik hem de hükümetsiz bir yaşam vardı. CNN Türk. https://www.cnnturk.com/haber/turkiye/catalhoyukte-9-bin-yil-once-hem-yerlesik-hem-de-hukumetsiz-esit-bir-yasam-vardi. Accessed 16 Apr 2022

ÇRP—Çatalhöyük Research Project. http://www.catalhoyuk.com. Accessed 11 Jan 2022

Deutsch BF (1965) Our lady of Ephesus. The Bruce Publishing Company, Milwaukee

Ebruli Tourism. Mother goddess tours. http://www.ebruliturizm.com/161/ancient-beliefs-and-mother-goddess/. Accessed 12 Jan 2022

Erdost MI (2016) Şemdinli Röportajı. Onur Yayınları, Ankara.

Ergener R (1988) Anatolia: land of mother goddess (trans: Isin M). Hitit Publications, Istanbul

Ergener R (2020) Religion and economics. Springer Nature, Cham, Switzerland

Farid S (2014) "10. From excavation to dissemination: breaking down the barriers." In: Peter G, Stone and Zhao H Sharing archaeology: Academe, practice and the public, pp 117–132. Oxford, Routledge

Foss C (1979) Ephesus after antiquity: a late antique, Byzantine, and Turkish city. Cambridge University Press, Cambridge and New York

Gallagher A (2016) Mary's house in Ephesus, Turkey: interfaith pilgrimage in the age of mass tourism. Vincentian Heritage J 33(2):Article 2. https://via.library.depaul.edu/vhj/vol33/iss2/2. Accessed 20 Jan 2022

Gimbutas M (1989) The language of the goddess: unearthing the hidden symbols of Western civilization. Harper & Row, San Francisco

Gimbutas M (1990) The world of the goddess. https://www.youtube.com/watch?v=yU1bEmq_pf0&t=4591s. Accessed 1 Feb 2022

Gimbutas M (1991) The civilization of the goddess: the world of old Europe. Harper, San Francisco

Gimbutas M (1999) The living goddesses (edited and supplemented by Dexter MR). University of California Press, Berkeley and Los Angeles, California

Hodder I (1999) The experience of the Çatalhöyük Web site. In: Archaeology and global information systems. Internet Archaeol 6(March). https://intarch.ac.uk/journal/issue6/hodder/part2.html. Accessed 14 Jan 2022

Hodder I (2000) Chapter 1: developing a reflexive method in archaeology. In: Hodder I (ed) Towards reflexive method in archaeology: the example at Çatalhöyük. British Institute of Archaeology, Ankara, pp 3–11

Hodder I (2003) Archaeological reflexivity and the 'local' voice. Anthropol Q 76(1):55–69.

Hodder I (2006) This old house. Natural History, June. http://www.naturalhistorymag.com/htmlsite/0606/0606_feature.html. Accessed 11 Jan 2022

Knott K (2010) Religion, space, and place: the spatial turn in research on religion. Relig Soc Adv Res 1:29–43

Ladstätter S, Büyükkolancı M, Topal C, Aktüre Z (2016) Ephesus. In: Ertürk N, Karakul Ö (eds) UNESCO world heritage in Turkey. Turkish National Commission for UNESCO, Ankara, pp 412–443.

Louise A, Hodder I (1998) Discussions with the Goddess community. http://www.catalhoyuk.com/library/goddess.html. No longer accessible

Lynch K (1989) Good city form. The MIT Press, Cambridge, Massachusetts and London, England

Marmara R (2021) La Maison de la Vierge Marie (Mont du Rossignol, Ephèse Smyrne). Izmir: Association de la Maison de la Vierge Marie

Meaden GT (ed) (2012) Archaeology of mother Earth sites and sanctuaries through the ages: rethinking symbols and images, art and artefacts from history and prehistory. BAR Publishing, Oxford

Meinardus OFA (1996) The Virgin Mary as mediatrix between Christians and Muslims in the Middle East. Marian Spirituality Interreligious Dialogue. Marian Stud 47(Article 10):88–101

Mellaart J (1964) A neolithic city in Turkey. Sci Am 210(4):94–104

Mellaart J (1967) Çatal Hüyük: a neolithic town in Anatolia. McGraw-Hill, New York

Polat G (2019) Batı Anadolu'daki Ana Tanrıça Kutsal Alanlarına Örnekler (Examples of sanctuaries of Mother Goddess in Western Anatolia). MSGSÜ Sosyal Bilimler Dergisi 3(19):154–169

Ponzo J (2020) The model of Mary between Islam and Catholicism: the figurativization of normative principles in the intercultural exchange. Int J Legal Discourse 5(2):301–315

Rountree K (2002) Goddess pilgrims as tourists: inscribing the body through sacred travel. Sociol Relig 63(4):475–496

Rountree K (2003) Reflexivity in practice. In: Çatalhöyük 2003 archive report. http://www.catalhoyuk.com/archive_reports/2003/ar03_20.html. Accessed 14 Jan 2022

Rountree K (2007) Archaeologists and Goddess feminists at Çatalhöyük: an experiment in multivocality. J Feminist Stud Relig 23(2):7–26

Rountree K (2015) The academy, the otherworld, and between. Pomegranate 17(1–2):155–169

Ruyle L. 9,000 years of the Goddess in Anatolia, July 24–25, 2005. https://www.academia.edu/11575268/9_000_Years_of_the_Goddess_in_Anatolia. Accessed 16 Jan 2022

Ruyle L. Goddess conversations at Çatalhoyuk, July 17, 2005. https://www.academia.edu/13023141/Goddess_Conversations_at_Catalhoyuk_2005. Accessed 12 Jan 2022

Ruyle L (2009) A (m)otherworld is possible: world icons of mothers and grandmothers. https://www.youtube.com/watch?v=zEGZh9SuhBg. Accessed 30 Jan 2022.

Selçuk Ticaret Odası. Selçuk'ta 15 Ağustos Meryem Ana Bayramı ayinlerle kutlandı. https://www.selcukticaretodasi.org.tr/haberler/selcukta-15-agustos-meryem-ana-bayrami-ayinlerle-kutlandi/609/. Accessed 30 Jan 2022

Selçuk Ticaret Odası. Turizm Sektör Raporu – 2016. https://www.selcukticaretodasi.org.tr/images/yuklenenler/MTViYTRkZDdmMTQ2MTQ.pdf. Accessed 14 Apr 2022

Senior (1953) Efes Panaya Kapılı Hazreti Meryem Ana Evi. Korsini Basımevi, İzmir

Shankland D (1996) Çatalhöyük: the anthropology of an archaeological presence. In: Hodder I (ed) On the surface: Çatalhöyük 1993–95. McDonald Institute for Archaeological Research & British Institute of Archaeology at Ankara, Cambridge, pp 349–358

St. Virgin Mary's House (2022) Restoration of the house. https://www.hzmeryemanaevi.com/en/marys-house-reconstruction/. Accessed 29 Jan 2022

Strocka VM (1979) Efes'teki Celsus Kitaplığı onarım çalışmaları. Belleten XLIII(172):809–831

Talisman Tours. Journey to the Anatolian Earth Goddesses. Turkey Goddess Tour. https://talismantours.com/tour-ideas/special-interest-tours/turkey-goddess-tour. Accessed 12 Jan 2022

T.C. İzmir Valiliği İl Kültür ve Turizm Müdürlüğü (2012) İzmir Taşınmaz Kültür Varlıkları Envanteri, vol. 3. İMAJ Basım Yayın Reklamcılık, Izmir

T.C. Kültür ve Turizm Bakanlığı, Döner Sermaye İşletmesi Merkez Müdürlüğü. 2019 Yılı Müze ve Örenyeri Ziyaretçi İstatistikleri. http://www.dosim.gov.tr/assets/documents/2019-ZIYARETCI-WEB-SITESI.pdf. Accessed 17 Jan 2022

Tringham R, Conkey M (1998) 1. rethinking figurines: a critical view from archaeology of Gimbutas, the "Goddess" and popular culture. In: Goodison L, Morris C (eds) Ancient Goddesses: the myth and the evidence. British Institute of Archaeology at Ankara, 2000. London: The British Museum Press, Ankara, pp 22–45, 197–202

Turner V, Turner E (1978) Image and pilgrimage in Christian culture: anthropological perspectives. Columbia University Press, New York

Türker N (2016) Chapter 10: religious tourism in Turkey. In: Egresi I (ed) Alternative tourism in Turkey: role, potential development and sustainability. GeoJournal Library, vol. 121. Springer, Cham

UNESCO WHC. Ephesus. https://whc.unesco.org/en/list/1018/. Accessed 29 Jan 2022

UNESCO WHC. Hattusha: the Hittite Capital. https://whc.unesco.org/en/list/377/. Accessed 7 March 2022

UNESCO WHC. List of world heritage in danger. https://whc.unesco.org/en/danger/. Accessed 13 Dec 2021

UNESCO WHC. Neolithic site of Çatalhöyük. https://whc.unesco.org/en/list/1405. Accessed 29 Jan 2022

UNESCO WHC. World heritage in danger. https://whc.unesco.org/en/158/. Accessed 13 Dec 2021

Ürüm AT (translation & commentary) (2014) John Turtle Wood'un Anıları: Efes Kazıları. Arkeoloji ve Sanat Yayınları, İstanbul

Wood JT (1987) Discoveries at Ephesus, Including the Site and Remains of the Great Temple of Diana. Boston, James R. Osgood and Company

Zwissler L (2011) Pagan pilgrimage: new religious movements research on sacred travel within Pagan and New Age communities. Relig Compass 5(7):326–342. https://doi.org/10.1111/j.1749-8171.2011.00282.x

Chapter 17
Cultural Landscape Modelling: A Practical Approach for Evaluating Sustainability. The Historic Village "Maymand" in Iran

Leila Kowkabi

Abstract The rural landscapes are particular due to cultural characteristics. As local residents in rural areas tend to use modern technology, the preservation of cultural values and natural attractions should not be underestimated. There are many criteria which must be considered to achieve sustainability. The research aims to answer the question: "How landscape sustainability is affected by culture, economy and qualities of natural resources?" This research will focus on rural landscapes investigating a sustainable approach to the relationships between people and nature. The case study—Maymand—has been continuously inhabited between 2000 and 3000 years which makes it one of Iran's oldest villages. It consists of several natural and man-made caves that are still used today for housing and sheltering. The rural landscape characteristics in this village were studied to find the signs of sustainability. Using Pasture landscape and agricultural landscape for livestock and planting simultaneously has led to ecological-economical sustainability in this rural landscape. Finally, the research revealed that rich cultures, the traditional manner of life and the biodiversity of the landscapes are the main causes of sustainability within this rural cultural landscape.

Keywords Cultural landscape · Rural landscape · Modelling · Sustainability · Meymand

17.1 Introduction

The concept of the landscape in which people live has been defined as the "cultural landscape". This is because human activity creates unique features in it (Frajer and Fiedor 2021). The cultural landscape is the continuous temporal variation of the material habitats of sedentary human societies, the intensity and diversity of the

L. Kowkabi (✉)
Faculty of Architecture and Urban Planning, University of Art, Tehran, Iran
e-mail: l.kowkabi@art.ac.ir

R. P. B. Singh et al. (eds.), *Placemaking and Cultural Landscapes*,
Advances in Geographical and Environmental Sciences,
https://doi.org/10.1007/978-981-19-6274-5_17

dynamic challenges posed by nature, the needs and desires of society itself, and the various. It corresponds to the historical realities of different eras in the region. This concept has evolved and spread into several fields such as environmental psychology and related fields such as landscape aesthetics and landscape ecology (Kowkabi and Akrami 2021; Conzen 2001).

The word "rural" comes from the Latin word "countryside". It is generally applied to agriculture as an economic activity and is contrasted with the term "urban" (Golley and Bellot 1999). People have survived by planting farms, burning forests, expanding farmlands through water management, collecting trees and grass for fertilizer, fuel and building materials and fishing in rivers and lakes. Rural landscapes are often used to study the whole system of human activity in agriculture as well as the natural world (Nassauer and Wascher 2007). It arose from the long-term mutual relationship between nature and humans. The rural landscape has become a space with a different function than before. In rural settings, it is no longer just major activities such as agriculture and forestry that affect landscape structure. Some rural areas have potential value due to their excellent nature and unique landscapes. In the past, soil fertility and ability to produce commodities were the basis for appreciating a landscape, but today it is the aesthetic, ecological and cultural qualities that make a landscape visible (Kowkabi and Badri 2020; Plieninger et al. 2016; Domona 2011; Banski and Wesolowska 2010).

Human intervention in nature, which is inherently due to socio-economic development, has increased worldwide since the industrial revolution. These have impacted rural ecosystem units by reducing biodiversity, endangering cultural diversity and destroying amenities (Benson and Roe 2007; Larcher et al. 2013; Kim et al. 2007; Lieskovsky et al. 2014). Local residents tend to be receptive to modern technology and development, but the emphasis should be on preserving traditions and cultural values, using local materials and construction methods and maintaining natural charm. "How is landscape sustainability affected by culture, economy, and the quality of natural resources?" It is to examine the criteria that affect local sustainability and how these phenomena challenge rural residents. Need for research: In recent years, modern technology has changed the lifestyles of most people in rural areas. This cultural change affected the shape and appearance of the village. Many villagers moved to larger towns and cities. The main problem is that change is accelerating rapidly. It is therefore important to look at the areas of least change to avoid further unnecessary changes and maintain sustainability in both the cultural and construction sectors.

17.2 Theoretical Frameworks

Some researchers study the concept of "sustainability in a rural area" from some perspectives but do not consider the variety of place-related facts. In this research, the different perspective to make an appropriate framework for analysing the rural landscape has been categorized. The framework consists of three parts:

1. Evaluating the theories and approaches which have underpinned rural landscape, ranging from natural beauty to cultural values. We investigate landscape sustainability as an official term from socio-economic sustainability to environmental sustainability.
2. Identifying how the cultural backgrounds of local residents have evolved.
3. Categorizing the criteria of rural landscape sustainability as the framework to analyse the rural landscape of Meymand.

The case study in this research has been chosen based on the specific construction of the instance to present a good example of rural cultural notion. Meymand—An old village—place in Kerman province, Iran was selected as a monographic case study. The lifestyle of people from past generations to the present has been brought into consideration to give a better understanding of rural cultural values.

17.2.1 Cultural Landscape Characters

The landscape is normally defined as an area of land with a certain use or function. It can be considered as a synthetic and integrating concept that refers to a material-physical reality, originating from a continuous dynamic interaction between natural processes and human activity, and to the immaterial existential values and symbols of which the landscape is the signifier (Antrop 2006). Landscape character is defined as a "distinct and recognizable pattern of elements that occur consistently in a particular type of landscape" (Countryside Agency and Scottish Natural Heritage 2002).

Cultural landscapes are the result of the consecutive reorganization of the land to adapt its use and spatial structure to the changing societal demands. Today, the changes are seen as a menace and as a negative evolution because they lead to the loss of diversity, coherence and identity, which were characteristic of the traditional cultural landscapes that are rapidly vanishing (Antrop 2005; Bolliger and Kienast 2010; Simensen et al. 2018).

Each part of the landscape has its distinctive character (Swanwick 2002), which offers a unique sense of place. The current condition of a landscape can be a reliable indicator of future trends and areas in need of further attention (Rippon 2012; Bell et al. 2009; Harvey 2012; Taylor and Lennon 2012). Moreover, there are specific cultural backgrounds, which have influenced the settlement patterns of regions and countries. This article reviews the backgrounds and meanings of "Persian culture" and shows that the landscape is seen here as a symbol of these concepts.

17.2.2 Cultural Landscape of Iran

The four major elements of "Persian culture" are "wind, water, earth and fire". These factors should be respected and everyone should strive to maintain them (Daneshdoust 2007).

Water—in Iran, water is the clearest symbol of life, and its abundance indicates abundant productivity and divine blessing (Kowkabi 2021; Martin 2005). Due to the purity and clarity of water, it has long been considered sacred in the religious beliefs of Iranians, so this element is sometimes considered a tribal symbol. Water is considered sacred to Iranians and has important spiritual values. Temples, ancient gardens and fire temples were built along the water with the utmost respect for its existence. The ancient Persians glorified the Iranian water goddess "Anahita" who protected water in Zoroastrianism.

Soil—Iranian architecture—anatomy is lifted from the soil and expected to return there (Nikpour et al. 2011; Parsi 2008). The human body, nature and mould are originally soil and return to the soil.

Wind—wind is the symbol of the purest air in nature. In Iranian culture, this element is a symbol of God's prayers, speeches and people's blessings. Wind is the clearest symbol of ether, the bearer of light and an indicator of the properties of heat and moisture in Persian culture.

Fire—in Iran, a country rich in sunlight, light has always been considered the most prominent aspect of fire (Ardalan 2000). In Zarathustra and Islamic religions, light is the symbol of divine wisdom, the source of all purity and good doing, and the symbol of man's emergence from the darkness of ignorance into the world of light (Avesta and Quran). Light, the most intangible and perceivable element of nature, has always been present in Iranian architecture and is indeed a sign of the higher worlds and spiritual spaces.

17.2.3 Cultural and Environmental Value in the Rural Landscape

Rural landscapes are multifunctional systems. These functions are influenced by both natural and cultural patterns. Some rural attractions are social and traditional values, culture, facilities and other infrastructures (Yoon 2002). For many years, scholars (such Selman 2009) have been valuing the natural beauty of landscapes. Selman and Swanwick (2010) claim that: "natural beauty relates to unspoilt rural areas free from large scale settlements or industry"; it includes rural landscapes which have been shaped by human activities, including fields and small settlements. Lafortezza and Brown (2004) have provided a tool, namely FDNP (Framework for the Design of New Patches), for incorporating structural and spatial attributes in the landscape planning and designing process. In another research, Talandier (2009) has investigated amenity indicators (i.e. cultural, traditional, naturalistic and economic indicators) for sustainable development in rural areas (Talandier 2009). Brown claims that "Beyond the traditional productive functions, rural landscapes are increasingly being recognized as complementary sources of biodiversity and places for cultural identification" (Brown et al. 2008, p. 395).

Due to significant changes in rural landscapes, several studies were performed by a wide range of scholars to help improve the environment through a sustainable rural landscape (Kowkabi 2020; Palang et al. 2005; Buhler-Natour and Herzog 1999). The Common Agricultural Policy (CAP) is mainly aimed at increasing agricultural productivity by promoting technical progress and stabilizing markets since 1962. The reformed CAP (1992) was for environmentally sustainable rural management, paying particular attention to the quality of food and the farmers' role (Banski and Wesolowska 2010). With the improvement of agricultural systems over the last few decades, the component biodiversity and specifically the floral diversity have decreased substantially (Waldhardt et al. 2003).

The configuration of rural landscapes is being re-drawn from 1999. The areas rich in visual quality are the object of strong development pressures. In the current approach, it is essential to view the amenity quality of landscape as the main resource for rural areas (Domona 2011). Considering the extent and rapid development in rural landscapes, the need for defining the cultural landscape pattern and process is increasing to manage and improve their functions.

17.2.4 Rural Landscape Sustainability

Several studies have looked into factors such as management, conservation and planning in rural landscapes employing a multifunctional perspective (Brandt et al. 2000; Jongman 2005; Pinto-Correia et al. 2006; Benson and Roe 2007; Brown et al. 2008; Selman 2009; Bolliger and Kienast 2010; Reed et al. 2017), integrated, multidisciplinary and transdisciplinary (Moss 2000; Tress et al. 2005) and systematic ones. Antrop claims, "Sustainability is not only for a particular landscape, it might be natural or cultural, traditional or contemporary, spectacular or ordinary" (Antrop 2005, p. 187). Therefore, policies for sustainable rural landscape should consider a range of perspectives including ecological, social, economic, historical and cultural. According to researchers in the field, this management system can lead to more policies from the perspective of both natural and cultural heritages for the development of rural areas and improvement of the environment (Ernoult et al. 2003; Pinto-Correia et al. 2006). Moreover, some scholars (Waldhardt et al. 2003; Palang et al. 2005) have emphasized the varieties of time lag and spatial scale in rural landscape changes due to urbanization. However, it did not pay attention to the variety of place-related approaches.

Evidence for the environmental sustainability of landscapes is often related to their multifunctionality services and resilience (Azari Dehkordi 2012). An ecological perspective defines a desired sustainable landscape as a landscape, which promotes viable and healthy populations through green infrastructure, biodiversity networks of habitat patches and corridors that offer appropriate quality, size and connectedness to support species' life cycles. In the same line of inquiry, Rosenberg suggests an "Emerging Paradigm for Landscape Architecture" that admires a landscape where "the human component can interrelate with water resources, wildlife habitats, edible landscapes, and urban woodlots" (Rosenberg 1986, p. 81). Selman (2009) claims, that most factors can be relevant to the sustainability of a landscape such as the conservation of soil, moisture and nutrient status, the influence of vegetation on the equability of microclimate and atmospheric carbon levels and the integrity of water quality and quantity in surface and subsurface hydrological systems.

Natural landscapes may have an additional cultural value, which enhances the heritage value (Mavromatidis 2012). Some researchers have focused on these challenges (Table 17.1). Nassauer, for instance, has documented one approach to achieving progress in a more sustainable direction through making positive changes attractive to the people with a say in the matter (Nassauer and Wascher 2007). She explains that many indigenous ecosystems and wildlife habitats violate cultural norms regarding tidiness and order when retained or introduced into the urban context.

Based on the aforementioned literature and experts' perspectives, sustainability criteria and characters in the rural cultural landscape can be classified as shown below (Table 17.2). According to the professional field and dominant attitude at the time, each of the researchers' views has been presented and compared in this Table 17.2.

Table 17.1 The diverse approaches to rural landscape sustainability

Approach	Scholars	Year
Multifunctional	Reed et al.	2017
	Lieskovsky et al.	2014
	Bolliger and Kienast	2010
	Selman	2009
	Brown et al.	2008
	Benson and Roe	2007
	Pinto-Correia et al.	2006
	Jongman	2005
	Brandt et al.	2000
Holistic	Palang et al.	2005
	Swanwick	2002
Integrated	Antrop	2006
	Tress et al.	2005
	Moss	2000
Total Human Ecosystems	Naveh	2001
Systematic	Grossmann and Bellot	1999

Table 17.2 Literature review of the rural landscape sustainability concept

Scholars	Year	Criteria/character
Zhao et al, Kowkabi and Badri Kowkabi	2021 2021 2020	Spatio-temporal connectivity Conservation of individual habitat patches Dynamic landscape
Simensen et al.	2018	Contemporary landscape characterization The distinction between biophysical and "holistic" approaches Methodological strategies
Reed et al.	2017	Landscape approaches' Attention to integrated land management
Plieninger	2016	Yield values
Lieskovsky et al. Kowkabi et al.,	2014, 2013	Scenic-aesthetic Structure: habitat mosaic, the Landscapes with high Ecological value
Domona	2011	Functions: (biodiversity, soil)
Banski and Wesolowska	2010	Amenity Quality of farmland Resources/abundance forest Quality of landscape
Bell et al.	2009	Communication patterns key features: churches, castles Historical events and persons Condition of the landscape Unique landform
Selman	2009	Valuing landscapes for "natural beauty" Natural environmental factors Soil moisture/nutrient status/vegetation Water quality and quantity Surface/subsurface hydrological systems Habitat fragmentation
Talandier	2009	Amenity indicators (cultural, traditional and naturalistic) Migration Inhabitant income/levels of income Employment in the service sector
Benson and Roe	2007	Socio-economic factors
Daneshdoust Nikpoor	2007 2007	Background of "Persian culture" Air, water, soil, fire
Antrop	2006 2005	Synthetic concept Immaterial existential values/symbols Interaction (between natural processes and human activity) Material-physical reality Social/historical/cultural Economic Ecology

(continued)

Table 17.2 (continued)

Scholars	Year	Criteria/character
Farina	2006	Maintaining healthy and viable populations/biodiversity/quality/facilitate species life cycles/spatial patterns and processes/green infrastructure/sufficient size/connectedness
Pinto-Correia et al.	2006	Cultural heritage Natural character
Martin	2005	Tribe's symbol
Antrop	2004	Societal demands Traditional/cultural landscapes Diversity Spatial structure
Lafortezza and Brown	2004	Land division and settlement Landscape aesthetic Microclimate Diversity Water and nutrient flows Spatial arrangement (Size/shape/core area)
Ernoult et al.	2003	Cultural heritage Natural character
Yoon	2002	Facilities Other infrastructures
Countryside Agency and Scottish Natural Heritage	2002	The pattern of elements that occur in a particular type of landscape
Conzen	2001	Landscape aesthetics Past/modern ways of life Living cultural landscapes Environmental psychology/landscape ecology
Ardalan	2000	Religions Symbol of divine wisdom and source of all purities Light/fire
Golley and Bellot	1999	Farming as an economic activity
Nassauer	1995 1997	Making positive changes attractive to the people
Nassauer	1995	Cultural norms Cultural framework for ecological quality
Forman, Godron	1986	Spatial patterns Processes
Ann Rosenberg	1986	Diversity and complexity human component Edible landscapes Water resources, wildlife habitats

17.3 Meymand, Kerman Province, Iran

The ancient village "Meymand" is one of the most amazing places in the world located in the Kerman province of Iran. It is 36 km from the historic town of Shahr Babak (or Town of Babak), said to be the birthplace of the founder of the Sassanid dynasty (Vaziri Kermani 1981). The village is located at 30°13' latitude and 55°25' longitude. The average altitude is 2240 m above sea level and the average annual rainfall is about 185 mm (National Geographical Organization 2002). Meymand has been inhabited for 2000–3000 years, making it one of the oldest surviving villages in Iran. The village consists of several natural and man-made caves, still used today as housing and shelter. The history of Meymand dates back to the days when the inhabitants of the Persian Plateau buried their dead in crypts dug into the mountains (Burke and Elliott 2008). Excavations have uncovered 10,000-year-old stone carvings and his 6000-year-old earthenware. The village's traditional houses are carved into the rock and feature corridors, pillars and stoves used to both cook and heat the homes during the frigid winter months. Inhabitants of the present village build cave dwellings called "kiche" by cutting horizontal cuts 6–9 m long into the soft sedimentary rock of the hill. Meymand has about 400 quiche (National Geographical Organization 2002). Each quiche covers an area of about 16–20 square metres and is almost 2 m high (Hanachi and Mehrdad 2007). Houses are built one above the other and house 130–150 people, many of whom live a nomadic lifestyle. Some of the lower units are grouped together, with up to five houses with entrance moats opening onto terraces called "big orchids". Villagers also use round sedimentary rocks to construct bulkheads and buildings in the valley floor.

17.4 Finding and Discussion

This paper is based on a literature review covering papers on landscape sustainability, including both suggested and empirically tested indicators. From the literature, some sustainability concepts were identified which together characterize the rural cultural landscape. These concepts are supported by different theories. While this paper focuses on landscape sustainability, characters and criteria have been developed to provide a framework for explaining what is important for rural cultural sustainability. This framework could thereby aid in identifying what characteristics of the rural landscape are important to describe. The results and dominant approaches of researchers in this field are classified in Table 17.3 and include culture, economy, ecology and spatial patterns.

To recognize the cultural pattern, the rural landscape components are identified concerning the two factors of relation and pattern. Some criteria which are important based on the sustainable cultural landscape are set as a framework to analyse the landscape components (Fig. 17.1).

Table 17.3 Categories of rural landscape sustainability and the criteria

Culture/tradition/social patterns	Diversity and complexity of the human component Cultural norms Positive changes attractive to the people Religions Symbol of divine wisdom and source of all purities The manner of life, traditional character Living cultural landscapes Societal demands Cultural heritage, tribe's symbol Immaterial existential values/symbols Viable populations The cultural background of Iranian Amenity indicators (cultural, traditional and naturalistic Migration Communication patterns Key features Historical events and person Land division and settlement Landscape aesthetic
Economy	Edible landscapes Farming as an economic activity Facilities The infrastructures Socio-economic factors Inhabitant income Levels of income Employment in the service sector Quality of farmland Diversity of incomes Having more than one job Yield values
Nature/environment	Natural resources (water, soil) Wildlife habitats Diversity Natural beauty Interaction between natural processes-and Human activity Biodiversity Species life cycles Natural environmental factors Nutrient status Hydrological systems Vegetation Natural beauty Condition of the landscape Abundance forest Quality of landscape Microclimate, temperature Water and nutrient flow Core area and connectedness

(continued)

Table 17.3 (continued)

Structural/spatial patterns	Spatial patterns and processes Green infrastructure Size of green patches The shape of green patches Habitant patches Habitat fragmentation Unique landform Spatial arrangement Habitat mosaic Traditional house type Settlement pattern

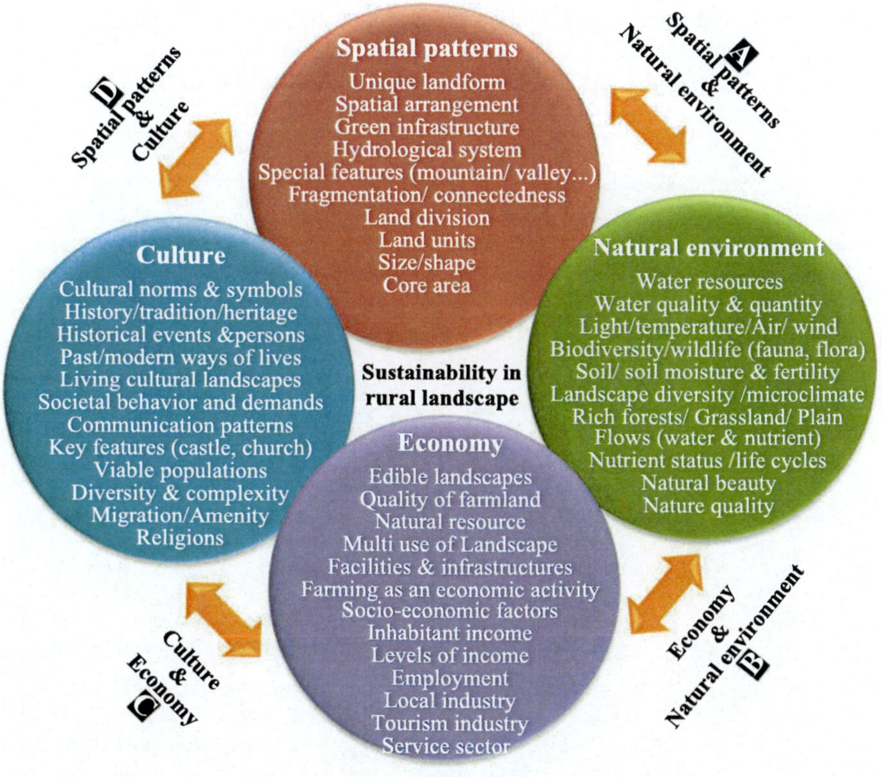

Fig. 17.1 Framework of sustainability in the rural landscape

Attempts are then made to analyse the village (Meymand) based on the criteria listed in this table. This research identified the main factors, which have influenced rural landscape sustainability in Meymand. Regarding rural landscape vitality, in this research the landscape component, the relations between them and the reasons for last living in the rural landscape are identified.

17.4.1 Spatial Patterns and Culture

There is a special manner of life in Meymand. A little over half of Meymand's population led semi-nomadic lives dwelling in the caves during the winter. They move with their herds of goats and sheep to the plains in the spring and then to higher pastures and cooler climes during the summer. Based on Fig. 17.2, the nomadic range is about 120 square kilometres with Meymand at its centre (official centre of Meymand site).

These nomads spend their winter months in Maymand, living in their cave homes with the women weaving carpets and the men tending to the herds. In the spring, they move to the lower plains as the grass in the pastures will have started to grow there first. The spring camps are designated areas called Aghols or Sar-aquls. Each residence is owned by a family. During the summer, the weather in the plains and Meymand gets quite hot and the nomadic herders find relief from the heat as well as fresh pastures in the higher altitudes around Meymand. They also harvest wild herbs and stay until early autumn.

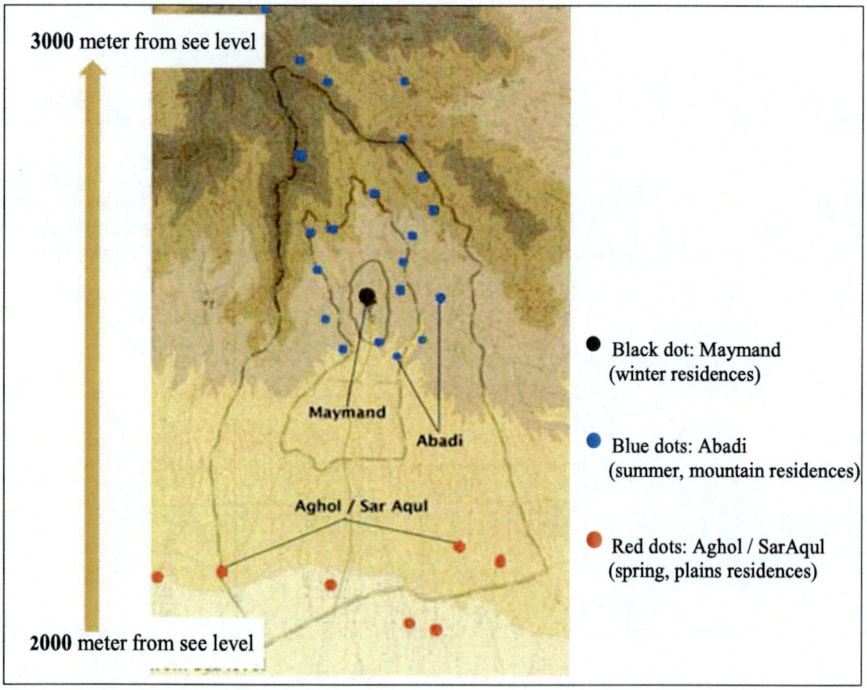

Fig. 17.2 Migration range of Maymand nomads

Table 17.4 The spatial manner of life in Meymand and the activities

Season	Residences	Place	Attitude	Weather	Activity
Winter	Maymand Cave	Mountain	2600–3000	Cold and harsh	Weaving carpets and Klim Animal breeding, Tending to herds
Summer	Abadi	Hillside	2500	Quite hot and dry	Picking wild herbs, nuts and seeds, orchards
Spring	Aghol/SarAghol	Plain residences	2200–2400	Warm	Sheep husbandry

Source The official site of Maymand

The mountain camp areas are called Abadis. 35–38 Abadi sites (Izadpanah, 2002) dotting the hillsides (Fig. 17.2). The various residences of Meymand in 3 seasons are illustrated in Table 17.4. Those residents that remain in the village occupy themselves with animal breeding and the tending of the orchards.

17.4.2 Paradox of Economy and Environment

The land cover of Meymand is diverse. There are wild herbs, nuts and seeds such as wild pistachio, almond, walnut, cumin seeds, black thyme, rosemary, yarrow, cumin, hollyhock, buttercup, fennel, peppermint, liquorice and astragalus, which have medicinal uses. Mulberry and blackberry trees can be found all around the village. The natural diversity in the region has helped people over the centuries to gain medicinal herbs, fruits and crops for themselves and their livestock. This variety of natural resources has also enabled the locals to meet their economic needs.

Water resources—seasonal rivers and springs provide villagers with rich sustainable agriculture. In Meymand basin, there is no constant river, only a few seasonal ones. These rivers include “Moorang”, “Lakhis” and “Lakhorin”. “Lakhorin River” passes through nearly 21 km of Meymand and is the main river of this village (Fig. 17.3). These rivers have water from early spring to April. In the past, Meymand basin had 72 wells but nowadays, there are only two springs of drinkable water: “Lay Ney Riz” and “KolKamar”. Also, two underground Kariz or Qanat (a water management system used to provide a reliable supply of water for human settlements and irrigation in hot, arid and semi-arid climates like Iran) aqueducts that bring water down from the water catchment areas in the upper slopes of the surrounding hills provide the water supply for this village So the village has two sources of water (2 Qanats). The orchards extend around the river.

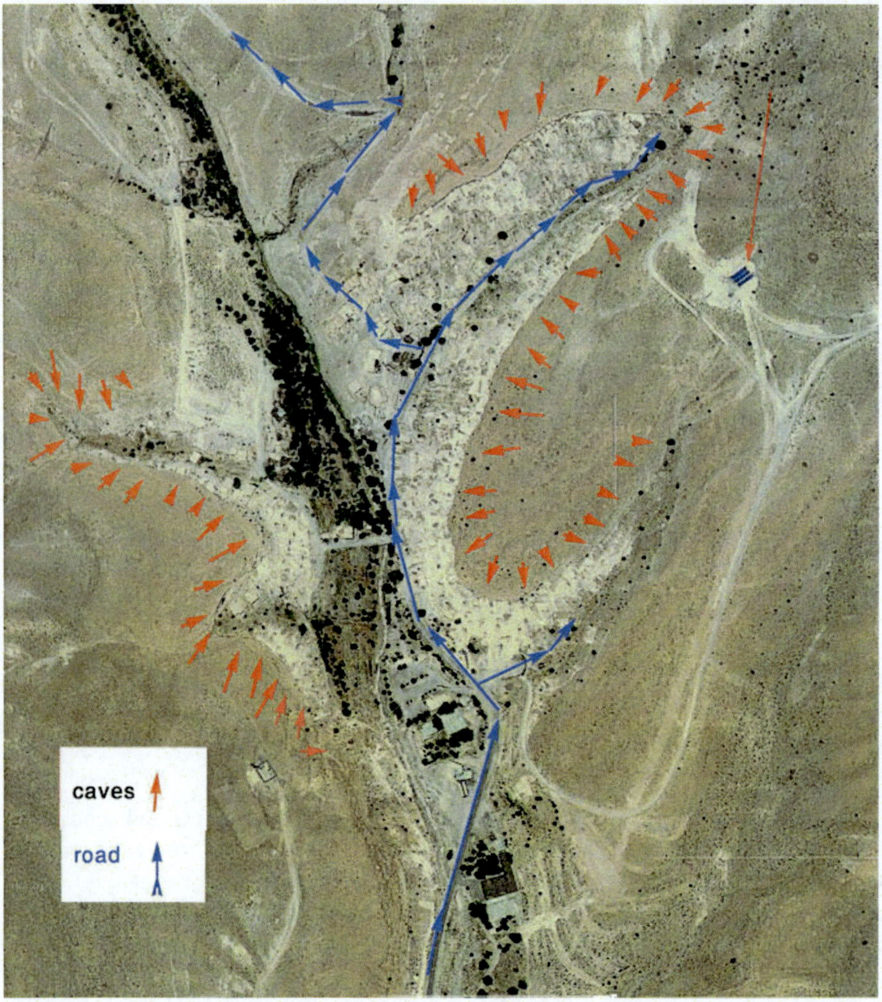

Fig. 17.3 Meymand Satellite image, the caves, surrounding mountains, river and main access roads are seen

17.4.3 Cultural and Environmental Value in Meymand

Situated between a desert and a mountain, Meymand enjoys a mountainous climate (Fig. 17.3). The winters in Meymand are cold and harsh. The summers are hot and dry. Some of Meymand's people are semi-nomadic shepherds who spend winters in the village and move with their herds of goats and sheep to the plains in the spring. They go to higher pastures and cooler climes during the summer. Villagers spend their summers in higher altitudes around Meymand picking wild herbs, nuts and seeds. So the diversity of climate and adaptation of people to the nature has helped them to easily live in the region for many years.

Air and Wind—the nomads of Meymand have a different manner of living. They make different types of shelters by using wood and stone outside the village. Those who spend summers in the village build special dwellings called "Kapars", which are made to allow the circulation of air to cool the Kapar temperature. Making a different kind of shelter and using them in different seasons help them to adapt more to nature.

Unique landform—the main factor to form this great housing collection is the mother material (Iravani & Pourkhosravani, 2011) which has made possible the synchronization between the houses and geomorphic properties of the region. It has provided the necessary bed to show creativity and manifest integrated spatial structure. It means that the form of the ground has created a kind of landscape within which rocky houses are most likely to appear.

Green infrastructure—there is many tiny oases in the ravines around the village. There is a good connection between vegetation and water resources, which underpin the natural functions of the settlement.

17.4.4 Spatial Patterns and Natural Environment

Though limitations of the nature have affected this structure, some sights of spatial structure are: restricted entrance to the village, location of the village and spatial arrangement of caves around the valley (Fig. 17.4); the summer camp area in the mountain and spring residences—Aghol—in plains.

Special feathers—there is a hill around the village which is well-known for its special feathers. The ancient hill called Markhazineh: is 3 km north of Meymand. A few charts of this hill are designed by the shape of goats, numeral ancient symbols like cross and also cravings and inscriptions craved of a few equestrians on the horse with a sword in the hand and picture of a mounted man and some animals that are most probably goats on the cliff.

Other key features are the ancient Fire Temple, Theatre of passion plays, Historical bath, old School and Mosque (Fig. 17.5). All the feathers verify the rich culture and long-time settlement in Meymand.

17.4.5 Economy

The economy of the villagers in Meymand is based on agriculture, herding, farming, animal husbandry, carpet weaving, making "Named" and tourism. The orchards consist of pistachio, almond, walnut, pomegranate and mulberry trees. Carpet weaving spawns supporting and ancillary crafts such as dyeing, felt-making, kilim weaving (a flat tapestry-woven carpet or rug) and crochet lace work. The tourism

Fig. 17.4 The houses (caves) and Dalans

Fig. 17.5 One of the caves as the mosque of Meymand

Table 17.5 Signs of landscape sustainability in Meymand according to sustainability indicators

Culture	Economy	Natural environment	Spatial patterns
Language Customs, traditions, ceremonies, Religion Traditional Food, Clothes Handicraft Use the spaces for social gathering Nomadic life Movement between mountain and Plain Life cycle Special ancient feathers	Gardening Multi-use of land Sheep Husbandry Agriculture/horticulture Carpet/Kilim weaving Making Namad Tourism Medical herbs	Mountainous features Unique landforms, soil Rocks, higher Attitude Underground water Surface water (River) Trees and vegetation Biodiversity Medicinal herbs Oasis/springs	The spatial arrangement of caves Restricted entrance Summer camp Spring residences Winter residences Underground water system (Qanats) Traditional house type Making Dalans for gathering

industry in Meymand is more cultural. Tourists visit the houses, the nature of Maymand, the lifestyle of the natives and how they adapt themselves to nature.

The village has maintained its original architecture and traditions, and the language, which has barely changed, still contains Sassanid and Pahlavi words. Villagers use their traditional materials for cooking, making clothes and building. They use the alleys as social public spaces. "Dalans" are used for family and social gatherings (Fig. 17.4).

Their lives are in adaptation and compliance with nature. This has made them resistant to natural hazards and unfavourable climate condition.

The particular location of the settlements especially the nature and the restricted entrance to the village has kept it safe and secure through adverse conditions of war, local conflicts and ethnic divisions. Another factor for the persistence of this rural community is being far from the major cities and current changes.

The results of landscape character surveys in Meymand are presented in the summary table (Table 17.5).

17.5 Conclusion

This chapter has tried to model the Cultural Landscape of "Maymand", the historic Village in Iran as a practical approach for evaluating sustainability. This research identified the main factors which have influenced rural landscape sustainability. Meymand's people change their place during the seasons to move from a bad situation to more suitable circumstances making their life more sustainable. The chapter findings help to promote the situation of Maymand village and the similar cases in the world according to the 8th goal of SDGs. This goal is about "Promoting sustained, inclusive and sustainable economic growth, full and productive employment and decent work for all".

By this practical model, the economic growth of such villages becomes more Sustain per capita following national circumstances and achieves higher levels of economic productivity through diversification and technological upgrading including through a focus on high-value added and labour-intensive sectors. Also, it promotes development-oriented policies that support decent job creation and entrepreneurship and encourages the growth of enterprises (micro, small and medium-sized) and productive activities, including access to financial services. All of these promote the sense of belonging to a place and decrease the rate of migration (due to lack of jobs and low incomes).

The 11th SDGs goal supports positive economic, social and environmental links between urban, peri-urban and rural areas by strengthening national and regional development planning. The sustainable model of the cultural landscape that has been discussed in this chapter and its effects on place-making can promote the Goal 11 of SDGs. This Goal is "Making cities and human settlements inclusive, safe, resilient and sustainable"; That By 2030, want to ensure access for all to adequate, safe and affordable housing and basic services and strengthen efforts to protect and safeguard the world's cultural and natural heritage.

This study calls for further research to establish stronger links between sustainable landscape theories and rural sustainability. Despite the limitations, we believe this framework is a useful approach to rural landscape sustainability.

Acknowledgements The author specially thanks Dr. Nasrin Kowkabi, from the Department of Language and Literacy Education at the University of British Columbia for her valuable comments on final editing of this chapter.

References

Antrop M (2005) Why landscapes of the past are important for the future? rural landscapes: past processes and future strategies. Landscape Urban Plann J 70(1–2):21–34

Antrop M (2006) Sustainable landscapes: contradiction, fiction or utopia? Landscape Urban Plann J 75(3–4):187–197

Ardalan N (2000) Sense of unity. Khak Press, Tehran

Azari Dehkordi F (2012) Landscape degradation modelling: an environmental impact assessment for rural landscape prioritisation. Landscape Res 37(5):613–634

Banski J, Wesolowska M (2010) Transformations in housing construction in rural areas of Poland's Lublin region—Influence on the spatial settlement structure and landscape aesthetics. Landscape Urban Plann J 94(2):116–126

Bell S et al (2009) Management of cultural landscapes: what does this mean in the former Soviet Union? A case study from Latvia. Landsc Res 34(4):425–455

Benson J, Roe M (2007) Landscape and sustainability. Taylor & Francis, London

Bolliger J, Kienast F (2010) Landscape functions in a changing environment. Landscape Online 21(1):1–5

Brandt J, et al (2000) Multifunctional landscapes: interdisciplinary approaches to landscape research and management. Conference material for the international conference on "Multifunctional Landscapes: Interdisciplinary Approaches to Landscape Research and Management". Centre for Landscape Research, Roskilde, Denmark.

Brown RD et al (2008) Cultural patterns as a component of environmental planning and design, Chapter 24. In: Hong K et al (eds) Landscape ecological applications in man influenced areas: linking man and nature systems. Springer, Dordrecht, pp 395–415

Buhler-Natour C, Herzog C (1999) Criteria for sustainability and their application at a regional level: the case of clearing islands in the Cubener Heide nature park (Eastern Germany). Landscape Urban Plann J 46(1–3):51–62

Burke A, Elliott M (2008) IRAN: Meymand. Lonely Planet, Iran, p 310

Conzen MP (2001) Cultural landscape in geography. In: Smelser N and Paul Baltes PB (eds) International encyclopedia of the social & behavioral sciences, vol. 5. Elsevier, Oxford, pp 3086–3092

Countryside Agency and Scottish Natural Heritage (2002) Landscape Character Assessment: Guidance for England and Wales. Countryside Agency, SNH, Battleby

Daneshdoust Y (2007) Islamic Gardens in Iran. Chapter 3 (Tehran)

Domona G (2011) Landscape as resource: consequences, challenges and opportunities for rural development. Landscape Urban Plann J 100(4):338–340

Ernoult A et al (2003) Patterns of organization in changing landscapes: implications for the management diversity. Landscape Ecol 18(3):239–251

Farina A (2006) Principles and methods in landscape ecology: towards a science of the landscape. Springer, Dordrecht

Forman RTT, Godron M (1986) Landscape ecology. Wiley, New York

Frajer J, Fiedor D (2021) A historical curiosity or a source of accurate spatial information on historical land use? Land Use Policy 100:1039. https://doi.org/10.1016/j.landusepol.2020.104937

Golley FB, Bellot J (1999) Systems analysis as a tool for rural planning. Rural Planning from an Environmental Systems Perspective. Springer, New York, pp 315–343

Hanachi P, Mehrdad M (2007) Politics, patterns and restoration samples in Meymand rural texture. Fine Arts Journal 30(1):53–62

Harvey D (2012) Culture, heritage and representation: perspectives on visuality and the past. Landsc Res 37(3):391–394

Iravani MR, Pourkhosravani M (2011) The role of geomorphic unites in forming tourism village of Meymand in Babak Town. Int J Bus Humanit Technol 1(2):150–155

Izadpanah F (2002) Architectural studies of Meymand historic village. Meymand cultural heritage site, pp 20–42

Jongman RHG (2005) Landscape ecology in land-use planning. In: Wiens J, Moss M (eds) Issues and perspectives in landscape ecology. Cambridge University Press, Cambridge, pp 316–328

Kim JE et al (2007) Landscape ecological applications in man influenced areas linking man and nature systems (chapter 29). In: Hong SK et al (eds) International trends of rural landscape researches for land management and policies. Springer Press, Dordrecht, pp 489–504

Kowkabi L et. al (2013) The Application of Fuzzy Multi-attribute Group Decision Making to Prioritize the Landscapes with high Ecological value: Khoshk River in Shiraz. International Journal of Environmental Research, 7(2): 423–434. https://doi.org/10.22059/ijer.2013.621

Kowkabi L (2020) The fundamental characteristics of the rural landscape through application of grounded theory (case study: historic village of Furg). J Environ Stud 45(4):693–709. https://doi.org/10.22059/jes.2020.295707.1007968

Kowkabi L (2021) Revitalization of bio infrastructure network, an approach to urban regeneration: a case study of the qanats of Tehran. Urban For Urban Greening 59. https://doi.org/10.1016/j.ufug.2021.127020

Kowkabi L, Akrami G (2021) The localization of sustainable landscape indicators in Iranian rural ecosystem based on the mental-evaluative images of stakeholders. J Rural Res 11(4):728–745. https://doi.org/10.22059/jrur.2020.299366.1478

Kowkabi L, Badri S (2020) Explaining sustainable landscape indicators in the rural area of Iran based on experts' and professionals' assessment. J Rural Res 11(1):50–65. https://doi.org/10.22059/jrur.2019.277942.1342

Lafortezza R, Brown RD (2004) A framework for landscape ecological design of new patches in the rural landscape. Environ Manage 34(4):461–473

Larcher F et al (2013) Planning rural landscapes: a participatory approach to analyse future scenarios in Monferrato Astigiano, Piedmont, Italy. Landscape Research 38(1):1–22

Lieskovsky J et al (2014) Factors affecting the persistence of traditional agricultural landscapes in Slovakia during the collectivization of agriculture. Landscape Ecol 29:867–877. https://doi.org/10.1007/s10980-014-0023-1

Martin L (2005) The symbol of water in Quran. Art, Tehran, p 28

Mavromatidis LE (2012) The aesthetic value of socio-cultural identities and the cultural dimension of the landscape. Hum Geogr 6(2):15–21

Moss MR (2000) Inter-disciplinarity, landscape ecology and the transformation of agricultural landscapes. Landscape Ecol 15(3):303–311

Nassauer JI, Wascher DM (2007) The globalized landscape: rural landscape change and policy in the United States and European Union. In: Wescoat J, Johnston D (eds) Political economies of landscape change. Springer, Dordrecht, pp 169–194

Nassauer JI (1997) Cultural sustainability: aligning aesthetics and ecology. In: Nassuer JI (ed) Placing nature: culture and landscape ecology. Island Press, Covelo, pp 65–83

National Geographical Organization (2002) Kerman-Shahr Babak. In: Kerman National Gazetteer of Iran (eds)

Naveh Z (2001) Ten major premises for a holistic conception of multifunctional landscapes. Landscape Urban Planning J 57(3–4):269–284

Nikpour M et al (2011) Investigating the role of nature elements in the central courtyards of traditional Iranian houses in hot and dry regions, recent researches in energy. Environ Landscape Archit 1:55–59

Official Center of Mymand site (2006) Handbook. Tehran

Palang H et al (2005) The forgotten rural landscapes of Central and Eastern Europe. Landscape Ecol 20(6):645–655

Parsi F (2008) Climatic areas in Iranian architecture. Architecture 48(1):112–117

Pinto-Correia T et al (2006) Bridging the gap between centrally defined policies and local decisions-towards more sensitive and creative rural landscape management. Landscape Ecol 21(3):333–346

Plieninger T et al (2016) The driving forces of landscape change in Europe: a systematic review of the evidence. Land Use Policy 57:204–214. https://doi.org/10.1016/j.landusepol.2016.04.040

Reed M et al (2017) A theory of participation: what makes stakeholder and public engagement in environmental management work? Restoration Ecology 26 (1): 7–17. https://doi.org/10.1111/rec.12541

Rippon S (2012) Historic landscape character and sense of place. Landsc Res 37:1–24

Rosenberg A (1986) An emerging paradigm for landscape architecture. Landsc J 5(2):75–82

Selman P, Swanwick C (2010) On the meaning of natural beauty in landscape legislation. Landsc Res 35(1):3–26

Selman P (2009) Landscape multi-functionality. Sustain Sci Pract Policy 5(2):45–52

Simensen T, Halvorsen R, Erikstad L (2018) Methods for landscape characterisation and mapping: a systematic review. Land Use Policy 75:557–569. https://doi.org/10.1016/j.landusepol.2018.04.022

Swanwick C (2002) Landscape character assessment: guidance for England and Scotland. The Countryside Agency and Scottish Natural Heritage, London

Talandier M (2009) Mesurer l'impact des aménités naturelles, culturelles et environnementales sur le développement économique local, XLVIe Colloque ASRDLF, Clermont-Ferrand. Available at http://www.asrdlf2009.com/propositions/final/upload-final/talandier.pdf.

Taylor K, Lennon J (2012) Managing cultural landscapes. Routledge, London

Tress B et al (2005) Integrative studies on rural landscapes: policy expectations and research practice. Landsc Urban Plan 70:177–191

Vaziri Kermani AA (1981) The history of Kerman. Elmi Press, Tehran

Waldhardt R et al (2003) Floristic diversity at the habitat scale in agricultural landscapes of Central Europe- Summary, conclusions and perspectives. Agric Ecosyst Environ J 98(1):79–85

Yoon Y (2002) Development of a structural model for tourism destination competitiveness from stakeholders' perspectives. Doctoral Dissertation in Hospitality and Tourism Management, Virginia Polytechnic Institute & State University, Blacksburg

Zhao HZ et al (2021) Incorporating spatio-temporal connectivity for prioritized conservation of individual habitat patches in a dynamic landscape. Ecol Ind 124:1029. https://doi.org/10.1016/j.ecolind.2021.107414

Chapter 18
Reuse of Unoccupied Religious Monuments for Tourist Accommodation: Santa Maria da Ínsua (North of Portugal)

Rolando Volzone, Soraya Genin, João Estêvão, and Mafalda Patuleia

Abstract This paper discusses the reuse of cultural heritage for tourist accommodation, a trend in Portuguese policies around tourism and culture. The convent (and fortress) of Santa Maria de Ínsua, in Caminha, south of the Minho river and close to the Spanish border, is a unique religious heritage site, due to its location on an islet and its diachronic evolution. After the dissolution of the religious orders in 1834, its abandonment and degradation led to the inclusion of this complex in a national programme aimed at conversion into a hotel. Starting from an international literature review, this case is evaluated through an integrated analysis, combining historical, architectural, and tourism data. The result shows the negative impact the programme has had for the protection of the site and the development of tourism without benefit for society. If sustainable tourism and conservation of cultural heritage are to be achieved, this will require an integrated framework for local and national government strategies for economic benefit through tourism and the occupation of abandoned buildings.

Keywords Rehabilitation of religious monuments · Tourism · Conservation of cultural heritage · Santa Maria da Ínsua · Portugal · Sustainability

---

The original version of this chapter was revised: The affiliation of the author has been amended. The correction to this chapter is available at https://doi.org/10.1007/978-981-19-6274-5_21

R. Volzone (✉) · S. Genin · J. Estêvão
DINÂMIA'CET-Iscte | Iscte-University Institute of Lisbon, 1649-026 Lisbon, Portugal
e-mail: rveoo@iscte-iul.pt

S. Genin
e-mail: soraya.genin@iscte-iul.pt

J. Estêvão
e-mail: joao.vaz.estevao@iscte-iul.pt

M. Patuleia
Lusófona University, Lisboa and GOVCOPP—Research Unit on Governance, Competitiveness and Public Policies, Aveiro University, Aveiro, Portugal
e-mail: p4998@ulusofona.pt

R. P. B. Singh et al. (eds.), *Placemaking and Cultural Landscapes*, Advances in Geographical and Environmental Sciences,
https://doi.org/10.1007/978-981-19-6274-5_18

18.1 Introduction

The sustainable reuse of unoccupied cultural heritage in line with international guidelines is a much discussed topic. This concern is even more relevant when it comes to religious heritage sites; as a result of continually decreasing entrances to the consecrated life, around 1% of old convents/monasteries in Portugal show functional continuity in their usage. The common practice in Portugal today is to transform this type of buildings into tourist accommodations. This phenomenon is clearly visible in the Revive programme, a government initiative for the rehabilitation of monuments for touristic purposes. It is urgent to rethink the refunctionalisation of these spaces in an integrated way that can bring benefits to both heritage and tourism.

When dealing with cultural heritage, which is part of a community's identity, previous research suggests that its sustainable use within the tourist industry should be carefully planned (Surugiu and Surugiu 2015). In Portugal, however, unoccupied cultural heritage sites (religious or otherwise) tend to be converted into accommodation facilities regardless of their previous function or geographical location—mainly into luxury hotels.

The ICOMOS International Cultural Tourism Committee (ICTC) pioneered the integration of heritage protection practice and tourism by adopting the first International Cultural Tourism Charter in 1976. The second Charter (ICOMOS—International Council of Monuments and Sites 1999), as well as the Florence Declaration on Cultural Heritage Conservation and Sustainable Tourism for Development (ICOMOS—International Council of Monuments and Sites 2014), is the reference for the international state of the art in this field. The first Charter refers to the positive and negative effects of cultural tourism on monuments and sites. Public knowledge of monuments and sites (the main purpose of cultural tourism) has a positive effect, as it contributes to their maintenance and protection. The Charter recommends respecting cultural heritage so that it continues to play its role as a touristic attraction.

This paper pursues the following questions: how is the rehabilitation of unoccupied heritage in Portugal being implemented? Is it possible to establish sustainable alternatives in reusing these buildings in an integrated way, taking a multicriteria analysis into account?

This chapter brings together a multidisciplinary study of architectural conservation and cultural tourism with the aim of understanding the strategies adopted by the Portuguese government for the valorisation of cultural heritage and tourism.

We begin by examining the international literature regarding cultural tourism, introducing the issues related to the allocation of cultural heritage for tourism purposes. We then delve deeper into the Portuguese *modus operandi* by analysing the Revive programme. The selected case study, the convent and fortress of Santa Maria da Ínsua, located in Caminha, is then analysed in terms of its diachronic evolution and in terms of material value and protection strategies. This case study is treated in depth within the JPI Cultural Heritage project "F-ATLAS. Franciscan Landscapes: Observance between Italy, Portugal and Spain", funded by the European Union's Horizon

2020 research and innovation programme under grant agreement n° 6995237 (https://www.f-atlas.eu/). Finally, the tourism indicators for Caminha municipality are taken into account.

Our study shows that the monument's outstanding value and the tourist indicators for the region are not being considered as criteria for rehabilitation. The methodology used in this research should be followed for an integrated workflow in the future programmes for heritage rehabilitation and tourism development.

18.2 Cultural Heritage and Tourism

In the public opinion and among decision-makers, there seems to be a broad consensus about the virtuous effects that the alliance between tourism and heritage brings to both, as well as to visitors and local communities in general (Butler 2018). Existing studies also highlight that a clash of interests between tourism and residents tends to occur in scenarios privileging a utilitarian use of heritage by tourism, insofar as these prioritize profit rather than the protection of a community's cultural legacy (Yang et al. 2021). In many cases, residents are deprived of the use of their own heritage for the sake of tourist development with questionable socioeconomic benefits to host communities (Akbar et al. 2019).

In some regions holding valuable state-owned heritage sites that no longer see day-to-day use, a pattern of utilitarian reuse through tourism seems to have become predominant, the virtues of which could be questioned (Bosone et al. 2021). In such contexts, instead of taking the tourism sector as a means to revitalise underused or even abandoned heritage sites by encouraging both residents and visitors to share their use, public administrations often convert them into accommodation units such as luxury hotels (Salim et al. 2021), privatising and limiting access.

Even when such conversions are carefully made so as to maintain the physical features of a given site, local communities lose touch with its symbolic value as well as become deprived of physical access to it (Salim et al. 2021). Moreover, the irony lies in the fact that such a pattern of converting monuments into hotels is also harmful to the tourism sector of a destination itself. As suggested by Benur and Bramwell (2015), the *raison d'être* of tourism destinations is the attractiveness of their natural and cultural features, which are usually called their primary resources. Even when access to such attractions requires the payment of a fee, they are often public and accessible to both visitors and residents. Tourism services such as accommodation units are usually the secondary or supporting resources that visitors require in order to temporarily inhabit the visited destination. Thus, it seems reasonable to posit that the conversion of heritage sites into accommodation units is often a lost opportunity to turn them into tourist attractions. In doing so, public administrations are single-handedly subtracting those same heritage sites from their array of attractions, jeopardising their symbolic value, and cutting their emotional and physical ties to local communities.

According to Hall (2008), four main approaches to tourism development at local, regional, and national levels may be identified: boosterism, economic, physical-spatial, and community based. Of these, the two which often lead to negative impacts in local communities are boosterism and the economic approach. In the case of boosterism, public authorities typically hold a simplistic and naïve understanding of the tourism sector as an automatic generator of positive impacts on residents' quality of life. Such a perspective posits that the tourism sector should grow as fast as possible, even if externally led, without the need for strategic planning (Hall 2008). Thus, given the fast pace of tourism growth in scenarios of boosterism, as well as the common mismatch between investments and the communities' knowledge and identity, residents have few chances to participate as key stakeholders (Hall 2008).

The economic approach differs from the previous because it does not disregard tourism planning, which is used to achieve and maintain the industry's performance levels (Bianchi and de Man 2021). However, such an approach fails to prioritise the needs and wants of local communities, considering natural and cultural heritage as resources that should be appropriated by tourism for the sake of its economic performance (Bianchi and de Man 2021).

We will analyse a national programme that, despite its stated good intentions, doesn't take into account factors that are essential for the sustainable reuse of cultural heritage and, consequently, for its potential to drive sustainable territorial development. With these concerns in mind, we shall examine the Revive programme as it is currently implemented in Portugal.

18.3 The Revive Programme

The Revive programme (https://REVIVE.turismodeportugal.pt/en) is a joint initiative of the Ministries of Economy, Culture, and Finance, aimed at promoting the preservation and monetisation of public property that is unoccupied and unused. The team responsible is made up of the General Directorates for Cultural Heritage, Treasury and Finance, National Defence Resources, and Tourism, together with the administrations of the municipalities in which the buildings are located. The team makes decisions about the safeguarding of classified heritage—or heritage in the process of classification—and about the adequacy of the refunctionalisation to the development needs of each region.

The programme intends to generate wealth and jobs as well as to strengthen the attractiveness of regional destinations. For this purpose, national heritage is opened up to private investment in order to develop tourism projects, by way of concession with a public tender.

Currently, 52 buildings are part of the programme. Since 2016, 24 of them have been the object of public tenders. Examination of these 24 properties shows that 15 buildings are related to religious heritage sites: 1 Jesuit College, 1 sanctuary, and 13 monasteries or convents. All the proposals of the 17 concluded public tenders planned the refunctionalisation of the properties into hotels, mostly in the four and five-star category (Fig. 18.1).

The relevance of this data should be appreciated. These properties have different characteristics regarding their original functions, architecture typology, urban context, locations, features, and they belong to completely different socio-political-economic conditions, yet they are being converted into the same function: tourist accommodations.

Moreover, potential benefits to local communities, especially in terms of job creation, are not directly linked. Indeed, four and five-star hotels tend to hire staff at national and international scales in order to meet the higher standards of service required in both categories (Hampton et al. 2018). Peripheral rural areas usually lack such manpower as well as training opportunities within the hotel industry; it is thus common that only a residual number of the hotel staff come from local communities (Hampton et al. 2018).

Cooper et al. (2008) suggest that a tourist destination is composed of four main components, namely (i) attractions; (ii) amenities; (iii) access; and (iv) ancillary services. As the designation indicates, attractions are the primary elements attracting visitors to a given destination. The remaining three components are relevant to ensure that visitors have a set of services at their disposal, such as hotels (amenities); that they can access and move comfortably in and around the destination (access); and that they are given proper support before and during their stays, namely through official destination websites or tourism information offices (ancillary services).

As argued by Jeon, Kang, and Desmarais (2016), cultural heritage tourism should be carefully planned to attend primarily to the needs and wants of local communities and to foster the preservation of their cultural assets. It is reasonable to ask if the top-down decision to integrate them into the Revive programme emerged, indeed, from the will of those same communities.

A rigorous and integrated analysis would be required in order to establish how the unoccupied religious cultural heritage sites function within the tourism industry, as well as to ascertain the characteristics of the private companies to whom the commercial use is conceded. This analysis is even more crucial with regard to the peripheral regions. In such areas, local communities are often in dire need of investments, demanding careful planning for the benefit of their fragile economies.

18.4 The Convent and Fortress of Santa Maria da Ínsua in Caminha

18.4.1 Historical Framework

The convent of Santa Maria da Ínsua (Fig. 18.2) belongs to the first foundations of the Franciscan Observant who reached Portugal in 1392 through the Galician group led by Gonçalo Mariño (Teixeira 2010; Fontes et al. 2020). In the same year, four

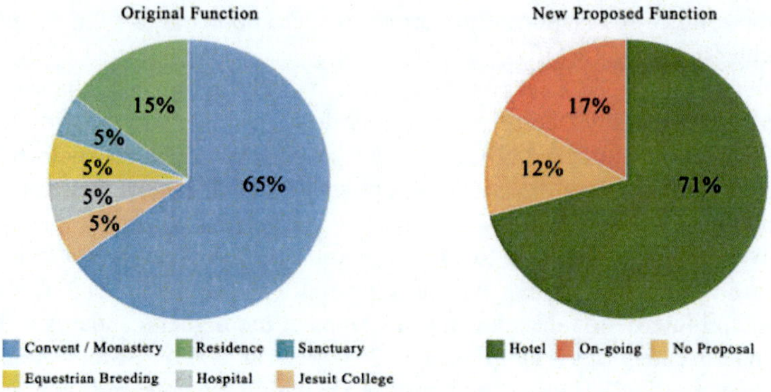

Name	Public tender	Original function	Proposed function
Convento de São Paulo	2016	Convent	Hotel (4 stars)
Pavilhões do Parque D. Carlos I	2017	Thermal hospital; Military barracks	Hotel (5 stars)
Convento do Carmo	2018	Convent	Hotel (5 stars)
Quartel do Carmo	2018	Convent	Hotel (5 stars)
Mosteiro de Arouca	2018	Monastery	Hotel (superior)
Convento de São Francisco	2018	Convent	No proposal
Convento de Santo António dos Capuchos	2018	Convent	Hotel (superior)
Convento de Santa Clara	2018	Convent	Hotel (5 stars)
Colégio de São Fiel	2018	Jesuit College	No proposal
Coudelaria de Alter	2018	Equestrian breeding	Hotel (4 stars)
Casa de Marrocos	2018	Residential	Ongoing
Mosteiro de Lorvão	2019	Monastery	Hotel
Palacete do Conde Dias Garcia	2019	Noble palace	Hotel (4 stars)
Mosteiro de São Salvador de Travanca	2019	Monastery	Hotel (4 stars)
Forte / Convento da Ínsua	2019	Convent / Military	Hotel (4 stars)
Quartel da Graça	2019	Convent	Hotel (5 stars)
Paço Real de Caxias	2019	Residence	Hotel
Castelo de Vila Nova de Cerveira	2019	Castle	Hotel (at least 4 stars)
Mosteiro de Santo André de Rendufe	2020	Monastery	Hotel
Forte da Barra de Aveiro	2020	Fortress	Hotel / Local Accomodation
Quinta do Paço de Valverde	2020	Convent	No proposal
Hotel Turismo da Guarda	2021	Hotel	Ongoing
Santuário de Cabo Espichel	2021	Sanctuary	Ongoing
Casa do Outeiro	2021	Rural residence	Ongoing

Fig. 18.1 List of the monuments included in the Revive programme: their original and proposed future functions. Data related to December of 2021

more settlements were founded on the northwest coast of Portugal: Santa Maria de Mosteiró (Valença), São Francisco de Viana (Viana do Castelo), São Paio do Monte (Vila Nova de Cerveira), and São Clemente das Penhas (Matosinhos).

It is located on an islet in the Minho River in front of the village of Caminha, 200 m from the Portuguese coast. This space represented the ideal *locus* for the settlement of Franciscans seeking a rigorous observance of the rules of humility, poverty, prayer, and contemplation from St. Francis of Assisi (Fig. 18.2). For this reason, the Franciscan Observants lived in authentic poverty in an eremitic setting isolated from cities and centres of power.

The convent of Santa Maria da Ínsua originated from an oratory founded at that time by Frei Diogo Árias (Sousa 2016) on the site of a pagan temple previously

Fig. 18.2 Map of the municipalities of NUT III Alto Minho, ©Rolando Volzone, 2021 (top). The convent and fortress of Santa Maria da Ínsua. Aerial picture by drone, ©Pietro Becherini, 2021

dedicated to Saturn (Cepa 1980). The settlement was restored and extended in the last quarter of the fifteenth century, developing into a conventual typology. A fortress was built in the mid-seventeenth century, as part of the reformation of the coastal line of defence on the banks of the Minho River, due to the Restoration War.

The complex is formed by the secularized convent of Santa Maria da Ínsua and the surrounding fortress.

The convent was altered over the centuries. The first built structures were the oratory, the refectory, and a few cells. In 1471, more cells were added and the oratory was expanded into a church. The sacristy and the chapel dedicated to Santa Maria Madalena were built in 1498. A second floor, including 5 cells, was built in 1675. In 1707, the chapter house was created and some interventions were carried out in the church. In 1717, the vault of the church was rebuilt, new cells were added, and the retable in the main chapel was introduced (Fig. 18.3).

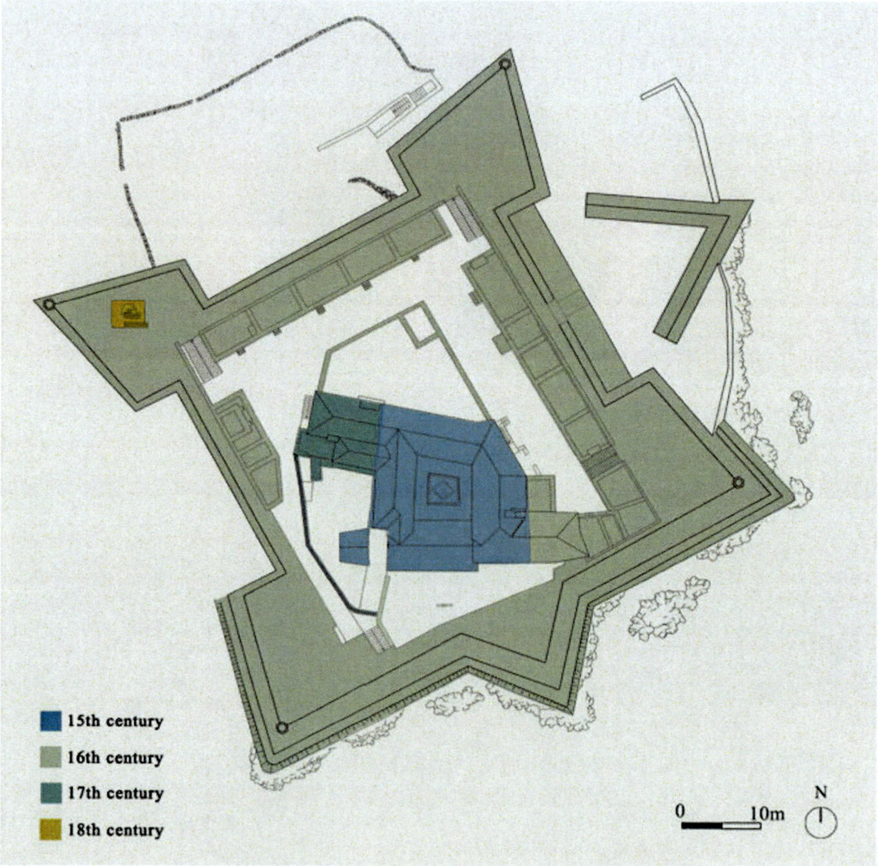

Fig. 18.3 Hypothesis of constructive evolution of the Convent and Fort of Ínsua, based on previous historical research, ©Rolando Volzone, 2021

The Franciscans were forced to leave the convent in 1834, with the Portuguese dissolution of the religious orders. The whole complex was managed by the Ministry of War until the last decade of that century, when it passed to the Navy Ministry and, finally, to the Finance Ministry in 1940. At this time, it was partially occupied by the general direction of lighthouse services, and a few rooms were used to store algae. Indeed, between the Middle Ages and the nineteenth century, the algae harvest has been a very important activity for the whole northwest coast of Portugal, both economically and socially. This was especially the case on this islet and was strictly linked to the Franciscan community, as attested by previous historical research (Cabral 2012). The islet, despite its small size, was rich in natural resources. This is also evident from the use of seashells for the production of lime, a necessity in mortars for the construction of buildings, including the convent of Santa Maria da Ínsua, as noted in previous studies. Whereas lime was produced from stone calcination in other regions, in areas with an abundant presence of shells the production of lime from their calcination is documented.

The abandoned complex today consists of the old convent and the surrounding fortress. The only way to access it is via the river. The singularity of this case study owes, firstly, to its strategic location, an islet in the Minho River at the Spanish border; secondly, to its double function, with the coexistence of a religious building—the convent—and a military one—the fortress.

18.4.2 Architecture and Outstanding Value

The convent has an irregular rectangular plan, formed by the church on two floors and an adjacent cloister which allows connection among the other conventual dependencies on the ground floor: the sacristy, chapter house, kitchen, refectory, and storage. The cells and other rooms—high choir and firehouse, among others—are located on the first floor. This nucleus is surrounded by an enclosure of granite masonry from materials available *in loco*, creating three different open spaces in which it is likely that kitchen gardens, orchards, and vineyards were cultivated.

The church (Fig. 18.4b), located at the south of the complex, is rectangular and has one only nave. This area is preceded by an endonarthex, under the high choir, and the main chapel is of reduced width. The old retable in the main chapel (*altar-mor* in Portuguese), as well as the lateral retables on the triumphal arch, which separated the nave from the main chapel, have disappeared. However, traces of mural painting on the church walls are still detectable.

The cloister (Fig. 18.4a), the distributing element for the whole complex, is quadrangular in shape, with four Tuscan columns on each side and a cross situated in the centre.

It is important to highlight the ground floor space called "De Profundis": a meeting area, adjacent to the refectory, where theological readings before and after meals occurred. Currently, mural decorations are still visible: a cross with symbols of the

Fig. 18.4 Outstanding components of the conventual complex. ©Rolando Volzone, 2021

Passion, painted, with sgraffito and *embrechados*, incrustations of crockery, glass, shells, and pebbles (Fig. 18.4c).

A final standout feature is the chapel of Santa Maria Madalena, connected to the enclosure, whose brick and stucco altar bears traces of polychrome (Fig. 18.4d).

The degradation of the construction is due to abandonment, as well as to a lack of management and maintenance of the site. Human causes have led to the exposure of the elements. As can be seen in Fig. 18.2, the original construction (the convent) is mostly protected by the roof, but the cloister has been left completely open to the air since the military constructions of the XVII century, which is why it's in a worse state of conservation. The partial protection, however, means that it is still possible to respect the integrity and the authenticity of the convent.

18.4.3 Strategies for Heritage Protection

As seen in the previous section, the convent of Santa Maria da Ínsua was the first nucleus built on the islet. Given its location, several strategies were implemented to safeguard the Franciscan convent, with its movable heritage, as well as to protect the northern border of Portugal. Potential threats were both natural and human: the ocean and its storms, as well as the continuous lootings, invasions, and attacks on the Portuguese territory.

These considerations first led to the construction of a wall in the fifteenth century and, in the seventeenth century, to that of a fortress. Moreover, over these centuries, the convent was rehabilitated, and some areas were at times rebuilt. This happened in 1717 with the reconstruction of the church. The fortress itself was rehabilitated in the last decade of the nineteenth century.

In 1910, the fortress, including the secularized convent, was classified as a National Monument through the decree of 16-06-1910 (DG nº 136 of 23 of June 1910).

Nevertheless, important movable assets have been lost, most of all since the 1940s. From that time until recently, various reports attest to the increasing state of degradation of the complex. The worsening situation led to the building's complete state of abandonment, the loss of movable heritage, including the retable—the main altarpiece—and the loss of the tiles that covered parts of the interior.

In 1979, a project for the adaption of the complex to a guesthouse was proposed; a project for a centre of marine research was developed in 1990 by the architect Fernando Távora. Neither has been implemented.

Since 2000, public access to the interior of the fort has been prohibited.

It has been abandoned and unused since then, representing an inactive endogenous resource of the territory in which it is located, one that is valuable due to its differentiating factors from the point of view of the landscape, of the movable and immovable components of its material heritage, of the know-how and the construction techniques that constitute its immaterial heritage, and of its exhibition of layers of historical stratification.

In 2016, the fortress and the convent were included in the list of properties to be leased by the Portuguese state to private individuals, through the Revive programme, with the aim of its conversion for tourist purposes. Through a public tender in 2019, the space has been leased for 50 years. The selected project foresees the installation of a lodging establishment (equivalent to a superior four-star touristic accommodation, including approximately 20 rooms) with tourist entertainment in the field of cultural and scenic touring, through experiences such as historical recreations, escape room activities, and augmented reality, complemented with tours along the Minho River (water/road/cycling). The adaptation work of the built construction is due to start soon.

Four levels of protection are defined in the programme brief: total protection, partial protection, no protection, and demolition (REVIVE. Rehabilitação, Património, Turismo 2019). The first level does not allow any demolition; the second is permissive in cases of non-structural walls without architectural or historical value, but volume and facades cannot be altered. The whole fortress and the main components of the conventual complex—enclosure, chapels, cross, church, and cloister—are safeguarded under the first level of protection. The second level includes the exterior of the fortress and the remaining conventual dependencies—sacristy, chapter house, refectory, and kitchen, among others. No components are included in the third level. Finally, consigned to the fourth level, the lighthouse is slated for demolition (Fig. 18.5).

Unfortunately, the programme does not valorise or even safeguard the material and immaterial values of the site that we noted in the previous section, and neither does it consider the state of conservation. The less protected construction includes part of the convent, some of the oldest parts of the complex that are still protected by an intact roof, now perhaps intended for hotel rooms. In contrast, the barracks of the fort, in a worse state of conservation, are classified with maximum protection.

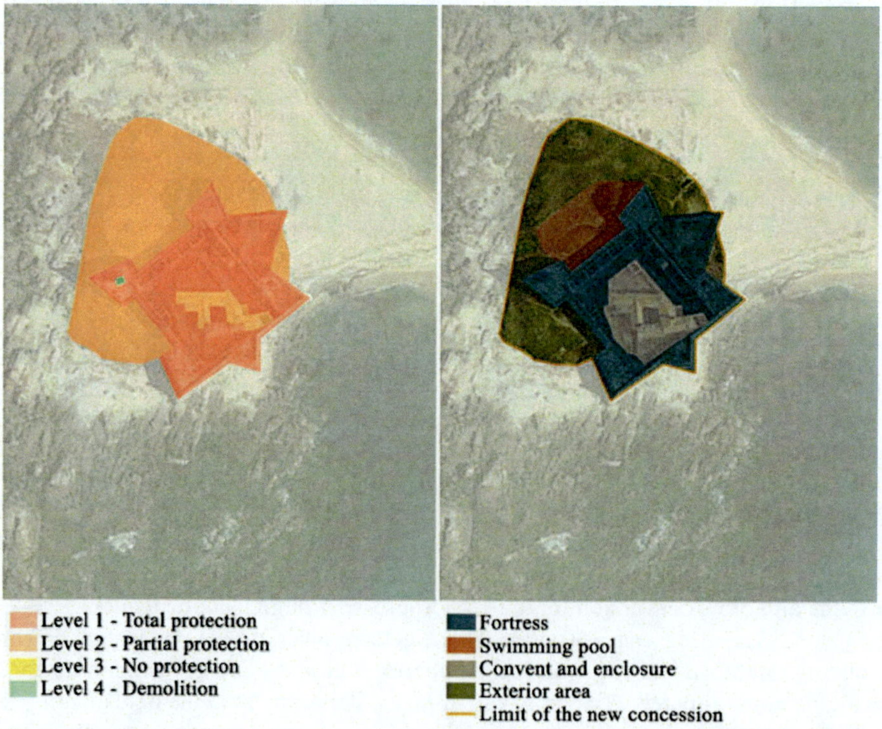

Fig. 18.5 The Revive programme for the rehabilitation of the monument: definition of four levels of protection (left); location of the swimming pool in the exterior (right) (Available at https://revive.turismodeportugal.pt/sites/default/files/Forte_insua_CE_Anexo_2.pdf)

Regarding the outstanding value of the cultural landscape, an important component of this site and, of course, of the Franciscan Observance architecture, the installation of a swimming pool is planned. This implies further mischaracterisation of the place, with a loss of its integrity and authenticity.

18.5 The Tourism-Related Indicators of Caminha

The rehabilitation project to turn the convent and fortress of Santa Maria da Ínsua into a hotel is a paradigmatic of the refunctionalisation practices for religious heritage sites in Portugal. Unfortunately, by converting a potential primary attraction of the municipality of Caminha into a hotel, the religious heritage site will be subtracted from visitors' experience. Indeed, consideration is given only to financial profit from this investment.

Taking into account the facts and figures of 2019, the best year ever for Portuguese tourism, from both the supply and demand sides, it is evident how the municipality of Caminha's tourism industry is not lacking hotel beds in order to become a more competitive tourism destination. The inclusion of data on tourism demand in 2019 is due to the subsequent exceptional and unprecedented decrease in tourism activity resulting from the COVID-19 pandemic. Referring to data from pandemic years could be misleading. Although located in the North of Portugal, less known for its beaches; Caminha has long been known as a sun-and-sea tourism destination (Panyik and Esteves 2013). In such destinations, the annual average stays of hotel guests tend to be longer than in most other types of tourism (Alegre et al. 2011). The average stay is usually a good indicator of the capacity of tourism destinations to retain their visitors and of the eventual need to create more attractions.

Surprisingly, in 2019, the annual average stay in Caminha's lodging units was only 1.7 nights, the second lowest of the ten municipalities which compose the NUTS (Nomenclature of Territorial Units for Statistics) III of Alto Minho (1.8 nights), where it is located, and well behind the national annual average (2.6 nights) (Instituto Nacional de Estatística 2020). Especially in a sun-and-sea tourism destination, this indicator alone seems to suggest that more attractions, such as Forte da Ínsua, should be developed in order to retain visitors for longer periods (Oklevik et al. 2019).

Another worrying indicator for the tourism activity of Caminha in 2019 is its annual accommodation occupancy rate of only 37%, lower than that of the NUTS II of Northern Portugal (42%) and of Portugal as a whole (47.3%). Such results from the supply side suggest that Caminha's tourism industry is much more in need of reasons to visit and to stay for longer periods than of more hotels.

Moreover, although Caminha is only the fifth most populated municipality ($n =$ 15.828) of the NUTS III Alto Minho (Fig. 18.2, top), in 2021 it was the one boasting the largest combined number of units of the three main commercial accommodation typologies in Portugal ($n = 467$), namely hotels ($n = 3$), rural tourism ($n = 12$), and short-term rentals ($n = 452$) (data available at https://rnt.turismodeportugal.pt/RNT/_default.aspx). This is much higher than the average number of accommodation units in the municipalities of the NUT III region ($n = 199$) and clearly above the national average number of units for Portugal's 308 municipalities in total ($n = 335$).

The above data strongly suggest that, at the time of writing, Caminha has an excessive number of beds in commercial accommodation units and a deficit in attractions, which are instrumental to increase occupancy rates and average stays of its guests. It is thus legitimate to question the conversion of Forte da Ínsua into accommodation units, as well as that of other public heritage sites—whether religious or not—in tourism destinations struggling with the same poor performance indicators.

18.6 Discussion

Following the advent of the concept of sustainable development, physical-spatial and community-based approaches consider tourism as a means to improve land use and residents' quality of life (Hall 2008). In such scenarios, tourism is often the object of participative planning processes in which communities are part of the decision-making, so that the selected tourism development models will safeguard their interests (Lalicic and Önder 2018).

Especially in the case of religious cultural heritage, there is a broad consensus between researchers that the main *raison d'être* of its use in the tourism sector should be the protection of its tangible and intangible elements, equally enjoyed by visitors and residents (Wang and Liu 2021). As argued by Li and Hunter (2015), use of cultural heritage for tourism that requires cutting its links to local communities is anything but sustainable.

According to the Portuguese General Secretariat for the Economy and Digital Transition, in 2018, Portugal was the 5th country in the world and the 2nd in Europe whose economy most heavily relied on tourism, accounting for 19% of its Gross Domestic Product (GDP). However, the usually enthusiastic views around the natural vocation of Portugal for the hospitality industry (Guizi et al. 2020) were, to some extent, replaced by narratives alerting the public sector to the need to diversify the Portuguese economy (Frey and Briviba 2021).

Unlike the tourism sector, the scenario for the cultural and creative industries is considerably humbler, being one of the poorest in the EU. While this sector accounted for 4.4% of the EU's GDP in 2019, the contribution to Portugal's GDP was a mere 3% (Laborinho 2020). One of the major clusters for this industry is that of craft arts, including rural traditional handicrafts. Being part of the cultural heritage of their communities, it seems adequate to ask if some of the reconversion projects intended to become hotels could be better revitalized through projects around cultural and creative industries. Indeed, previous research suggests that cultural projects involving local communities, with no a priori intents to become tourist products, often become primary attractions for visitors (Mısırlısoy and Günçe 2016; Cavalheiro et al. 2020).

These are an alternative to the reuse of cultural heritage. Indeed, it is at times not possible to maintain the function of monumental buildings due to cultural, social, and physical conditions. However, the adaptive reuse of historic buildings poses challenges for research to decide on the new use with appropriate functions, considering the different dimensions of heritage values, spatial characteristics, socioeconomic impact, and policy guidance. Successful adaptive reuse should respect the existing building and its environment for sustainable contemporary uses (Royal Australian Institute of Architects, Australia, and Department of the Environment and Heritage 2004; Wang and Liu 2021).

Adaptive reuse has recently become prevalent worldwide, employing a flexible approach to reusing historical buildings. When multicriteria decision-making methods are used (Chen et al. 2018) in order to establish appropriate new uses (Vehbi et al. 2021; Vizzarri et al. 2021), these buildings represent resources for sustainable urban development, a critical issue in the UN 2030 Agenda for Sustainable Development.

In architecture, adaptive reuse refers to the repurposing of an existing structure for new use (Foster and Saleh 2021). It should be considered as a strategy for the conservation of cultural heritage, and consequently for its utilisation and reuse, with the potential for economic and social benefits, especially in the tourism sector.

18.7 Conclusion

Religious placemaking contributed, over the centuries, to the sustainable development of remote locations. Indeed, friars most often occupied abandoned lands, contributing to their cultivation and to the definition of centres of production and the application of know-how linked to the built construction and to agricultural techniques. They provided assistance to the local communities and disseminated knowledge through the colleges and seminars in all the Portuguese territory. However, the dissolution of religious orders in 1834 led to the loss of material and immaterial values linked to the friars. Almost two centuries later, various spaces have been privatised or transformed by the local authorities without taking into account the principles of conservation of cultural heritage, including the involvement of local communities.

Decision-making processes that exclude broad discussion within local communities often prove to be harmful to the tourism sector's own interests, especially when dealing with cultural heritage. This appears to be paradigmatic with regard to the refunctionalisation of unoccupied religious heritage sites. However, when reusing tangible and intangible elements which have been part of a community's heritage and identity for centuries, the primary goal of the decision-making process should be that of sustainable rehabilitation. By taking the Forte da Ínsua as a case study, this chapter has sought to highlight the importance for decision-makers to carefully weigh which types of uses of the heritage sites being intervened upon would best serve the needs of the tourist destination.

In addition, given the composite nature of the tourism product and the consequent diversity of its components, it seems paramount that decision-makers adopt a holistic view of a tourism destination's needs before choosing the future use of a heritage site. The conversion of the Forte da Ínsua into a hotel is likely to become a lost opportunity to enrich the set of attractions of the municipality, only adding to the current surplus of hotel beds.

The case of Caminha is not an isolated event in which there is a mismatch between decisions concerning the use of heritage within tourism and the needs of the destination. As previously mentioned, there seems to be a pattern in the Revive programme of privileging the conversion of cultural heritage sites into hotel units, thus impoverishing the possibilities of turning them into heritage tourism attractions accessible to visitors and residents alike. This pattern might be explained by the absence of a theoretical framework for the decision-making process within this programme, which is today apparently committed to the belief that hotels alone will bring visitors to a given destination. A sustainable approach to regional planning in these cultural landscapes is still lacking. Understanding the sacred heritage and religious built space could stimulate the liveability and the development of remote areas, as well as peripheral or urban areas, in line with the UN-SDGs. The big challenge, due to the continuous evolution of these religious cultural landscapes, is to define sustainable alternatives that could bring benefits to all of the actors, including the cultural heritage itself. A balance needs to be achieved between preservation and changes, between environmental preservation, social equity, and economic growth. Future studies should therefore address the description and critical analysis of methodologies aimed at reutilising cultural heritage through tourism, as well as examining the level of strategic vision at work in the corresponding decision-making process. Future studies should also make use of applied methodologies such as Design Science Research (Hevner 2007) in order to propose a framework that ensures a decision-making process based on a sound knowledge base that considers the interests of the diverse actors in the environment that will be affected.

References

Akbar I, Zhaoping Y, Fang H, Gulnar K (2019) The influence of negative political environment on sustainable tourism: a study of Aksu-Jabagly world heritage site, Kazakhstan. Sustain 12(1):143. https://doi.org/10.3390/su12010143

Alegre J, Mateo S, Pou L (2011) A latent class approach to tourists length of stay. Tour Manage 32(3):555–563. https://doi.org/10.1016/j.tourman.2010.05.003

Benur AM, Bramwell B (2015) Tourism product development and product diversification in destinations. Tour Manage 50(October):213–224. https://doi.org/10.1016/j.tourman.2015.02.005

Bianchi RV, Man FD (2021) Tourism, inclusive growth and decent work: a political economy critique. J Sustain Tour 29(2–3):353–371. https://doi.org/10.1080/09669582.2020.1730862

Bosone M, De Toro P, Girard LG, Gravagnuolo A, Iodice S (2021) Indicators for ex-post evaluation of cultural heritage adaptive reuse impacts in the perspective of the circular economy. Sustain 13(9):4759. https://doi.org/10.3390/su13094759

Butler R (2018) Sustainable tourism in sensitive environments: a wolf in sheep's clothing? Sustain 10(6):1789. https://doi.org/10.3390/su10061789

Cabral JP (2012 [2005]) A apanha de algas na ilha da Ínsua (Caminha) nos séculos XVII - XIX. Singularidades e conflitos. Finisterra 40(80). https://doi.org/10.18055/FINIS1475

Cavalheiro MB, Joia LA, Cavalheiro GMC (2020) Towards a smart tourism destination development model: promoting environmental, economic, socio-cultural and political values. Tourism Plann Dev 17(3):237–259. https://doi.org/10.1080/21568316.2019.1597763

Cepa A (1980) Caminiana: Revista de Cultura Histórica, Literária, Artística, Etnográfica e Numismática 2(June):148–59

Chen CS, Chiu YH, Tsai L (2018) Evaluating the adaptive reuse of historic buildings through multicriteria decision-making. Habitat Int 81(November):12–23. https://doi.org/10.1016/j.habitatint.2018.09.003

Cooper C, Fletcher J, Fyall A, Gilbert D, Wanhill SRC (2008) Tourism: principles and practice, 4th edn. Prentice Hall Financial Times, Harlow, England

Fontes JL, Rodrigues AMSA, Andrade MF (2020) La(s) Reformas En El Franciscanismo Português En La Edad Media. Hispania Sacra LXXII(145):51–63. https://doi.org/10.3989/hs.2020.004

Foster G, Saleh R (2021) The adaptive reuse of cultural heritage in European circular city plans: a systematic review. Sustain 13(5):2889. https://doi.org/10.3390/su13052889

Frey BS, Briviba A (2021) A policy proposal to deal with excessive cultural tourism. Eur Plan Stud 29(4):601–618. https://doi.org/10.1080/09654313.2021.1903841

General Secretariat for the Economy and Digital Transition (2021). Portugal é o 5º país com maior contributo do turismo para o PIB [Portugal is the country with the 5th highest contribution of tourism to its GDP]. Available online: https://www.sgeconomia.gov.pt/noticias/portugal-e-o-5-pais-com-mais-forte-contributo-do-turismo-para-o-pib.aspx

Guizi A, Breda Z, Costa R (2020) How are overtourism and host-guest relationships portrayed by the Portuguese print media? International Journal of Tourism Cities 6(1):215–232. https://doi.org/10.1108/IJTC-06-2019-0081

Hall CM (2008) Tourism planning: policies, processes and relationships, 2nd edn. Pearson/Prentice Hall, Harlow, England; New York

Hampton MP, Jeyacheya J, Long PH (2018) Can tourism promote inclusive growth? supply chains, ownership and employment in Ha Long Bay, Vietnam. J Dev Stud 54(2):359–376. https://doi.org/10.1080/00220388.2017.1296572

Hevner AR (2007) A three cycle view of design science research. Scand J Inf Syst 19(2):87–92

ICOMOS—International Council of Monuments and Sites (1999) International cultural tourism charter: managing tourism at places of heritage significance. https://www.icomos.org/charters/tourism_e.pdf

ICOMOS—International Council of Monuments and Sites (2014) The Florence Declaration on heritage and landscape as human values. Declaration of the principles and recommendations on the value of cultural heritage and landscapes for promoting peaceful and democratic societies. https://www.icomos.org/images/DOCUMENTS/Secretariat/2015/GA_2014_results/GA2014_Symposium_FlorenceDeclaration_EN_final_20150318.pdf

Instituto Nacional de Estatística (2020) Statistical yearbook of Portugal 2019. Instituto Nacional de Estatística, Lisbon

Jeon MM, Kang M, Desmarais E (2016) Residents perceived quality of life in a cultural-heritage tourism destination. Appl Res Qual Life 11(1):105–123. https://doi.org/10.1007/s11482-014-9357-8

Laborinho AP (2020) Cultura é (também) economia. Diário de Notícias, Agosto 2020. https://www.dn.pt/opiniao/opiniao-dn/convidados/cultura-e-tambem-economia-12515728.html

Lalicic L, Önder I (2018) Residents involvement in urban tourism planning: opportunities from a smart city perspective. Sustain 10(6):1852. https://doi.org/10.3390/su10061852

Li Y, Hunter C (2015) Community involvement for sustainable heritage tourism: a conceptual model. J Cult Heritage Manage Sustain Dev 5(3):248–262. https://doi.org/10.1108/JCHMSD-08-2014-0027

Mısırlısoy D, Günçe K (2016) Adaptive reuse strategies for heritage buildings: a holistic approach. Sustain Cities Soc 26(October):91–98. https://doi.org/10.1016/j.scs.2016.05.017

Oklevik O, Gössling S, Hall CM, Jacobsen JKS, Grøtte IP, McCabe S (2019) Overtourism, optimisation, and destination performance indicators: a case study of activities in Fjord Norway. J Sustain Tour 27(12):1804–1824. https://doi.org/10.1080/09669582.2018.1533020

Panyik E, Esteves A (2013) Evolução das funcionalidades das praias: uma análise do turismo balnear no Alto Minho. Interconexões (July):15–32. https://doi.org/10.34632/INTERCONEXOES.2013.8567

REVIVE. Rehabilitação, Património, Turismo (2019) Forte e convento de Santa Maria da Ínsua. Instalação de um estabelecimento hoteleiro. Caderno de encargos - Termos de Referência para o Projeto. https://revive.turismodeportugal.pt/sites/default/files/Forte_insua_CE_Anexo_2.pdf

Royal Australian Institute of Architects, Australia, and Department of the Environment and Heritage (2004) Adaptive reuse: preserving our past, building our future. Dept. of the Environment and Heritage, Canberra

Salim N, Mohamed B, Marzuki A (2021) Commodification stage of George Town Historic Waterfront: an assessment. J Tour Cult Chang 19(1):38–58. https://doi.org/10.1080/14766825.2020.1849243

Sousa BV (2016) Ordens Religiosas Em Portugal Das Origens a Trento - Guia Histórico. 3rd edn. Livros Horizonte, Horizonte Histórico. Lisboa

Surugiu MR, Surugiu C (2015) Heritage tourism entrepreneurship and social media: opportunities and challenges. Procedia Soc Behav Sci 188(May):74–81. https://doi.org/10.1016/j.sbspro.2015.03.340

Teixeira VG (2010) *O Movimento Da Observância Franciscana Em Portugal, 1392–1517: História, Património e Cultura de Uma Experiência de Reforma Religiosa.* Editorial Franciscana, Porto

Vehbi BO, Günçe K, Iranmanesh A (2021) Multi-criteria assessment for defining compatible new use: old administrative hospital, Kyrenia, Cyprus. Sustain 13(4):1922. https://doi.org/10.3390/su13041922

Vizzarri C, Sangiorgio V, Fatiguso F, Calderazzi A (2021) A holistic approach for the adaptive reuse project selection: the case of the former Enel power station in Bari. Land Use Policy 111(December):105709. https://doi.org/10.1016/j.landusepol.2021.105709

Wang G, Liu S (2021) Adaptability evaluation of historic buildings as an approach to propose adaptive reuse strategies based on complex adaptive system theory. J Cult Herit 52(November):134–145. https://doi.org/10.1016/j.culher.2021.09.009

Yang H, Qiu L, Fu X (2021) Toward cultural heritage sustainability through participatory planning based on investigation of the value perceptions and preservation attitudes: Qing Mu Chuan, China. Sustain 13(3):1171. https://doi.org/10.3390/su13031171

Chapter 19
Survival of Heritage in Majuli in the Context of Climate Change

M. Satish Kumar

Abstract Globalisation and its attendant consequences have had a major impact on the traditions and cultural heritage of communities across the world. The flow of global and regional capital into redevelopment projects and the inevitable turn of climate-related disasters and natural hazards including the increased political instability have all impacted the preservation and conservation of cultural heritage. At the same time, gradual erosion and destruction of heritage have far-reaching consequences on the sustainability of communities both in the developed and developing world. Culture, therefore, matters when engaging with international development. In this chapter, an attempt will be made to interrogate the ideas and perspectives emanating from cultural heritage in meeting the global demands of sustainable development goals (SDGs). This will help enable a perspective towards shared prosperity, humanity, respect, and dignity in a rapidly polarised multipolar world. Drawing upon data from the field studies conducted in Majuli this paper makes a case for decolonising heritage management and highlights the challenges faced by cultural heritage centres as a result of the inexorable march of climate change.

Keywords Tangible and intangible cultural heritage · SDGs · Climate change · Cultural rights · Landscape

Embankments have come
The wild floods have receded
Now our children don't know how to swim in the river
While floods have disappeared due to huge embankments,
Our tradition and culture,
Our ability to survive floods
have been diminishing day after the day.

M. S. Kumar (✉)
Department of Geography, Archaeology & Palaeoecology, School of Natural and Built Environment, Faculty of Engineering and Physical Sciences, Queen's University Belfast, Belfast BT7 1NN, Northern Ireland, UK
e-mail: s.kumar@qub.ac.uk

R. P. B. Singh et al. (eds.), *Placemaking and Cultural Landscapes*,
Advances in Geographical and Environmental Sciences,
https://doi.org/10.1007/978-981-19-6274-5_19

It's good not to have floods
But at the same time,
We don't know what to do with the water now
We don't have water buffaloes,
as now we can't take them to water
We don't have access to that river anymore but through floods.
(Kumar 2017, 2018)

19.1 Questioning the Survival and the Place Personality

Globalisation and its attendant consequences have had a major impact on the traditions and cultural heritage of communities across the world. The flow of global and regional capital into redevelopment projects and the inevitable turn of climate-related disasters and natural hazards including the increased political instability have all impacted the preservation and conservation of cultural heritage. At the same time, gradual erosion and destruction of heritage have a far-reaching impact on the sustainability of communities both in the developed and developing world. Culture, therefore, matters when engaging with international development. The World Commission on Culture and Development notes that it is not culture that is inserted in development, it is development that is embedded in the culture. In this respect, the protection, preservation, and conservation of cultural heritage are of utmost importance across all communities and regions of the world.

This paper reiterates the danger of using a normalised yardstick by UNESCO (2005) and World Heritage Sites to advance a Eurocentric agenda in the fields of conservation and preservation for India. At the same time, it also seeks to identify the unique neo-Vaishnavite placemaking landscape, which is a product of the natural, ecological, and cultural, informed by the sixteenth-century teachings and experimentation of Srimanta Sankardeva. The paper positions its arguments in the context of heritage and the dangers posed by climate change. It seeks to engage with questions of resilience and resourcefulness in the landscape. The next section highlights the role of the mighty river Brahmaputra in the formation of Majuli's unique landscape. The next section elaborates on the heritage of Majuli and its role in the placemaking landscape over time. The observation from the fieldwork is then presented to bring together the highlights from the island. The concluding remarks draw together an analysis of the key debates on climate change and its impact on cultural heritage and the future of placemaking.

Majuli is the world's largest inhabited riverine island located on the Brahmaputra River in Assam. With a population of 168,000, it covers an area of 580 sq. km and has been designated as a carbon–neutral, biodiverse heritage site in the world. The location of the island on the Brahmaputra is fraught with challenges being situated in the largest sediment-bearing river in the world. This has resulted in the regular shrinking of the island from 1200 sq. km to just about 400 sq. km over the last 100 years due to annual floods and related morphotectonic activity. Majuli has been able to dominate this unique cultural heritage landscape for over 450 years due to the resilience demonstrated by the islander.

One will need to be mindful of the need to distinguish between the impacts of climate change on both slow as well the fast degradation of tangible and intangible cultural heritage across countries. While extremities of changes in the climate regimes and unabated natural disasters are becoming a norm with extreme drought to unseasonal rainfall, these will impact differentially on the heritage sites and practices. This calls for better anticipation of climate impact on cultural and natural heritage sites (Lefèvre 2014). Here, the building of community resilience becomes as important as providing strategies for adaptation and mitigation (Longworth 2014). One of the exciting developments in heritage research is that of 'relational thinking' which enables us to move beyond the binary dualistic thinking of developed and underdeveloped nations or rural and urban or society and nature (Massey 2009; Anderson 2000). In this paper, I invoke the rights of the people of Majuli to belong to a unique space, to exist as a community with a clear ideology, and philosophy, where rights are defined by culture and tradition and tested over time without being boxed into majority versus minority concerns. It is the cultural contract that adds value to the place of belonging and goes beyond mere coexistence.

19.2 Heritage

In India, there has been a need to expand the concept of heritage to include not only monuments and buildings of exceptional cultural and architectural value but also traditional and vernacular establishments and other built forms. Likewise, the concept of conservation includes not only the protection but also the revitalisation, regeneration, renewal, and sustainable development of both tangible and intangible cultural resources. Rapid economic development in India while focusing significantly on infrastructure provisions has tended to ignore the historical layering of the landscape. A regional landscape approach to identifying, conserving, and managing heritage sites involves appreciating the inter-linkages of their physical form, their spatial organisation, their regional-cultural specificities, and indeed their social-cultural and economic values becoming a main point of departure in this paper.

Cultural heritage plays a significant role in promoting climate-resilient strategies thereby promoting the advancement of sustainable development goals (SDGs) and at the same time ensuring cultural rights and transformative climate action. This includes above all recognising equity and justice as key to the understanding of and addressing the imminent dangers emanating from climate change. Here, active promotion of solidarity with the indigenous and vulnerable communities becomes the key to future climate action (ICOMOS Resolution 20GA/15 declared a Climate and Ecological Emergency).

The promotion of a people-centred approach to cultural heritage in particular the connection of people with heritage and places, engaged intercultural dialogues to enable understanding of sustainability and well-being within the context of heritage policies and practices becomes an imperative (ICOMOS Resolution 20GA/19). Thus, cultural heritage is closely affiliated to their ecology and landscape. Thus, according to the recent ICOMOS report, states, "a rights-based and a people-centred heritage management" (2022, 8), is the way by which cultural heritage can deliver on climate-resilient pathways. This will also help to promote zero-carbon-low carbon footprints and futures. Thus, taking terms such as 'cultural extinction', cultural rights, and cultural impacts seriously will ensure that policies relating to economic development are aligned to SDGs in India.

19.3 Resilience to Foster Cultural Rights and Justice

Developing resilience to external threats such as climate change, natural disasters, political strife, and economic transformations is critical for communities in marginal ecological zones. Such forms of resilience will also engender cultural rights to be protected and thereby ensure justice and dignity for all. According to Adger (2003), resilience is about the ability of communities to withstand external shocks to their social infrastructure. More recently, there is a suggestion that 'resilience' as a concept can be substituted into 'resourcefulness' to capture the importance of a just and progressive social relation (MacKinnon and Derickson 2012), largely seen in traditional communities as in Majuli. As Spivak (2013) notes, the immediate and most pressing task is to 'cultivate the will to social justice among everyday people'. There is an added advantage of using 'resourcefulness' in that it considers the uneven distribution of resources within and between communities in India, which remains a complex civilisational society layered by differences in caste, religion, ethnicities, languages, and culture. Such an approach opens space for cultural self-determination by reiterating the significance of local skills and traditional ecological knowledge systems. The fostering of translocal collaboration among the communities and across cultural spaces and institutions becomes the key to securing cultural justice and rights. The fact that Majuli as a riverine community has adapted to the vagaries of morphotectonic and fluvial transformations over the last 450 years is a standing testimony to the resilience, resourcefulness, and adaptability of the community. The unique cultural heritage landscape of Majuli reflects dynamism over time and acquires a symbolic identity.

Majuli island is located on the Brahmaputra River (or 'son of Brahma') is the only Hindu masculine river deity and is the world's youngest river with a catchment of 712,035 sq. km and has the highest sediment load and erosional deposits. The island extends for a length of about 80 km, 10–15 km north–south direction with a total area of about 875 sq. km and at an elevation of 85–90 m above the mean sea level (see Fig. 19.1). Today, it supports a population of c.168,000 (2011 Census of India) and is an important international site of ancient cultural heritage, wetland ecologies, and migrating birdlife. Historical reports suggest Majuli has been settled more than four and a half centuries ago. In CE 1188, Sri Gauri Narayana of the Great Bodo Tribes is reputed to have founded a dynasty with Ratanpur as its capital (Hazarika

2004). Ratanpur was situated near 'Kherkatia Suti' on the southern bank of the Lohit River and disappeared due to the raging floods in the region. As two branches of the river flowed on either side of Ratanpur, Majuli was originally known as 'Majali' (Hazarika 2004). Majuli is an island mentioned in the *Satsari Asam Buranji* and the *Deodhai Asam Buranji* chronicles of Assam in the sixteenth century. The term Majuli and Majali or Mojali are casually used to mean the island (Fig. 19.1). Erosion as a natural phenomenon has major implications for the survival of the cultural heritage of Majuli.

19.4 Heritage of Majuli

Majuli is distinguished for its unique riverine geography and culture where Vaishnavism flourished uninterrupted since the sixteenth century. The legendary Sattras of Assam provided a unique cultural landscape for crafting a composite socio-religious culture that continues to influence art, literature, music, and dance forms, including mask-making tradition and most significantly the neo-Vaishnavite adaptation of the annual four-day Krishna Leela Festival during the first full moon after Diwali (the festival of lights). The island of Majuli as a living testimony of the 450-year-old cultural heritage houses 243 small and large villages whose population has demonstrated incredible ability to adapt their existence to the changing dynamics of its ecosystem over the centuries. The island is of the outstanding universal value of cross-cultural diversity, unique continuous sixteenth century Vaishnavite religious and cultural tradition in parallel to the Renaissance period in Europe. Once there were 65 Sattras (monasteries), but erosion forced 28 to move off the island; many have been damaged and have relocated elsewhere on the island. Its people have learned to adapt their lifestyle to suit the river, building their houses on stilts and moving when required. There used to be around 70 monasteries called 'Sattras' in Majuli but due to the shrinking of the island, there are now only 22 in existence and under constant threat. Mainly built on the higher stable areas of the island, the Sattras are believed to have been in existence since the mid-sixteenth century and so have weathered over four centuries of devastating floods and earthquakes and have relocated from their original sites many times over in the past (Fig. 19.2).

In the context of Majuli, this notion of culture remains a 'living heritage' whose everyday performance remains deeply embedded in the quotidian lifestyles of its inhabitants. Its heritage acquires value and meaning from its long-standing shared past as well as from its unique physical landscape. Here, the preservation and conservation of its cultural identity, its values, and its ethics remain the core of existential principles as informed by the neo-Vaishnavite philosophy of Saint Srimanta Sankardeva. Therefore, conservation follows the eternal philosophy of preservation which has remained in force over the years on the island, presenting a unique approach towards heritage sustainability. Cultural heritage in Majuli's context includes both the tangible, the *Sattras* Institutions (Vaishnavite monasteries), the *Namghars*, the Ponds, the *Mukuts* (Masks), and intangible, the *Bhaonas, Raas* festivals, etc. It is a

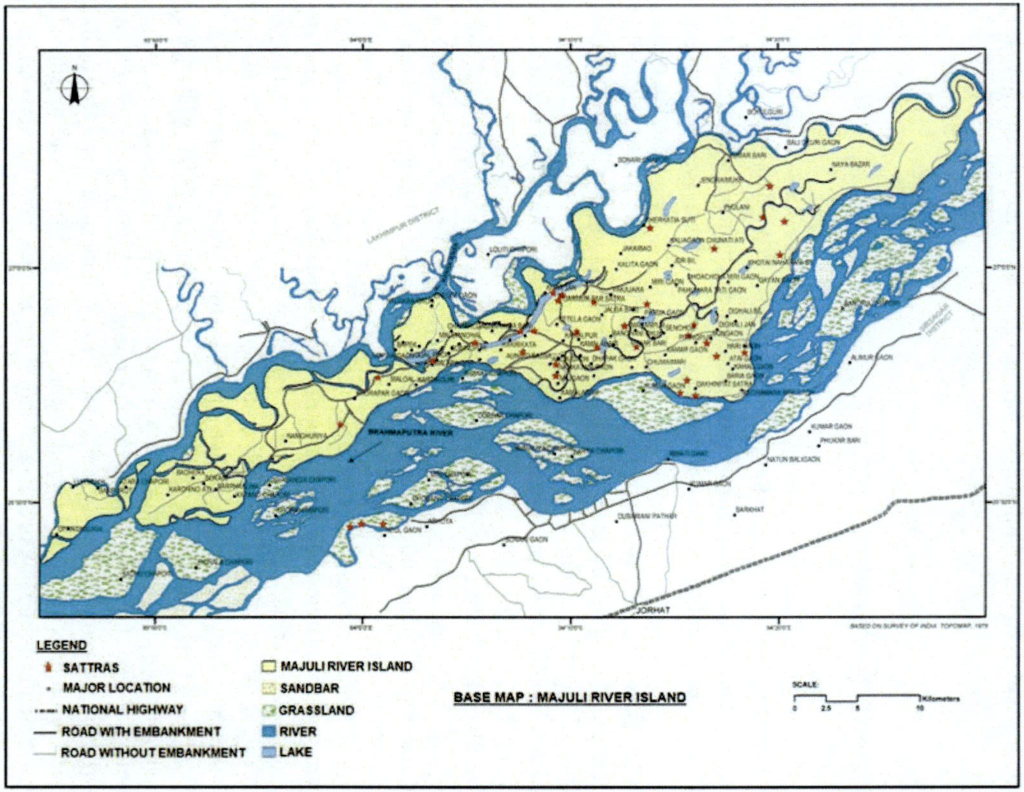

Fig. 19.1 Location of *Chaporis* (sandbars) and interconnected channels and cluster of islands (*Source* River Island of Majuli: Cultural landscape and Living Tradition in the Midstream of Brahmaputra River in Assam, India for UNESCO/WHS, ASI, 2012)

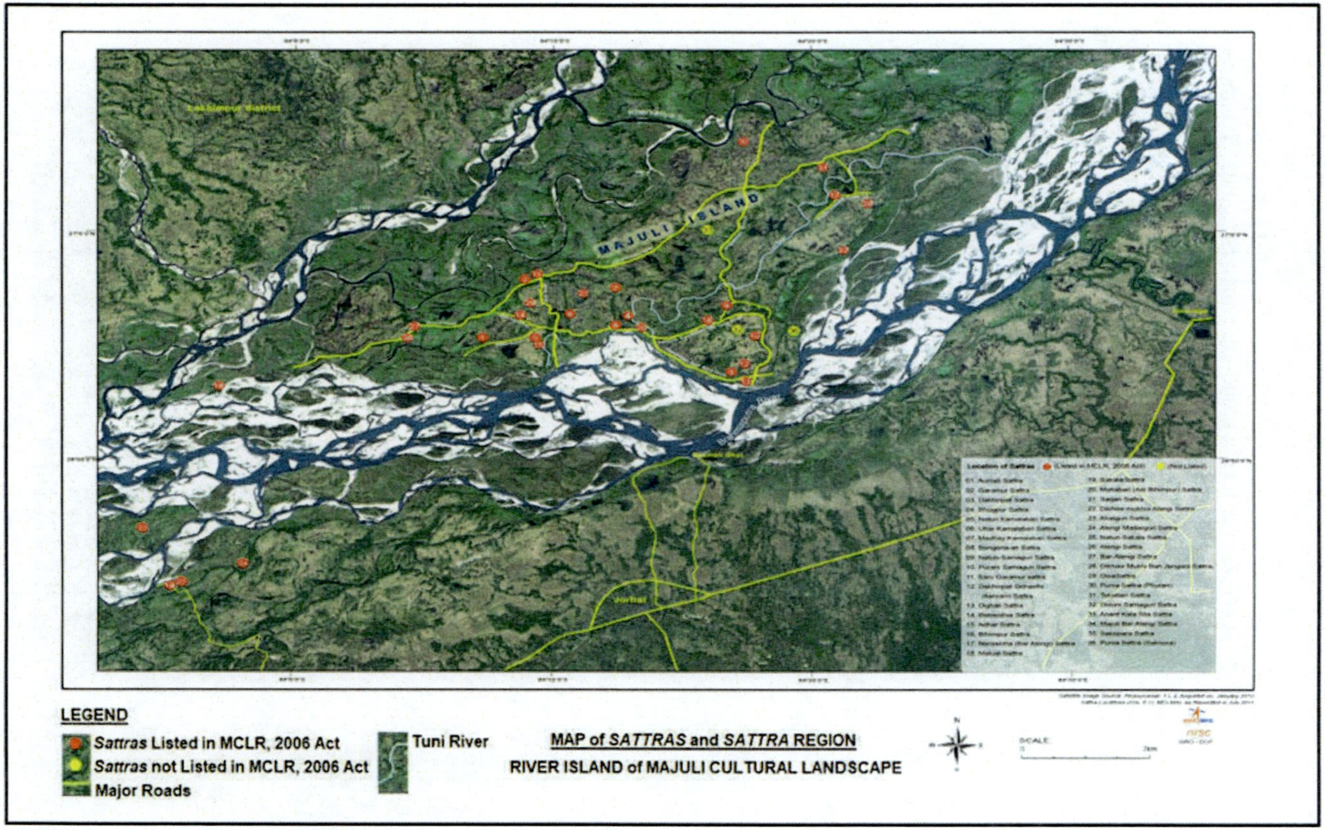

Fig. 19.2 Locational Sattras (monasteries) are under severe threat from the raging Brahmaputra (*Source* River Island of Majuli: Cultural landscape and Living Tradition in the Midstream of Brahmaputra River in Assam, India for UNESCO/WHS, ASI, 2012)

philosophy that infuses the philosophy of coexistence, irrespective of castes, tribes, and class. Such a tradition of inclusivity is arranged around the Sacred Book-the *Bhagavata Purana* instead of the numerous gods and goddesses within the Hindu pantheon (Fig. 19.3).

It was the pioneering Vaishnavite teacher and reformer Srimanta Sankardeva and the practitioners who established the Sattras traditions in Majuli. The first Sattra commenced at Belaguri Dhuwahat in West Majuli and hosted the prestigious '*Manikanchan Sanjog*' (Neog 1965). These simple teachings of Vaishnavism heralded the socio-cultural reformation of Assamese society. This Bhakti movement

Fig. 19.3 *Bhagavata Purana,* Sacred Book as prescribed by Srimanta Sankardeva (*Source* Picture by Parasher Barua, taken during the fieldwork, 13 February 2019)

of Assam instigated the spiritual renaissance and ran parallel to the golden Ahom dynastic rule. This movement was inclusive and bypassed caste, creed, and ethnicity and incorporated indigenous tribes including the Mishmi, Deoris, and Sonowal among others. Public spaces for performance became the heart of the movement and included the *Sattras, Namghars*, and *Kirtan Ghars* as prayer halls. These spaces also reinforced the 'power geometries' (Massey 2009) across the Sattras vying for patronage and resources.

At the same time, these became democratic institutions to foster social changes and developments in medieval Assam. In Majuli, the predominant deity, Lord Krishna presided as the saviour with supreme power. The Bhakti movement inculcated a rare call for a "Single God and humankind" as the prime objective behind the whole ideology (Das 1953). Such a religious and cultural awakening played a critical role in promoting incredible changes which were reflected in the cultural heritage landscape of Majuli in particular and Assam in general. It was indeed a socio-religious and more importantly a cultural reformation within the Hindu society. Indeed, cultural experimentation introduced was manifested in unique practices such as congregational prayers, theatre, music, dance, painting, and others largely defined and developed by Srimanta Sankardeva making it inclusive across the society (Neog 1965). It created a highly spiritual, egalitarian civil society that respected collective values of fraternity, justness, humanism, and enlightenment. This unique philosophy also fostered literature art and culture as a guide for generations.

The neo-Vaishnava Bhakti tradition established a belief in Majuli referred to as the '*Eka Sarana Nam Dharma*' (faith in the single God), i.e. Lord Krishna (*Eka Sarana*). The *Bhagavata Purana* adopted by Srimanta Sankardeva proscribed different modes of Bhakti or *devotion* such as *Sravana* and *Kirtana,* accessible to the society, irrespective of caste and creed without any rigid adherence to ritualistic commandments (Borkakoti 2017). He also organised dramatic performances, or *Cihna-Yatra*, which included dance accompanied by music and covered the sacredness of the pristine environment of Majuli. The use of locally produced unique *Mridangas* (conical drums) made by the native potters and the cobblers is significant in the upholding of this intangible cultural tradition since the sixteenth century (Fig. 19.4). In addition, a variety of Cymbals *(manjira, khutital, bartal),* an instrument of copper, and bell metal were also made as instructed by Sankardeva (Neog 1965). Majuli also has a repository of paintings of Sankardeva on cotton-based paper called *'Tulapat'* representing *Vaikuntha* (the abode of Lord Vishnu). Thus, Bhakti was propagated via literature, drama, painting, music, or *Bargeet*. The evolution of Sattras or monasteries in Majuli followed the framework outlined by the vision of Sankardeva and his disciples over time. He initiated the 'One Act Plays' known as *Ankiya-nat* and composed the *Six Ankiya Nats* in *Brajabuli* (combination of Maithili and Assamese) language and also developed the technique and methodology of theatre-based performance (Pathak 2008; Fig. 19.5). Thus, it was only through devotion and meditation to Lord Krishna that can one expect to attain absolute knowledge or pure liberation (Nath 2009). The unique dance form of *Satriya* in Majuli's neo-Vaishnavite monasteries

Fig. 19.4 Unique form of Dance and Music using traditional Mridangas and Cymbals (*Source* Pictures by the author, undertaken during the fieldwork, 4 November 2017)

was the *Bhaona* depicting events from the ancient sacred texts, of *Mahabharata*, the *Ramayana*, and the *Bhagavata Puran.*

19.5 Observations from the Field

Majuli's landscape-its physicality is defined by the built environment and the layered cultural history since antiquity. The island and its people instil energy, commitment, beauty, and elegance. At the same time, the existence of such a living culture and heritage reiterated the challenges of preservation and conservation of both tangible and intangible heritage in the wake of recurrent changes to the climate. The management of the cultural landscape of Majuli is therefore guided by a deep appreciation of its tradition and its cultural heritage over time. Regular invocation of their traditional ecological knowledge systems helped to counter the contingencies imposed by recurrent annual floods which remain an existential threat to this unique landscape. A unique aspect of organic *Samuguri* Mask-making and Boat-building tradition, the vernacular architecture, their traditional *Apsara* dance forms and *'bhaonas'* reiterates the challenges imposed in terms of conservation and preservation of tangible and intangible heritage. *Srimanta Sankaradeva's* philosophy became a precursor for

Fig. 19.5 Theatrical performance and the use of traditional crafts and masks (*Source* Pictures by the author, undertaken during the fieldwork, 23 November 2018)

an ecologically balanced approach towards meeting sustainable development goals in the region. The focus is on a form of circular economy wherein the use of natural resources is within the region's ecological limits and carrying capacity. The inhabitants of Majuli have since built their economic, social, and cultural traditions around these natural settings and continue to carry the legacy of its cosmic Living Traditions. The message is that for communities to remain connected to the places they value, the management of their unique cultural heritage landscape remains imperative.

19.6 Concluding Remarks

Placemaking in Majuli emerges in a unique natural, ecological and cultural landscape where the imagined and the real coalesce. Majuli as a place is as much a product of the fertile imagination of Saint Sankardeva and his associates, rooted in deep neo-Vaishnavite Bhakti philosophy and the inhabitants continue to enact the virtues of its ideals in space and time. It provides a "symbolic landscape" (Meinig 1979, 164) that is striving to remain intact in the wake of significant climate change and at the same time to remain relevant despite the inexorable march of globalisation and development. It remains an iconography for the people of Assam and India, sharing the ideals, memories, beliefs, and feelings thereby binding all in an inclusive 'embrace'.

Indeed, heritage conservation in the Indian context demands a non-standardised approach away from the Western ethics and practices in vogue in the global South (Tandon 2015). The dominance of politics of World Heritage entities subverts the uniqueness of indigenous and innovative conservation and preservation practices and norms in existence in Majuli since the sixteenth century. Decolonising such practices will enable the inhabitants of Majuli and the Government of Assam and India to reinforce their commitment to sustainability. The local conservationist in Majuli/Assam needs to play a far more significant role in their heritage than paid advisors parachuting into the island without an understanding of its historicity and culture. There is a need to go beyond a tangible, object-centred approach to cultural heritage and should include the intangible when discussing the impact of climate change (Kim 2011). In addition, the significant role of multi-generational cultural engagement in Majuli remains a key factor that needs acknowledgement both nationally, locally, and globally. Understanding the cultural, historical, and geographical contexts of riverine Majuli in fragile ecological zones remains critical in advancing the imperatives of conservation and preservation. Here, the sacredness and exceptionalism of unique neo-Vaishnav philosophy centred around Sri Krishna as an avatar of Lord Vishnu remains deeply rooted in the island of Majuli's interpretation of cultural continuity and identity. The tangible-material forms emerge as a by-product of the intangibles of values, ethics, and aesthetics thereby transforming the very semiotics of expression in the cultural landscape of Majuli. We see here the social and cultural values being reiterated since the sixteenth century via festivals such as the Raslila, Palnam, Janmashtami, Barsik Bhaona, or Bar-Sabha and Bihu. This is very much reminiscent of the Javanese context where cultural adaptation and resilience have emerged as a by-product of long-established traditions (Cowherd 2013). The physical spaces of Majuli, therefore, embody meaning and values to the island. Thus, the physical landscape provides the space for the expression of rituals, experiences, and interactions to ensure the continuity of traditions in terms of preservation and conservation. This is seen during annual festivals when scattered diaspora undertakes a pilgrimage to their roots to reinforce their commitment to the sacredness of their culture and heritage. In particular, the imminent challenges posed by climate change on the heritage are unavoidable thereby reiterating the impermanence and unpredictability faced by places such as Majuli. Indeed, history bears witness to the resilience of Majuli as the centre for neo-Vaishnavism that has tackled key challenges to its existence through the centuries.

There is a need to design short and long-term sustainable adaptation planning policies that help to mitigate damages and enhance the preservation of cultural heritage. This calls for an interdisciplinary approach to the cultural heritage landscape management of Majuli. The troika of stabilisation, adaptation, and mitigation becomes the way forward in dealing with the detrimental impact of climate change on the ancient heritage of Majuli. Majuli has demonstrated the way this island community has evolved from the unique attributes of climatic, fluvio-morphotectonic, and geomorphological actions, and culturally informed human activities imprinted on the landscape over the last 450 years. Indeed, our study has shown that Majuli cultural heritage landscape presents a record of past and present environmental and climate-related actions which in effect demonstrates traditional ecological knowledge systems in operation, which are overlooked by present policies.

Without community engagement and recognition of the unique values and norms of the Majulian community, planning for the preservation and conservation of cultural heritage in riverine communities will not succeed. Traditional and time-tested knowledge about conservation and preservation of tangible and intangible cultural heritage needs to be integrated into local and national policies rather than import policies of World Heritage Sites to be fitted into local policies. Without incorporating community values and traditions, commitment to the conservation of cultural heritage will remain on paper only (Kato 2008; Mitchell and Barrett 2015). Incorporating these values and knowledge will enable a realistic risk assessment of heritage being impacted by climate change. It will also help to align scientific data analysis with the local traditional ecological knowledge systems and will ensure a balanced approach towards stabilisation, adaptation, and mitigation for climate-proofing in Majuli (Carmichael 2015; Leon et al. 2015). Involving local communities in the decision-making towards preserving and conserving their heritage is the key (Brabec and Chilton 2015). As historical custodians of their unique cultural heritage, their ethical needs should take precedence over the political imperatives and aspirations of WHS, UNESCO, ICOMOS, and other global bodies (Walter and Hamilton 2014). Climate justice for whom is a key question in the context of Majuli as it dilutes the obligations of the global North and the governments in the global South to effectively deal with the enormous challenges faced by marginal communities seeking to protect their heritage in the wake of climate change (Maus 2014).

Cultural rights are an expression of and a prerequisite for human dignity. They are core to the human experience, and essential for implementing other universal human rights and achieving the sustainable development goals. They are not a luxury, even during a global health crisis. These goals are key to the implementation of cultural rights and vice versa. Progress on human rights obligations and the SDGs are therefore two sides of the same coin. Moreover, safeguarding and promoting culture in all its diversities contributes directly to achieving many of these goals (Hosagrahar 2021). Indigenous heritage management in Majuli island calls for greater support and oversight towards their protection. The lack of community veto over unplanned development is a key issue that limits the protection of ancient traditions and practices. The livid experience of communities remains critical in any heritage planning and they should not be forced to choose between caring for their culture and heritage versus having to adjust to greedy developers both from within and outside the State of Assam. The sacred has to go hand in hand with the modern. Being culturally aware is far more significant than being a Wharton graduate. The impact of the loss of cultural heritage due to climate change and recurrent natural disasters on the individual and the community's collective identity and well-being remains significant and real. Conservation leading to commodification must be avoided at all costs to preserve the values of cultural and social capital in Majuli. This will also help to achieve the key globally mandated SDG-15, of strengthening resilience and integrating climate change measures in planning and development. SDG-13 supports heritage-related livelihoods; SDG-8 promotes entrepreneurship, creativity, and innovation, and SDG-10, promotes social, economic, political, and cultural inclusion of all stakeholders. It will also help support a zero-carbon management heritage cultural landscape in Majuli.

Indeed, there is a clear nexus between climate change and cultural rights in which culture has a close affinity to specific ecosystems, in particular for the people of Majuli which includes the indigenous population (Hosagrahar 2021). Suffice it to add cultures are located in a particular environment and are invariably place-based. Culture, therefore, influences our understanding of the environment and promotes placemaking at the most fundamental level. Therefore, concerns for the welfare of future generations in Majuli are both explicitly environmental as well as cultural. Managing the cultural heritage landscapes also implies protecting and securing natural and public spaces for interactions towards maintaining continuity of ways of life. Placemaking has enabled the diverse communities in Majuli to see what they need to see about themselves, about their authentic self, rooted in ancient tradition. There is an assurance that their past is stable, and their presence is assured. However, the challenge for the future remains tenuous due to the changing environment. This study of placemaking has highlighted the continuity and purity of Majuli since the sixteenth century, where myth and reality are enacted daily. A strong sense of place is established thereby providing both an experiential and associational engagement between the people and the natural and cultural landscape, which remains a product of placemaking. Thus, placemaking becomes a unique cultural activity as demonstrated by the Sattra tradition and mores. It also provides the much-needed socio-cultural opportunities to engage with climate action, emanating from actions associated with *adaptation,* i.e. *preservation* and *conservation*, i.e. *stabilisation* and *mitigation* in the wake of imminent upheavals ushered by global climate change triggered by the Anthropocene activities.

Funding Funding Support was provided by UKRI/AHRC/QUB/DfE-UK-2017–2019, for the Field Visits to Majuli.

References

Adger WN (2003) Social capital, collective action, and adaptation to climate change. Econ Geogr 79(4):387–404

Anderson K (2000) "Thinking 'Postnationally': Dialogue across Multicultural, Indigenous and Settler Spaces". Ann Assoc Am Geogr 90(2):381–391.

Borkakoti SK (2017) Majuli, resources, and challenges. Authors' Press, Guwahati

Brabec E, Chilton E (2015) Towards an ecology of cultural heritage. Change over time. Int J Conserv Built Environ 5 (2): 266–285

Carmichael B (2015) Supporting indigenous rangers' management of climate-change impacts on heritage sites: developing an effective planning tool and assessing its value. Rangeland J 37:597–607

Cowherd R (2013) The cultural reconstruction of Surakarta Java. In: Bharne V (ed) The emerging Asian City: Concomitant urbanities and urbanisms. Routledge, Oxford, pp 66–74

Das HM (1953) Contribution of Srimanta Sankardeva. Sri Gouranga Press. Government of Assam, Majuli District, Calcutta (Kolkata), Accessed 3 May 2022 at 15.57. https://majuli.gov.in/information-services/geography

Hazarika N (2004) Majuli: A unique ecological entity. In: Grover N, Singh KN (eds) Cultural geography: form and process. Essays in honour of Prof. A.B. Mukherji. Concept Publishing, New Delhi, p. 447–454

Hosagrahar J (2021) Annexe: International legal framework for report by the special rapporteur in the field of cultural rights on a human rights approach to cultural mixing, mixed cultural identities, and syncretism. Available at https://www.ohchr.org/sites/default/files/Documents/Issues/CulturalRights/A_76_178_Annex.docx. Accessed 5 May 2022

ICOMOS (2019a) The future of our pasts: Engaging cultural heritage in climate action. ICOMOS Climate Change and Heritage Working Group. International Council on Monuments and Sites, Paris, France

ICOMOS (2019b) People-Centred Approaches to Cultural Heritage Resolution 20GA/19. Available at https://www.icomos.org/en/77-articles-en-francais/59522-icomos-releases-future-of-our-pasts-report-to-increase-engagement-of-cultural-heritage-in-climate-action. Accessed 5 May 2022

ICOMOS (2020) Agenda item 6–6–1 Resolution 20GA/15–Cultural Heritage and the Climate Emergency Ver. 05/12/2020. Available Accessed 5 May 2022

ICOMOS (2022) Heritage, climate justice, and equity-climate action working group. Available at https://www.icomos.org/images/DOCUMENTS/Working_Groups/CCH/Heritage_Climate_Justice_and_Equity-English.pdf. Accessed 5 May 2022

Kato K (2008) Addressing global responsibility for conservation through cross-cultural collaboration: Kodama Forest, a forest of tree spirits. Environmentalist 28:148–154

Kim HE (2011) Changing climate, changing culture: Adding the climate change dimension to the protection of intangible cultural heritage. Int J Cult Prop 18:259–290

Kumar MS (2017, 2018) Climate change and hidden heritage: Lessons from Majuli. Queen's University Belfast. Available at https://youtu.be/7JNQKDCTc-U (Grateful to Minal Sagare, for the translation- Climate Change and Learnings from Traditional Land-Water Practices of Majuli Island, Assam, 2021, INTACH-Mimeo)

Kumar MS (2017–2019) Based on UKRI/GCRF supported grants to conduct fieldwork in Majuli- "Climate Change and Hidden Heritage: Lessons from Majuli" (2017); "Transforming heritage- An International Workshop for Enhancing UK-South Asia Networks" (2017); "Role of heritage in zero carbon footprint: Majuli" (2018–19)

Lefèvre R-A (2014) The impact of climate change on slow degradation of monuments in contrasts to extreme events. In: Von Schorlemer S, Maus S (eds) Climate Change as a threat to peace: Impacts on cultural heritage and cultural diversity. Peter Lang AG, New York, pp 83–100

Leon JX, Hardcastle J, James R, Albert S, Kereseka J, Woodroffe CD (2015) Supporting local and traditional knowledge with science for adaptation to climate change: lessons learned from participatory three-dimensional modelling in BoeBoe, Solomon Islands. Coast Manag 43(4):424–438

Longworth E (2014) The culture of prevention: Heritage and resilience. In: Von Schorlemer S, Maus S (eds) Climate change as a threat to peace. Peter Lang AG, New York, pp 119–126

MacKinnon D, Derickson KD (2012) From resilience to resourcefulness: A critique of resilience policy and activism. Prog Hum Geogr 37(2):253–270

Massey D (2009) Concepts of space and power in theory and political practice. Documents D'anàlisi Geogràfica 55:15–26

Maus S (2014) Hand in hand against climate change cultural human rights and the protection of cultural heritage. Camb Rev Int Aff 27(4):699–716

Meinig DW (1979) Symbolic landscapes: Models of American Community. In: Meinig DW (ed) The interpretation of ordinary landscapes: Geographical essays. Oxford University Press, New York

Mitchell NJ, Barrett B (2015) Heritage values and agricultural landscapes towards a new synthesis. Landsc Res 40(6):701–716

Nath D (2009) The Majuli Island Society, economy, and culture. Anshah Publishing House, Guwahati

Neog M (1965) Sankardev and his times: the early history of the Vaisnava faith and movement in Assam. Nabajiban Press, Calcutta (Kolkata)

Pathak NC (2008) Assam's history and its graphics. Mittal Publications, New Delhi

Spivak G (2013) Scattered speculations on geography. The 2012 Antipode AAG Lecture. Antipode 46 (1): 1–12

Tandon A (2015) First aid to cultural heritage in times of crisis. Smithsonian Institution/UNESCO/ICCROM

UNESCO (2005) Convention Concerning the Protection of the World Cultural and Natural Heritage. WHC-document 29COM 7B. a,2005, Nr.5- Available at, https://whc.unesco.org/archive/2005/whc05-29com-07BReve.pdf. Accessed 5 May 2022.

Walter RK, Hamilton RJ (2014) A cultural landscape approach to community-based conservation in the Solomon Islands. Ecological Society 19(4):41–50

Chapter 20
'*Pūch*' as an Institution for Maintaining the Cultural Landscape of the Kullu Valley

Ashwini Pethe and Kiran Shinde

Abstract The Kullu Valley in Himachal Pradesh in India, known as "valley of gods," represents a distinct cultural landscape where the natural landscape is imbued with sacred values and beliefs. The inhabitants of this landscape reinforce and perpetuate the sanctity of this landscape through cultural and religious practices. '*Pūch*' (a shamanistic ritual) is one such socio-religious practice, believing that the *debatās* (*devatās* in Sanskrit, means gods) communicate with people through a medium. *Gūr* (shaman) is believed to possess the divine powers to communicate with the *debatās* with his religious authority. This chapter discusses the role of *pūch* in maintaining the fragile ecosystem and natural landscape of the valley. Based on the fieldwork conducted in sixteen villages of the Kullu Valley (interviews, oral histories), and literature review, this study examines how the verdict from the divine received during *pūch* affects the collective decisions of communities for the collective good. At various instances, *pūch* has demonstrated its power to mobilize communities in order to build up a pressure on policy decisions. This chapter argues that the ritual of *pūch* plays a significant role in the decision-making processes around new developments that tend to threaten the culture and ecosystem of the valley.

Keywords Cultural landscape · *Pūch* · Shamanism · Community mobilization

20.1 Introduction

In many cultures around the world, shamanistic practices are commonly used for communicating with the divine or with an unknown energy. This is particularly the case where people want to appease pagan deities representing forces of nature. For

A. Pethe (✉)
School of Architecture, MIT ADT University, Pune MR 412201, India
e-mail: ashwini.pethe@mituniversity.edu.in

K. Shinde
Department of Social Inquiry, La Trobe University, Melbourne, VIC VIC-3552, Australia
e-mail: K.shinde@latrobe.edu.au

R. P. B. Singh et al. (eds.), *Placemaking and Cultural Landscapes*,
Advances in Geographical and Environmental Sciences,
https://doi.org/10.1007/978-981-19-6274-5_20

instance, in Africa, such practices framed as religious rituals and beliefs are often used by religious leaders in conserving sacred sites and protecting nature (Byers et al. 2001). These processes are not uncommon among people who live off natural resources and within or in the proximity of places such as mountainous terrains, river deltas, mangroves, forests, and coastal areas (Drew 2017; Bande 2006). Through their social-religious and cultural practices, such geographical territories are transformed into cultural landscapes (Singh 2017). Such landscapes have many kinds of shamanistic practices; some more prevalent than the other and more explicit concerning environmental conservation and environmental management (Conklin 2002; Naomi 1996). This research aims to add an empirical study to the discourse on the centrality of such practices in conserving the fragile ecosystems that characterize mountainous regions. In this chapter, we discuss *pūch*—a shamanistic ritual that reinforces a strong belief in the divinity of nature and has been successfully invoked time and again for addressing environmental change in the Kullu Valley (Fig. 20.1), also known as the valley of gods, situated in the lower Himalayan belt in India.

The cultural landscape of the Kullu Valley is characterized by a mountainous rural terrain dotted with hundreds of sacred sites (Fig. 20.2). These sacred sites are indispensable parts of the socio-cultural life of the people and anchor considerable tangible and intangible cultural heritage that revolves around religious beliefs, religious rituals, and religious institutions (see Singh et al. 2022, p. 14). Using these as key conceptual categories, this chapter explains the intersections and interdependence of religious practices with the land, the landscape, and the administrative structures in the valley. These are particularly pronounced when development agendas seem to cross with environmental issues. It is through *pūch* that questions regarding impacts and dilemmas of certain actions are raised, and answers are sought.

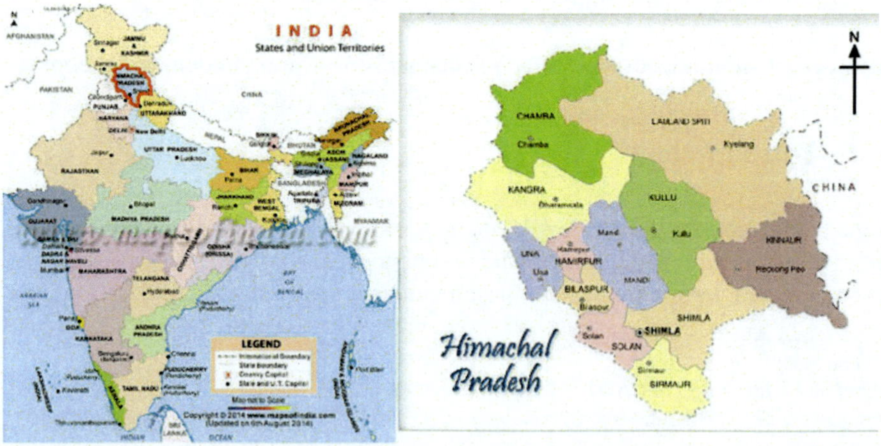

Fig. 20.1 Map of India showing the state of Himachal Pradesh and Kullu District (*Source* www.mapsofindia.com)

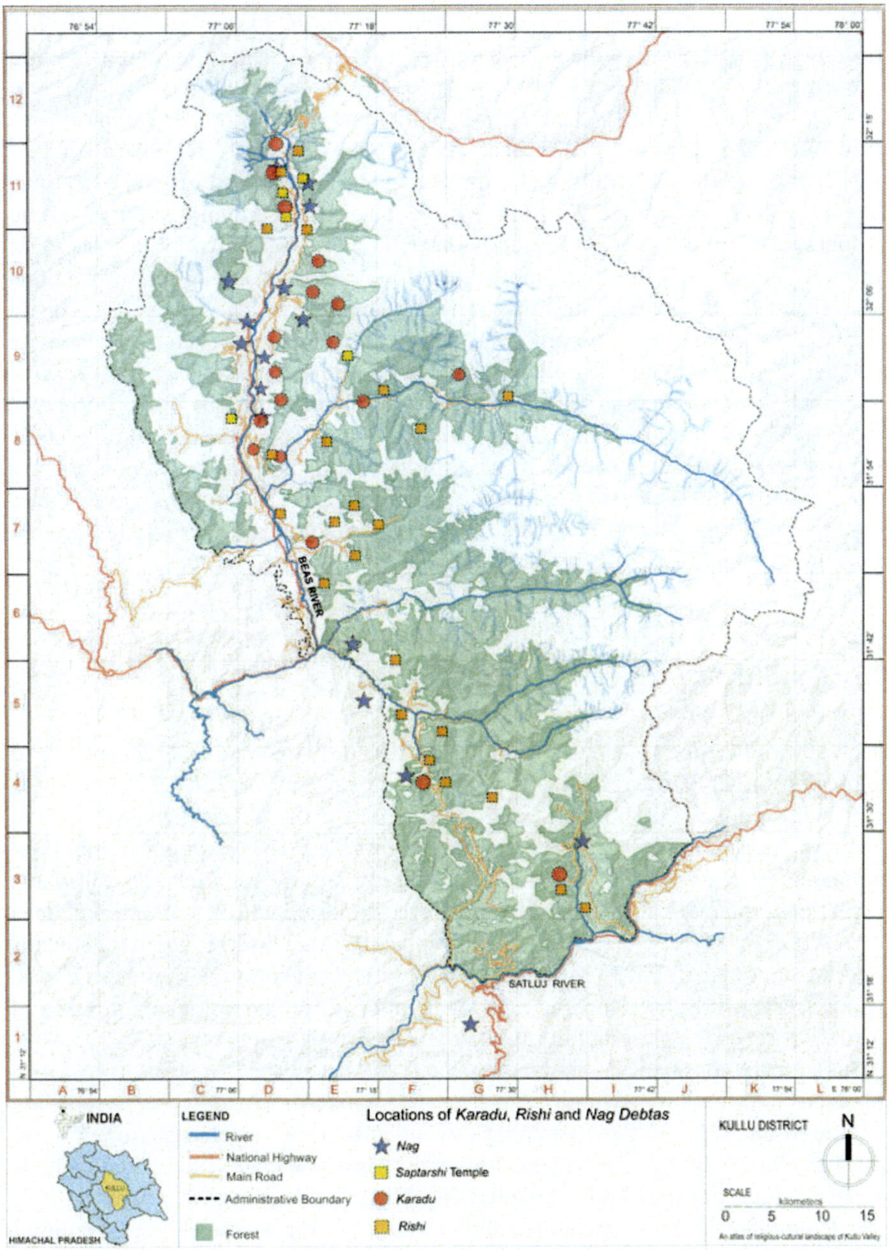

Fig. 20.2 Map showing the locations of 3 different cults of gods in the Kullu valley (*Source* Drawn by Author)

This study is based on the doctoral research done by the first author from 2012 to 2017 under the guidance of the second author. The study focused on understanding the making of cultural landscapes in the valley of gods and involved physical mapping exercises, conducting in-depth interviews, participant observation, and photo-video documentation. The findings discussed here are derived from one aspect of that study. Following the brief introduction, the chapter is organized into five sections. The next section presents a review of the literature—global and indigenous—on the key concepts of cultural landscape and shamanism. Based on this review, a conceptual framework comprising religious beliefs, religious rituals, and religious institutions is developed for the study. The third section discusses the *pūch* as a shamanistic practice in the context of 'the land, the gods, and the people' which is central to the understanding of the cultural landscape. The fourth section discusses how *pūch* is established as environmental management mechanism and has been invoked at various spatial scales to influence decision-making in recent development projects in the valley. The concluding section highlights the central role of *pūch* in maintaining the sacred and cultural landscape of the Kullu Valley and situates it within the broader literature around indigenous practices that rely on religious systems to conserve the environment.

20.2 Literature Review

Cultural Landscape: As a concept, cultural landscape has gained wide acceptance to explore the human-nature relationships in many meaningful ways (Atha 2012). There is a growing emphasis on both the natural formations in the physical and geographic sense and the human agency in altering and modifying those (Rapoport 1992). The cultural landscape approach has been particularly significant for areas that have special geographical features or areas that have supported unique cultural practices or twinned divinity (Meyhew 2000). Nature and culture are both regarded together as co-productions (Singh 2013). Moreover, "In the process of Nature-Culture interfaces, humans create places by imposing their inherent or possessed unique manifestation to certain parts of the geographic space and its environs" (Singh et al. 2022, p. 4).

This study builds on the understanding of cultural landscape from Belsare's (1962) indigenous interpretations where culture means '*Saṁskritī*' *(in Sanskrit)*. According to Belsare, *saṁskritī* (culture) comprises two parts—'*saṁ*' (to cleanse, to purify, to shape) and '*kritī*' (to do). Put together, this means '*samyak kritī.*' He further explains "When a child is born, by body and mind the child is equivalent to any natural living thing. Cultural dimensions are added and given to developing the child as a cultural identity" (Belsare 1962). Thus, *samskrutī* broadly refers to the continuing of tradition over generations while each generation overlaying its own ideas. The traditions that continue themselves mean they are still relevant to a certain portion of society. In Belsare's conceptualization cultural landscape means '*Saṁskṛtirūpa Sṛṣṭi*' (Belsare 1962, p. 19), which is manifested as '*Antaraṅga*' (inner) and '*Bahiraṅga*' (outer).

The '*Antaraṅga*' means the philosophical base or value system while '*Bahiraṅga*' means the physical, social, and cultural manifestation of this *Sṛṣṭi.*

'*Saṁskṛti parikṣetra*' (cultural landscape) is a similar term used by Rana Singh that involves both a natural and cultural milieu and refers to the ideas, beliefs, and ways of life that people value and use when faced with change—above all the link to an intimate relationship between the human psyche (humanity) and the mystery of nature (divinity) (Singh 2011a, p. 6). Through religious beliefs, religious rituals, and religious institutions, religion is manifested in the sacred territory and the cultural landscape. Further, provoked that "all aspects of landscape are regarded as representational, thus cultural signifiers, the interpretation of which reveals social attitudes and material processes" (Singh 2010, p. 28). This is validated in this study.

Religious beliefs are closely tied with the notion of fear of, and from nature. Some of these may align with cosmic principles, aspects of astronomy and astrology, but largely they are channelized through folklore and oral histories. However, the most significant way of transmission is through religious rituals inscribed in religious practice. Rituals may be a means to achieve some goals but they provide a deeper meaning of connecting with belief in the supernatural. Rituals, particularly in rural societies, are seen as effective in invoking divine powers to deal with calamities and adverse impacts (Damodaran 2012, p. 323). Often in rituals, cosmology is imitated and reproduced, and a setting is created where the performers can perform activities using natural elements and symbolically connect with the divine. Correct performance of these acts is a must to achieve the desired results. To ensure this, certain institutional roles are established with a religious authority that may be in a personified form such as priest, shaman, monk, or be scaled to the level of organization where many people collectively perform these roles. Thus, a connected framework is established to ensure that the divine, religious, and cultural aspects of society are maintained.

As will be shown in this chapter, all three components come together in the practice of shamanism, which is simultaneously a belief, a ritual, and an institution. This understanding is particularly important because that is how indigenous communities connect with their cultural landscapes.

Shamanism: Shamanism is one of the oldest practices of communication between humans and the divine forces. As per Hamayon (Hamayon and Paris 1993/2007), 'Shamanism is a symbolic system … which implies that mankind entertains relations of alliance and exchange with supernatural beings that are thought to govern natural beings' (Walraven 2009). Shamans, the main actors, invoke their spiritual self and spirits of nature through various rituals in highly charged natural settings. They undertake a journey to the unknown reality so that they can provide some answers to those seeking their assistance (Coleman 2004, p. 201). Using their supernatural abilities, they determine the basic principles for maintaining the spiritual connection with nature by prescribing what should be done and what should be avoided. These are then taken as doctrines of religious-cultural practice in establishing and maintaining the ecological and social relationships of the local communities with their surroundings. The idea of stewardship toward ecology by institutionalizing the religious practices, especially shamanism, is observed globally and resonates with the modern moral agendas of conservation (Verschuuren et al. 2010; Bhagwat et al. 2011). Such ideas of using a religious basis for environmental protection and management become central to the discourse of "religious ecology" (Maarif 2015). The religious ideas and practices of forest conservation have been ably demonstrated using in several examples including sacred forests in Zimbabwe (Byres et al. 2001); sacred groves in India (Malhotra et al. 2001); Sarana Mata Sacred Forest in Central India (Borde and Jackman 2010).

This study does not intend to argue whether shamanistic practices are true or imaginary or hypnotic, but rather acknowledges the fact that they take place in the Kullu Valley and have something important to contribute in terms of maintaining its cultural landscape.

20.3 The Institution of *Pūch* in the Kullu Valley: The Land, the *Debatās,* and the People

The Kullu Valley refers to the landscape formed within the valleys of the river Beas, and its tributaries Parbati and Tirthan in the lower Himalayan ranges in Kullu, in Himachal Pradesh in India. It is worth noting that this landscape represents more of a culturally coherent area than the administrative boundary of the Kullu District (although there are overlaps). In the valley, one finds shrines dedicated to several local cults that are represented through deities called *debatās*. These include *Athārā Karaḍu* (name of local cult), *Athārā Nāg* (snake), *Ṛṣi* (sage), *Vīra* (warrior), *Asura* (evil being), *Joginī* (sacred feminine deity), *Jamalu* (name of local cult), etc. Each cult has its own mythological narratives of origins, evolution, and manifestations, and are known through oral histories and vernacular texts. These cults generate strong religious-cultural identities and interdependencies among the inhabitants of the valley. The mapping of these cults using their catchments and areas of influence show the extent of this landscape spreading beyond the administrative boundaries of the Kullu district (Pethe and Shinde 2016). To be remembered that "the mythological literatures are the basic source in this context as one mythology is interconnected to the other" (Singh 2011b, p. 11), as illustrated in the present study.

Local communities revere '*debatās*' as a savior of their land. *Debatās* manifest through many forms such as *deba-bana/narāyaṇḍī* (*deva-āraṇya* in Sanskrit, means sacred grove), sacred tree/s, sacred stones, sacred water bodies, sacred mountain, *bhāṇḍāra* (temple), *ḍerī* (*sthān* in Sanskrit, means shrine), *modh* (*yajñaśālā* in Sanskrit, means a place for fire), a tall wooden shaft, a (vertical or horizontal) stone slab, sacred precinct, or a sacred place in someone's residence (Pethe and Shinde 2020). Instead of the common practice of people visiting temples of deities, in Kullu, it is the *debatās* that are taken outside into villages and surroundings where people receive them and seek blessings through the act of "*darśana*" (the act of seeing and being seen by a deity). The physical manifestation of *debatā* is a *moharā* (a pressed face image in gold/silver kept in a basket). It is not consecrated traditionally in a temple as its location is not fixed. But it is aesthetically arranged in a basket, along with *moharās* of other *debatās*, and kept in the attic of a structure called *bhāṇḍāra* that occupies a significant location in a village. People believe that *debatās* communicate with them and therefore seek their permission before they take major decisions: both at the individual and household level and the community level in villages. These communications take place through many kinds of rituals; *pūch* being the most prominent. The practice of *pūch* is like a mechanism of religious governance institutionalized to bring law and order to the village and the wider cultural landscape.

Before turning to a discussion of *pūch*, it is necessary to introduce the structure through which villages and different jurisdictions in the valley are governed because that provides the context for the performance of *pūch*. Based on the spatial hierarchy, a four-tiered structure exists for local-self-government. Most matters at the level of the village are governed by the institution of *grāma-pañcāyata* (village council): a collective of villagers elected by the residents of the village. Several villages brought together under a shared boundary generate an administrative unit called a tehsil (a district administration in India); several *tehsīls* are combined to form a *district*, and districts are brought together to form the jurisdiction of the state. Administrative responsibilities for development are distributed across these tiers.

However, another hierarchy is more important in the Kullu Valley for maintaining its cultural landscape. This is the hierarchy of religious institutions associated with different *debatās* and how they operationalize everyday religious and cultural and social life of the valley dwellers, the people in the service of *debatās*. While mapping the spatiality of three cults in the Kullu Valley, the *Atharā Karadū, the Atharā Nāg*, and *ṛishis*, a three-tiered religious administrative structure was noticed that included *Grāma debatā* (village deity), *Baḍā debatā* (deity commanding a group of village deities), and Lord Raghunathji at Kullu (the highly revered, commanding all deities in Kullu Valley). This structure appears to be well-knit in the fabric of villages.

The villagers in a village revere the *grama debatā* which includes local deities such as *nāga, vīra, asura, ṛiṣi, and joginī.* They believe that their existence depends on the land blessed by these *debatās.* They pay homage to the *debatā* by offering a part of what they produce, as it is believed that the earth/land is divine and owned by *debatā.* These offerings are used by the village authorities for the welfare of villagers during the times of festivals, gatherings, and calamities like drought, disease, etc. To serve the *Debatā*, villagers form a local institution called '*deolī*' (the committee for the *deity*) that is seen as a representative of the deity and has significant religious authority to decide on village matters. This committee includes representation from all classes and castes in the village. While some roles in the committee are hereditary, some are appointed by the *debatā* through the process of *pūch.* A shaman is known as *gūr* in Kullu and is an integral part of the *deolī*: he acts as the medium of communication between the villagers, *deolī*, and the *debatā.* It is believed that the *gūr* possesses supernatural powers and therefore is to be appointed by the *debatā.* Through the *gūr* and *pūch*, villagers invoke the local *debatā* whenever a family has a problem, any land dispute, or a village-level issue.

The *badā debatā*, as the name suggests, commands a much larger area—often a group of villages; the number ranges from two to fourteen (as found during fieldwork). This territory of the *badā debatā* is called '*Debatākṣetra*' (a territory under command of a particular deity) (Fig. 20.3). The villages in the *debatākṣetra* present part of their produce to this *badā debatā* as homage and in return, this *debatā* takes care of them, their settlements, and also the natural forces. *Badā debatā* is generally invoked during larger calamities in the region. For instance, in 2015, the incident of villagers seeking divine intervention to protect the apple crop from hailstorms was widely reported (Bisht 2015). During *yātrās* (pilgrimages in the region), the *debatā* visit their neighboring *debatās*.

All *baḍā debatās* along with *grama debatās* in their *debatākshetra* visit Raghunathji during the annual festival of *Daśaharā*. They pay their homage in terms of grains, offerings, etc., to Raghunathji patronized by the king of Kullu (refer to Berti 2015 for a discussion on the politics of why a classical god-like Raghunathji is placed higher in the hierarchical order although he is not from this area). It is important to note that the presence of Raghunathji in the hierarchy does not undermine the importance of the local *debatās* as they are routinely invoked by villagers for a variety of reasons using the performance of *pūch*.

In a village, there are many occasions during which the *grām debatā* is invoked by the *gūr* for the purpose of *pūch*. One such occasion that was observed and documented at village Jana during fieldwork is illustrated below (see Fig. 20.4).

One young girl from a family of the *Luhāra* (blacksmith) caste was suffering from an ailment for a long time. Hence, the family wanted to consult *debatā* regarding her health issues. The family contacted the *pujari* (priest) and *kārdār* (the executive head of *deolī*) and requested a *pūch*. The *pujarī* checked the *panchāng* (Indian religious calendar) and fixed an auspicious day for the *pūch* and that was communicated publicly in the village. On the day of the *pūch*, the *bājantrī* (local musicians who belong to the *Luhāra* community) collected the musical instruments from the store next to *modh* and paraded through the village playing devotional music and arrived at a designated place in the sacred grove where others had already gathered. Soon the *gūr* entered the temple near the sacred grove and brought the *nishān* of *debatā*, i.e., a *ghaṇṭā* (bell), a silver plate, and the *dhūpa* (incense) from the temple. *Gūr* brought the leaves of medicinal plants that are used as offerings in the sacrificial fire (*āhūti* in the *dhūpa*). He had procured those from forests in the upper Himalayas. This arrangement was placed on a natural stone platform that became the altar near the *ḍerī* (a small shrine). The chanting, music of *bājantrīs*, lighting of *dhūpa*, burning of the medicinal leaves—all this creates an aura that facilitates the *gūr* to move into the state of trance. Now, he is assumed to be communicating with the *debatā* or the supernatural energies and answers the questions asked by the *Luhāra* family in the local Kulbi language. The leaves having medicinal values are given to the family as a *Prasad* (a religious offering) for burning in their home for purification of air. After the *gūr* regained normal calm, the meeting with the *debatā* ends with a final ritual.

The practice of *pūch* is invoked and performed for various reasons. In the next section, we focus on how it is conducted for development-related matters that influence the cultural landscape of the valley.

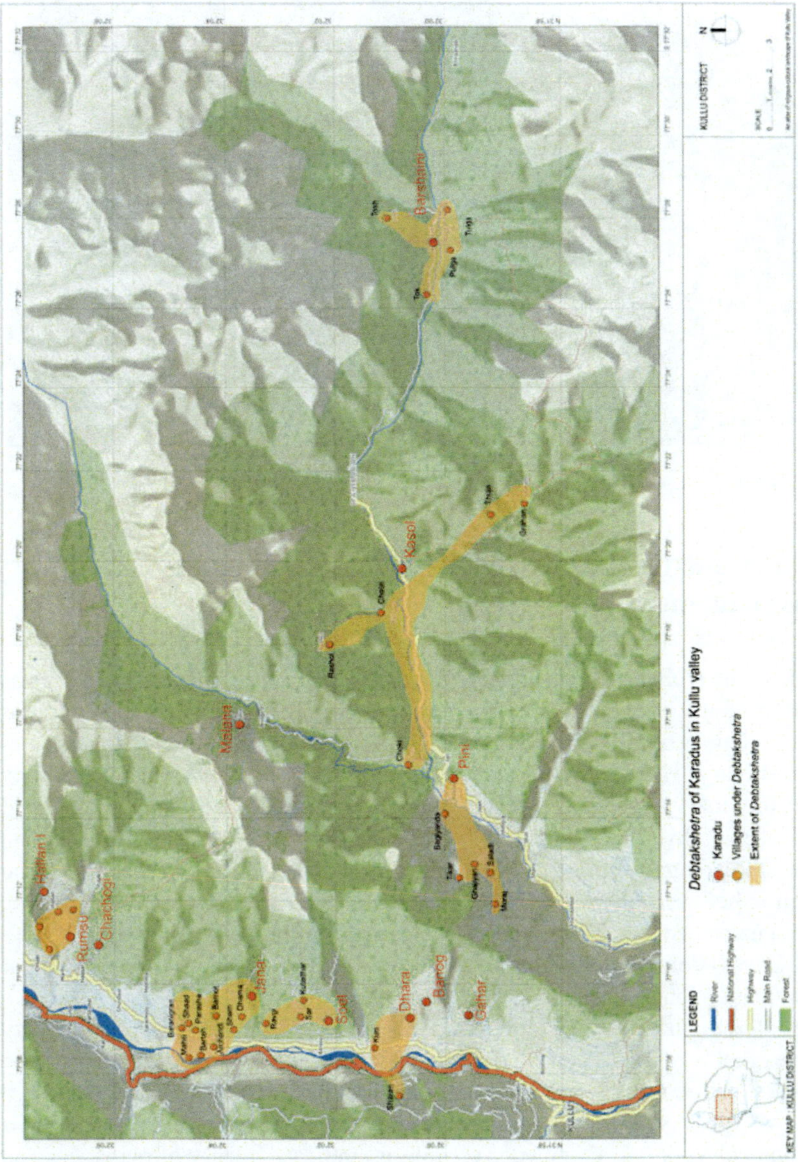

Fig. 20.3 Map showing the Debtākshetra of cult of Athārā Karaḍū in the Kullu valley (*Source* Drawing by Author)

Fig. 20.4 *Pūch* being carried out in the sacred grove at village Jānā

20.4 Shamanistic Rituals, *Pūch*, and Environmental Management

Village life in the Kullu Valley is replete with rituals that are shamanistic in nature as they relate to natural forces that affect their agriculture and livelihood and predictions of the future. This is best illustrated in case of the annual festival of Fāglī that is celebrated in February. There are many narratives regarding its celebrations: mythology states that this marks the victory of gods over *asuras* and is an announcement of the return of gods from Indrapurī (city of lord Indra in heaven). From an agrarian point, this is to celebrate spring and the beginning of the cropping season, and the celebration is believed to remove all evil from the landscape. This is the occasion for the declaration of fortune by *debatās*. An example of Fāglī celebrations in the village of Goshal demonstrates how shamanism is subtly woven into many kinds of rituals that mark such celebrations.

In Goshal, the *g͟ūr*s enter the temple of Gautam Ṝishi and bring a spherical clay lump made up of wet and sieved soil (prepared 21 days ahead) to predict the fortune of the village or the region. In one ritual, the clay lump is broken and distributed among the villagers. Villagers crush the clay on palms and try to make sense of their find. Finding *kumkum* (sacred powder vermillion) indicates plenty of marriages and sacred rituals; finding stones indicate a forecast of drought; finding seeds is good for agriculture while apple leaves mean apple plantations will bring in revenue. Thus, the religious beliefs in the power of nature are strongly imbued in these religious rituals (Fig. 20.5).

Another well-known example of invoking of divine powers of *debatā* is found with local communities that grow apples as a cash crop for income. Fearing hailstorms, they protect their crop with measures including covering apple trees with anti-hail nets. However, it is their resolve in praying to their *debatās* for saving them from the wrath of weather gods that is unwavering. A village headman in Bondra mentions villagers going up "in high hills and mak[ing] different kinds of offerings to the deity, seeking protection from the hailstorm that has damaged apple crops in the area" (Bhagat Singh Thakur, head of Bondra Temple Committee). Media reports are quick to observe how "The villagers, particularly the fruit growers, are turning toward the local deities to protect the apple crop from hailstorm" (Bisht 2015).

Religious practices like holding prayers, making offerings to appease the deities, and connecting to the *debatās* through *gūrs* are the common ways of approaching *debatās* to safeguard during calamities. We found another interesting incident that highlights the belief in the authority of *debatā*. This was noted by a villager in her diary in 2010 and shared with us. It is of merit to reproduce as written in her diary. It says,

Fig. 20.5 Religious rituals by the *Gūrs* during Fāglī at village Goshal in the Kullu Valley

Jap by nāg-debatās at Nagoni:

On 28th February, Dumal debatā from Halan 2, came along with all the villagers and bajantri to visit Basuki-nāg in Halan 1, in order to relieve the villagers by bringing in rains. Dhumal-nag, who is the son of Basuki-nāg and one of the Atharā-nāg, is a powerful debatā in the region … Gūr of Vasuki was possessed and asked first about the 'issue of rain'. This act was followed by the gūr of Dhumal. Villagers and deolī took tirtha snan - holi bath at the tirtha sthan and then return to bhandaar of Vasuki-nag. The villagers, bajantri and debatā Dhumal returned back after 'brahma-bhoj' (divine-food) offered by the villagers of Halan 1, at Bhandara of debatā Vasuki. Dhumal-nag had visited Basuki after six long years in order to bring in rains. (Diary of Sushma Thakur, 2010)

This note in the diary clearly emphasizes these beliefs: (1) *Nāg-debatās* are associated with water and sources of water and are revered to bring in rains, (2) A visit to the original *sthāna* (place) of *debatā* brings in prosperity, (3) The movement of *debatās* (with people) in the landscape is an important practice, (4) Villagers have strong faith in worshipping nature, e.g., *tīrtha-snāna* (holy bath), and (5) *Debatā* is the savior of nature and mankind. Such beliefs reinforce the power of rituals and shamanistic practices.

The following paragraphs focus on how the institution of *pūch* is invoked to address environmental issues at different spatial hierarchies in the valley.

At the village level, there are many examples where *debatā,* during *pūch,* seem to give a direct command regarding scrutiny of constructions of houses. This is best illustrated during the movement of *debatā* Vasiṣṭha (deified sage) in a *yātrā* as narrated by Luchesi:

… in Vashishtha, unauthorized stalls and markets were demolished by the rath of angry debatā on this way to scared yatra on the occasion of festival in April 2001… (Luchesi 2006, p. 76)

The unauthorized construction is noticed by *debatā.* This power is evident in the example where *debatā* instructed a petitioner during *pūch* to demolish part of their house in Manali (Luchesi 2006). Reoccurrence of mishaps in a family made them consult the *gūr* of Vasiṣṭha and Hiḍimba. During the *pūch,* the *gūr* confirmed the reason for the mishap was that they had built their house by encroaching on other's land not owned by them. During this *pūch,* angry *debatā* showed his resentment for this unlawful act by the family and gave them two years' time to demolish the same. This incidence exemplifies the religious authority through the act of *pūch.* Using several examples, Luchesi (2006) argues that the *debatā provides* an unbiased and fair verdict, particularly in case of encroachments and unlawful house extensions.

The institution of '*jagati pūch*' is a powerful mechanism within religious governance in the valley. *Jagati pūch* literally means a *jāgatika saṁsada* (global assembly) of *debatās.* As a tradition, the call for collective *pūch* is raised by the Raja of Kullu, who is believed to be ruling the throne on behalf of Lord Raghunathji. When an issue involves larger or regional concerns, the Raja invites all village deities to Jagati-Pat (place of gathering) at Naggar for the *jagati pūch.* There is a long history of significant outcomes from such gatherings and the power of the *debatās* was once again felt when the high profile 'Himalayan Ski Village Project' was knocked down at a recent *jagati pūch.*

As per the detailed study by Asher (2008), a private company proposed to develop a ski resort on 133 acres of land near Kullu in 2005. Like most resorts, it comprised hotels and recreational facilities, but most importantly could potentially open access to more than 3000 acres of forest areas. Estimated at Rs 2200 crores (in USD), the project claimed to provide a much-needed economic boost to tourism and to the development of the region. However, it soon started to receive criticism for its significant negative impacts on the pristine environment of the valley and the project threatened "sacred natural sites, traditional temples, and broadly to the existing culture" (Asher 2008, p. 6).

The unpopularity of the proposal united the local community, and they appealed the then king, Maheshwar Singh, to organize *jagati pūch*. He obliged and at the *jagati pūch*, the power of this institution was at full display.

> *During 'jagati-puchch', the representatives were seated in rows and the King went around asking each to speak his or her mind. This prompted the representatives to go into a trance, ... god spoke and made it clear that they did not want the ski village in the area since it would desecrate holy places. They warned the people that if the project was approved, the gods would leave and destroy the area...* (Asher 2008, p. 16).

The wide reporting of the event in the media mobilized more community support. Besides several protests, the project was fought in a legal battle as a hotelier from Vasiṣṭha village and a community organization filed Public Interest Litigation (PIL) in the High Court at Shimla. The prevailing sentiment was that the project would "ruin the local culture by subjecting people to Western influence" (interview with a *gūr*). Another community leader opined that "since the land belonged to the *debatā*, no one else had the authority to change its status and its use."

The reverence for nature and the environment was at the core of this struggle and the *debatās* and the communities were together in this. Thus, *debatās* have a significant influence in the public sphere, 'especially in the context of the *jagati pūch*, where they may operate as counter powers or as an arena for contesting government decisions' (Berti 2015, p. 117). The successful rebuttal of the Ski Village Project inspired a series of such large-scale community-driven *pūch* in the following years because it ascertained itself as a mechanism to "safeguard the environmental interests against the non-congruent development decisions" (Damodaran 2012, p. 323). A notable example is the consternation of a new construction project in Jogini forest at Vasiṣṭha village in 2011. A private water supply company began to build a water tank in a forest near the village of Vashishtha. Villagers protested because they believed that this area was inhabited by Joginis (powerful feminine deity) (Berti 2015) and conducted a *pūch.* The *debatās* gave a verdict against the project and yet the company continued with the project. Finally, the village communities as a collective decided to seek advice at the larger court of *debatās* called *jagati pūch.* However, by 2012, given the community mobilization and political pressures, the government stopped the company from the construction of the project. Berti remarks that this case reiterated that 'nature is governed by village deities' and that such decision-making mechanisms need to be integrated into the ways state institutions decide on the future of communities (Berti 2015, p. 126).

20.5 Concluding Remarks

In Kullu, people consider their land as the valley of gods and that they are a part of the divine nature contributes to the imagination of the Kullu Valley as a cultural landscape. The beliefs, rituals, performances, and actions of villagers are all acts of placemaking through reverence: they "know that their activities and destinies are governed by these gods [*debatās*] and that they cannot afford to disobey them at any cost" (Sharma 2007). There are several ways in which they look after this cultural landscape, but this chapter focused on the shamanistic practice of *pūch* that intimately ties the people of the valley with their *debatās*. Such indigenous ways of placemaking in sacred landscapes is not unique to Kullu valley but is widespread among communities that live close to nature and their livelihood depends directly on natural resources. The ideas of nature-culture interdependence presented in this chapter are germane to many situations where indigenous communities rely on supernatural and divine connections with the landscape when they have to protect their environments from different kinds of forces. For instance, in some forested areas in East-Central India, women have invoked a deity named Sarnā-mātā—the earth goddess in their struggle for the conservation and regeneration of forests (Borde and Jackman 2010).

The chapter has illustrated how shamanistic practices can also be seen as instruments of placemaking because of their multifaceted nature as belief, ritual, and institution as epitomized in the practice of *pūch*. Such practices that emerge from the environment and the perception of the ecological systems form a part of traditional wisdom (Kaushal 1998). Moreover, they reinforce the deep connections between the land and divinity. "Unlike sacred space and objects, whose boundaries are clearly defined, conceptions of sacred land are typically abstract and abstract notions of sacred land are indivisible and must remain intact" (Sosis 2011, p. 17). They also serve to awaken the communities, bring them together, and empower them to raise their voice, and ultimately influence decisions about changes to the very environment that supports them and need to be protected. The centrality of belief and institution is reinforced by Byres in the conservation efforts at Zambery, Zimbabwe (Byers et al. 2001). Byers et al. show that forests have better chances of protection when traditional spiritual leaders continue to perform religious activities in those forests and add to their image as sacred.

As shown in this chapter, invoking sacred and cultural values through beliefs and practices in development vs environment debates offer a more sustained action toward environmental conservation. This is also demonstrated in case of the contests against three hydroelectric projects on the Ganga River where local communities have asserted their traditional wisdom and sacred values of in their (Drew 2017). Similar renewed enthusiasm for emphasizing sacred powers of nature through their religious and faith-based practices in community action is evident in recent studies and projects in mountainous areas (Rao 2010; Rescia et al. 2008; Sood 2006). A notable example is the Kailash Sacred Landscape Conservation and Development Initiative (KSLCDI) being undertaken at the Kailash Mansarovar landscape. This initiative focuses on employing the sacred values of this landscape and cultural linkages among local populations to achieve the long-term conservation of ecosystems. Like this project, we hope that someday the cultural landscape of Kullu can become exemplary for demonstrating an intrinsic role of religious activities in the stewardship of natural conservation.

Finally, at a more conceptual level, the discussion of *pūch* as a shamanistic practice provides more rounded support to Belsare's appeal of using indigenous approach to understand cultural landscapes as a totality of the inner and outer expressions. The inner values are deeply rooted in the belief that nature is sacred and can be maintained and reinforced by performing religious and cultural rituals in specific settings. Mediated by institutions, they can help in transmitting the traditional wisdom about being sustainable with natural resources. To maintain the deep connections, it is necessary that the outer physical, social, and cultural relationships are nurtured and nourished rather than sacrificing them in the name of development. These ideas have been passed through generations via Vedic literature. For instance, *Bhūmisūkta* (a Vedic treatise) advocates the theory of stewardship, eco-spirituality, and conservation of the environment, not only at a physical level, but also at the religious, spiritual, and cosmic level, by building the attitudes and perceptions of the people toward nature (Pant 2007, p. 198). These sentiments are at the core of religious ecology and conservation in cultural landscapes (Marton-Lefevre 2010). The philosophical or inner values and the outer physical, social, and cultural expressions are interdependent on each other. Change in one will affect the other. Thus, understanding of Indian cultural landscape is incomplete without indulging in the metaphysical aspects of the religious rituals. With this approach exploring the role of the mind and ultimately collective minds in the Indian cultural landscape in future studies may give more insights.

References

Asher M (2008) Dream destination for world class tourists... nightmare for the Himalayas! Impacts of the proposed Himalayan ski-village project in Kullu, Himachal Pradesh. Him Niti Campaign, Jan Jagran Evam Vikas Samiti and EQUATIONS, Bangalore, India

Atha M (2012) A neglected heritage: towards a fuller appreciation of the landscapes and lifeways of Hong Kong's rice farming. Asian Anthropol 11(1):129–156. https://doi.org/10.1080/1683478X.2012.10600860

Bande U (2006) Folklore and ecology: the connecting Links. In: Bande U (ed) Folk traditional & ecology in Himachal Pradesh. Indus Publication Company, Shimla, Himachal Pradesh

Belsare K (1962) Adhyatma Darshana. Manohar Yashwant Joshi, Pune

Berti D (2015) Gods' rights vs hydroelectric projects: environmental conflicts and the judicialization of nature in India. Hum Person Nat Class Mod India (Rivista Degli Studi Orientali Nuova Serie, ISSN 0392-4866) LXXXVIII:111–129

Bhagwat SA, Dudley N, Harrop SR (2011) Religious following in biodiversity hotspots: challenges and opportunities for conservation and development. Conserv Lett: J Soc Conserv Biol 4(3):234–240

Bisht G (2015, May 15) Villagers seek divine intervention to protect apple from hailstorm. Hindustan Times. Himachal Pradesh

Borde R, Jackman AJ (2010) The Devi as ecofeminist warrior: reclaiming the role of sacred natural sites in East Central India. In: Verscheuren B, Wild R, McNeely J, Oviedo G (eds), Sacred natural sites: conserving nature and culture. Earthscan, Taylor & Francis, London, Washington, DC, pp 272–279

Byers B, Cunliffe R, Hudak A (2001) Linking the conservation of culture and nature: a case study of sacred forests in Zimbabwe. Hum Ecol 29(2):187–218

Coleman SM (2004) Pilgrimage and shamanism. In: Walter MN, Fridman EJ (ed), Shamanism an encyclopedia of world beliefs, practices, and culture, vol. 1. ABC-CLIO, Inc., Santa Barbara, California, pp 197–200

Conklin B (2002) Shamans versus pirates in the Amazonian treasure chest. Am Anthropol 104(4):1050–1061

Damodaran V (2012) Globalisation and sacred *adivasi* landscapes in Eastern India. In: Daly P, Winter T (eds) Handbook of heritage in Asia. Routledge, London, pp 311–327

Drew G (2017) River dialogues: Hindu faith and the political ecology of dams on the sacred Ganga. University of Arizona Press, Tucson, United States of America

Hamayon R, Paris E (1993/2007) Are "trance," "ecstasy" and similar concepts appropriate in the study of shamanism? Shaman Int J Shamanistic Res 1(1–2):17–40

Kane P (1941) History of Dharmasastra (ancient and medieval religious and civil Law)—Vol II, Part II. Bhandarkar Oriental Research Institute, Poona

Kaushal M (1998) Sacred response to environment: the Gaddi and his Mountain. In: Saraswati B (ed), The cultural dimensions of ecology. IGNCA and D. K. Printworld Pvt. Ltd., New Delhi, pp 51–55

Luchesi B (2006) Fighting enemies and protecting territory: deities as local rulers in Kullu, Himachal Pradesh. Eur Bull Himalayan Res 29–30

Maarif S (2015) Ammatoan Indigenous Religion and Forest Conservation. Worldviews Global Religions, Cult Ecol 19(2):144–160. https://doi.org/10.1163/15685357-01902005

Malhotra KC, Gokhale Y, Chatterjee S, Srivastava S (2001) Cultural and ecological dimensions of sacred groves in India. Indian National Science Academy and Development Alliance. http://wgbis.ces.iisc.ernet.in/biodiversity/sahyadri_enews/newsletter/issue4/Yogesh_CEdimensions.pdf

Marton-Lefevre J (2010) Foreword. In: Verschuuren B, Wild R, McNeely J, Oviedo G (eds), Sacred natural sites: conserving nature and culture. Earthscan, Switzerland

Meyhew RJ (2000) Geography is twinned with divinity: The Laudian Geography of Peter Heylyn. Geogr Rev 90(1):18–34

Naomi R-A (1996) Of seeds and shamans: the appropriation of the scientific and technical knowledge of indigenous and local communities. 17 Mich. J. Int'l L

Pant U (2007) Ecology and conservation in the Bhumisukta of Atharvaveda. 2nd World Congress on Vedic Sciences

Pethe A, Shinde K (2016) A religious-cultural Atlas of the Kullu Valley. Intach Heritage Academy, New Delhi.

Pethe A, Shinde K (2020) Debta and the religious-cultural landscape of rural settlements in the Kullu Valley. INTACH J Heritage Stud 4(1):1–12

Rao MS (2010) Landscape urbanism: decoding traditional knowledge through the sustainable water management. J Landscape Archit 28:80–84

Rapoport A (1992) On cultural landscapes. Tradit Dwellings Settlements Rev 3(2):33–47

Rescia AJ, Pons A, Lomba I, Esteban C, Dover JW (2008) Reformulating the social–ecological system in a cultural rural mountain landscape in the Picos de Europa region (northern Spain). Landsc Urban Plan 88(1):23–33

Roerich N (2014) Gods of Kuluta Kullu, the land of Gods and Goddesses, a reference book on Kullu Devtas. In D. D. Committee (ed), A reference book on Kullu Devtas. District Dussehara Committee, Kullu, p 7

Sharma BR (2007) Gods of Himachal Pradesh. Indus Publishing Company, Shimla

Sinha A (2006) Cultural landscape of pavagadh: the abode of Mother Goddess Kalika. J Cult Geogr 23(2):89–103
Singh RPB (ed) (2010) Heritagescapes and cultural landscapes. Planet Earth & cultural understanding Series, Pub. 6. Subhi Publications, New Delhi
Singh RPB (2011a) Rural cultural landscapes: Asian vision of man-nature interrelatedness and sustainability. International Symposium on Sustainable Rural Landscape and Planning in Asia Pacific Region, Korea
Singh RPB (ed) (2011b) Sacredscapes and pilgrimage systems. Planet Earth & cultural understanding series, Pub. 7. Subhi Publications, New Delhi
Singh RPB (2013) Indian cultural landscape vis-à-vis ecological cosmology: a vision for the 21st century. Ann Nat Assoc Geogr India 33(2):36–62
Singh RPB (2017) Appraising the Indian cultural landscape: envisioning ecological cosmology in the 21st century. North Eastern Geogr 39(1–2):3–28
Singh RPB, Olimpia N, Rana PS (2022) Introduction: interfaces among placemaking and cultural landscapes: review and appraisal. In: Singh RPB, Olimpia N, Rana PS (Eds), Placemaking and cultural landscapes. Springer Nature Singapore Pte Ltd, Singapore, ca. pp 1–18
Sood KK (2006) The influence of household economics and farming aspects on adoption of traditional agroforestry in Western Himalaya. Mt Res Dev 26(2):124–130
Sosis R (2011) Why sacred lands are not indivisible: the cognitive foundations of sacralising land. J Terrorism Res Cent Study Terrorism Polit Violence 2(1):17–44
Taylor K (2012) Heritage challenges in Asian urban cultural landscape settings. In: Daly P, Winter T (eds), Handbook of heritage in Asia. Routledge, London
Verschuuren B, Wild R, McNeely J, Oviedo G (2010) Introduction: sacred natural sites the foundation of conservation. In: Verschuuren B, Wild R, McNeely J, Oviedo G (eds), Sacred natural sites: conserving nature & culture. Earthscan, Switzerland, pp 1–13
Walraven B (2009) National pantheon, regional deities, personal spirits? Mushindo, songsu, and the nature of Korean Shamanism. Asian Ethnology 68(1):55–80

Correction to: Reuse of Unoccupied Religious Monuments for Tourist Accommodation: Santa Maria da Ínsua (North of Portugal)

Rolando Volzone, Soraya Genin, João Estêvão, and Mafalda Patuleia

Correction to:
Chapter 18 in: R. P. B. Singh et al. (eds.), *Placemaking and Cultural Landscapes*, Advances in Geographical and Environmental Sciences,
https://doi.org/10.1007/978-981-19-6274-5_18

The original version of the book was inadvertently published with incomplete affiliation of the author in Chapter 18, which has now been corrected. The book has been updated with the change.

The updated version of this chapter can be found at https://doi.org/10.1007/978-981-19-6274-5_18

R. P. B. Singh et al. (eds.), *Placemaking and Cultural Landscapes*,
Advances in Geographical and Environmental Sciences,
https://doi.org/10.1007/978-981-19-6274-5_21